KLIMA POLIS Vol. 1   City, Climate, and Architecture

City, Climate, and Architecture

A Theory of Collective Practice

Second and Revised Edition

Sascha Roesler

KLIMA POLIS Vol. 1

Birkhäuser

Table of Contents   Preface   7

I   European Developments before 1945:
           The Field of Knowlegde

Urban Studies of Man-made Climates   11
An Introduction

1 Thermal Geographies of the European City   17
Urban Hygiene and the Heat Economy

2 Man-made Climate by Design   43
Urban Climatology and Modern Architecture

3 Democratizing Urban Nature   75
Infrastructures of Climate Control at Large Scales

II   Global Adaptations after 1945:
           Three Design Metaphors

Thought Styles of Man-made Climates   115
An Introduction

4 Thermal Heritage   121
Collective Memory and the Natural History of Cities

5 Microclimatic Islands   155
Thermal Spectacles and Closed-system Imaginaries

6 Energy-Synergy   179
Energy Ecologies and Settlement Structures

III   Epilogue

7 Singapore as a Model?   213
Urban Climate Control in Practice

           Appendices

Endnotes   228
References   256
Index   265
About the Author   272
Acknowledgments   273
Illustration Credits   274

Preface to the second and revised edition

It is not surprising that climate change and rising temperatures in cities have sparked a renewed public interest in concepts of urban climatology. However, the current debate on urban climate largely excludes the design issues that are so important. Currently, it is science journalists, rather than architectural journalists that are setting the tone in the public sphere. So far, there is hardly any talk of architectural solutions in a more comprehensive sense, although the urban climate is in a fundamental way the product of the design—the form, arrangement and material—of urban buildings, and the cooling and heating requirements inside buildings depend to a large extent on the climatic conditions outside.

How can we translate insights from urban climatology into design? This methodological question was the starting point of a long-term research project. I initiated this project in 2013 at the Future Cities Laboratory of ETH in Singapore and it was continued in Switzerland as part of a six-year research grant on "Architecture and Urban Climates". The research was conducted at the Academy of Architecture in Mendrisio and at ETH Zurich in Switzerland thanks to generous funding by the Swiss National Science Foundation providing a grant of SFr. 2.2 million between 2015 and 2021.

Our conclusions have now taken the form of two major publications. These two books entitled *City, Climate, and Architecture* (Vol. 1) and *Coping with Urban Climates* (Vol. 2) launch a new international series entitled KLIMA POLIS, published by Birkhäuser.

The two first volumes of this series aim at rethinking climate control—a key concern of the discipline of architecture—through the lens of urban climate phenomena. They aim to stimulate new ways of thinking about the spatial order of cities by complying with the potentials of climate control at the scale of groups of buildings and their surrounding (urban microclimates). The two books clearly question whether the energy-source supply of urban architecture can still be taken as a private matter. Vol. 1 is an intellectual history, tracing the emergence of modern urban climatology and its adaptation by architecture and urban design. Vol. 2 is a cross-cultural study of four cities around the world, exploring the manifold relations between urban climates, architecture and society both within and beyond buildings. Each volume is self-contained, but they are complementary in their assessment of architecture and urban climates from a historical (Vol. 1) and a contemporary (Vol. 2) perspective.

We are delighted that the first edition is already sold out after just two years. It can be seen as a sign that interest in this topic is continuing. For the second edition, minor corrections and additions were made and the bibliography was expanded.

Sascha Roesler, December 2024

European Developments before 1945:
The Field of Knowledge

I

Thermal Geographies of the European City
Man-made Climate by Design
Democratizing Urban Nature

Urban Studies of Man-made Climates

An Introduction

The primary aim of this publication is to *urbanize* the thinking on climate in architecture. For this, an understanding of the practice of climate control is to be developed in light of an architectural history and theory of urban climates. My central thesis is that certain discourses and built projects of the 20th century contain the elements for a new urban theory of climate control. The reconstruction of urban climatological knowledge in 20th-century architecture results in a different understanding of climate than that which was brought to fruition by the research guided by a notion of *comfort*. Most importantly: the practice of climate control is understood as being a collective practice—rather than a practice of the individual.

Today, the American lines of thought on comfort must be subjected to a radical revision through the perspective of urban environments. What is commonly understood by *climate control* did not emerge by considering such environments: there is a prevailing suburban context in the theory of climate control, and in sustainability at large. Even the so-called passive solar movement remained steeped in the comfort thinking of *air conditioning*, and thus focused on a practical methodology that was entirely committed to the individual building and individualism.[1] Today, there is an urgent need to relate the theory of climate control to the particular conditions of cities and to identify the relevant urban agencies of climate control. The consideration of groups of buildings, rather than isolated individual buildings, as being the fundamental "climatic unit"[2] also challenges basic assumptions of current architectural theory.[3]

The Notion of Man-made Climate

The notion of *man-made climate* forms the theoretical pivot of this publication. It is the starting point for the elaboration of an urban theory of climate control. By referring to groups of buildings and to *thermal difference* (rather than to climatic stabilization), the concept of man-made climate opens up new ways of thinking about climate control in, through and of cities. This notion (in German: *künstliches Klima*) was widely used in Europe—in particular, in Germany and Austria—in the interwar period. It was equally applied to outdoor and indoor environmental conditions and not just to mechanically conditioned interiors.[4] Beyond this, the examination of urban climates—one could say *the urban studies of man-made climates*—promoted a new modern sensitivity for the interdependence of the scale of the building and the scale of the city.[5] In the urban climatological thinking of the interwar period, the topological interconnection of interior and exterior perspectives—of the apartment and the city—was particularly present. As such, the imperative of *integrating indoors and outdoors* in order to reduce environmental and thermal loads in cities is the critical legacy of interwar urban climatology. By promoting a systemic and multi-scalar understanding of heat management throughout the city, the notion of man-made climate can be viewed as being proto-ecological.

Historically, the notion of man-made climate has been the decisive theoretical interface along which the mutual appropriation of modern urban climatology and modern architecture and urban planning has taken place. Berlin and Vienna in particular were laboratories and intellectual centers of urban climatology in the 1920s and 30s, which can be explained not least by the political conditions in both cities at the time—German social democracy and Austro-Marxism. The interest in the living conditions of the many and not—as in the case of the American ecology movement—of the individual may go some way to explaining why urban climatology originally developed under politically left-wing conditions. From its inception, urban climate research appeared to be both an applied and interdisciplinary endeavor aimed at improving the built environment of broad segments of the population. The interdisciplinary project of applied urban climatology, as I will show, was equally driven by climatologists, physicians, architects and urban planners; it developed in the wake of a discourse on urban hygiene that increasingly took into account the urban climate.

Interwar urban climatology had a twofold interest: on the one hand, it posed the comprehensive question of society's dependency on the climate ("the great role climate plays in people's health and in their economic and cultural achievements"[6]); on the other, however, it also raised the considerably novel question of the influence of the city—as a human artifact—on the climate ("the manner in which these great concentrations of human beings influence their climate"[7]). This approach gave architects and urban planners a completely new scope for thinking about the relationship between architecture and climate. The urban climate was now no longer merely a driving force of building design but, just the other way around, rather the *result* of it. Climate in cities thus came to be seen not only as a naturally given influencing factor but also as a result of urban configurations.

This publication aims to make this important intellectual heritage visible and to present its relevance for today's architecture and urban design. With the growing awareness of the Anthropocene and the Earth as a whole as a "world ecology",[8] the notion of the man-made climate has become more relevant still. An enlightened political ecology provides the further theoretical framework for this. Referring to Karl Marx, Maria Kaika and Erik Swyngedouw proposed the metaphor of "metabolism"[9] to do justice to the mutual correspondences between nature and society, in which "non-human 'actants' play an active role in mobilizing socio-natural circulatory and metabolic processes".[10] This publication relies on this tradition of political ecology to build bridges within a highly relevant epistemic field.[11] Seen from the perspective of man-made climate, urban microclimates appear as man-made artifacts and thus as the result of consciously designed buildings.

European Developments before 1945

The Architectural Historiography of Urban Climatology

The central intuitions and insights of the German-language urban climatology of the interwar period have still not yet been properly received in architectural history and theory.[12] The reasons for this are primarily linguistic: the majority of source texts have not yet been translated from German into English. As such, the insights into urban climate made by meteorologists and architects are made accessible to a global readership probably for the first time through this publication. In addition, there might be a certain subconscious aversion (in the Anglo-Saxon world) to research that was published only after the National Socialists came to power in 1933: basic writings of modern urban climatology, such as Albert Kratzer's *Das Stadtklima* and Brezina and Schmidt's *Das Künstliche Klima*, were both published in 1937. This makes it all the more difficult for today's readers to separate the (bio)political echo chambers of this research from the findings that continue to be relevant (today).

Even more importantly, the urban climate, by being reduced to a phenomenon of outdoor space—without any relation to indoor space—has stood in the slipstream of the controlled indoor environment, as it spread with air conditioning, central heating and insulation in the West and beyond. Part of this one-sided fixation on outdoor space is revealed by the fact that the notion of "urban heat island" was elevated to the central paradigm of urban climatology in the late 20th century. By focusing on the relationship between "meteorology and urban design", a one-sided focus on urban outdoor spaces was established—one that threatened to make the central epistemological implications of urban climatology obsolete.[13] The narrowing of the notion of "man-made climate" to mechanical air conditioning inside buildings gave urban climate the status of a niche topic in the discipline of architecture and exorcized a deeper sense for climate control in urban environments. It is precisely the mutual reaction between the exterior and the interior that has not yet been adequately taken into account by the historiography of urban climatology. Such an integration forms, in essence, the intellectual program of an urban theory of climate control.[14]

The rhetoric of "light, air and sun," as promoted by members of the Bauhaus and CIAM reveals another meaning in the context of urban climatology: It also appeared as part of a scientific, ultimately empirical, evidence-based project to which the Bauhaus, for one, was committed. Its directors, Walter Gropius and Hannes Meyer, were particularly convinced of the interdisciplinary character of their school. However, this rhetoric—widespread among modern architects of the interwar period— has so far hardly been seen in the context of the emerging urban climate discourse of the interwar period.[15] Although the Modern Movement widely addressed air and sun, "the evidence basis remained sketchy, since designers had little empirical understanding of the complex nature of urban microclimates".[16] The predominant helio-therapeutic meaning given to complex climatic phenomena by early CIAM members obscured the scope of architectural reflection on the man-made climate in the 1920s and 30s. This is exemplified by the schematic and ideological debates about the proper orientation of buildings and streets in the big city.[17]

Architectural historiography followed the intellectual guidelines of modern architects without examining the scientific contexts more closely. In the context of this publication, historiographical corrections will be made, illuminating the knowledge of urban climate in the context of wind- and sun-oriented modern architecture. Beyond that, however, the work of those modern architects will be discussed for whom an interest in urban climate research was made *expressis verbis*—such as Gustav Hassenpflug, who was trained at the Bauhaus, and Ludwig Hilberseimer, who taught there. Hilberseimer was the protagonist of the modern architectural movement who most comprehensively attempted to translate the new findings of urban climatology into new principles of planning and design; accordingly, he is the secret hero of this publication, whose far-reaching considerations repeatedly appear throughout the book.[18] While the first part provides a systematic overview of urban climatological knowledge of architecture and urban design, as it emerged in the interwar period in Germany and Austria, the second part follows developments as they took place after 1945 under globalized conditions. In both the first and the second part, Berlin plays a central role in this publication. With its buildings and landscapes, forests and lakes, it is a critical agency and enabler for a new kind of discourse that connects human with non-human actors. Berlin shows the possibilities and the limits of a design-driven appropriation of urban climatological knowledge through architecture and urban design.

The Translation of Science into Design

Today, the vast quantity of scientific studies on urban climate stands in sharp contrast to the lack of architectural methodologies to apply the insights of these studies. The rudimentary awareness of the *translation* needed between urban climatological knowledge and architectural practice manifests itself in a fundamental lack of design methodologies. This is also revealed by the fact that Baruch Givoni is still considered the central representative of a transfer of urban climatology into architecture, although his work is exemplary for its disregard of the *agency* of design.[19] In his case, urban climatology appears as a field of applied natural laws, with marginal reference to architectural history and theory—and thus to design and culture. Accordingly, the actual aim of this study is to provide theoretical and historical foundations for a *methodological* transfer between urban climatology and architecture. The direction taken by this publication is towards the fragmentary appropriation of urban climatology by 20th-century architecture and the ideas and methods developed along the way. Two remarks on the main approach of this book:

Discourses

This publication reconstructs discursive links between hygienists, climatologists and architects in order to highlight architectural dimensions of urban climate. The emergence of the new science of urban climatology is described as being parallel to that of modern architecture and urban planning, which at the time was also in its nascent stages. In this publication, urban climatological texts are examined for their references to models of urban design and, vice versa, urban projects by architects for their climatological implications. Discourses played a unifying role: linking the practices of architects with those of scientists; civil engineering and meteorology came together in the discourse on hygiene.[20] Thus, following

→ fig. 1 European Developments before 1945

Michel Foucault, the aim of this study is to find the common denominator of *les mots et les choses*, as developed in his "archaeology of the human sciences", pursuing an archaeology that examines the epistemological transformations of a scientific object—the urban climate—into a design artifact. The book presents manifold transformations of the concept of the urban climate through architecture in the 20th century. Transmission and sublimation play an important role in the incorporation of scientific knowledge by design disciplines; the theoretical focus is accordingly on processes of exchange, appropriation and transformation. This publication insists on an asymmetrical relationship between urban climatology and architecture and it reveals fluid transitions and hybrids consisting of architecture and landscape architecture, architectural projects and research interests.

Buildings

By combining a history of science with a history of architecture, the publication brings together the "imaginary and the real".[21] In this context, one can speak of a "discrepancy between the many new urban planning ideas and proposals and their actual realization";[22] many urban climate-related-ideas failed due to prevailing social and spatial conditions. However, the aim of this study is to highlight an architectural knowledge of urban-climatology insights as components of a new theory of practice. Such a theory of collective practice combines a "history of ideas" with a "practical science" of architecture.[23] Crucial to this is a discourse history that negotiates interior and exterior conditions as thermally interdependent and conceives the control of climate as a collective endeavor. This is the reason why buildings—as the interface between inside and outside—remain epistemologically at the center. The building at the center of this book is not a self-sufficient entity but one that is closely interconnected with its urban environment and embedded in its city quarter. Urban buildings are found in a cultural, social, and ecological exchange with their surroundings. Accordingly, these investigations are located in the transitional area between architecture and urban design. The main focus is on the question of how a building is thermally influenced by adjoining buildings, streets, parks and winds.

Climatic Determinism, Revisited

Unsurprisingly, the debates on global warming and the Anthropocene have promoted a new architectural interest in concepts of *climatic determinism*.[24] In the history of the architecture–environment relationship, the city is the exemplary artifact where subordination gradually turns into a "progressive ability to control nature".[25] Clarence J. Glacken's classic 1967 study *Traces on the Rhodian Shore* demonstrates the importance of ancient and early modern theorists of architecture, such as Vitruvius and Leon Battista Alberti, in theorizing the complex relationship between nature and culture, and, more specifically, between climate and urbanization. Urban climate is a prominent field of application of an intellectual interaction that Glacken sees as being at work in Europe since antiquity. He points to the constancy of a certain way of thinking, which underwent a fundamental transformation only in the 19th century. "Buffon, Kant, or Montesquieu, I think, would have found the classical world strange, but the gulf between their times and classical times would have been less than that between 1800 and 1900."[26] The empirical approach to urban

Urban Studies of Man-made Climates

climate—which forms the center of my study—has experienced a strong developmental push, especially since the early 20th century, which can be described as the *scientification* of urban climates. In this process, the atmospheres of the evolving modern city were subjected to investigations based on the methods of meteorology and thermodynamics.

Urban climatology as a science emerged at the beginning of the 20th century with the development of the modern European city. In the future, however, its integration into architecture and urban design will have to draw increasingly upon the manifold manifestations of planetary urbanization.[27] A future architectural theory of urban climate must further elaborate the modes of thought in terms of three forms of appropriation: 1. As a *transnational metabolism* between West and East, North and South. Here, the focus is on urban landscapes as cross-cultural phenomena and under reciprocal transcontinental influences; 2. As a *scientific metabolism* between meteorology and the building sector; this is applied research in architecture and urban design; and 3. As a *political–regulatory metabolism* between law and architecture. In the context of the latter, the approaches strive for a regulation of urban climates; they aim at a new kind of thermal governance that exceeds (if not replaces) the concept of climate control. For the kind of architectural theory of the urban climate aimed at here, the inclusion of these three lines of thought plays an important role, in which they are brought together as urban studies of man-made climate.

1
Urban climates from the perspective of the architect.

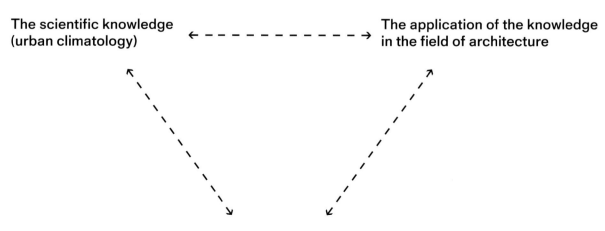

European Developments before 1945

1 Thermal Geographies of the European City

Urban Hygiene and the Heat Economy

The urbanization that emerged in parallel with the industrialization of Europe, was characterized by an ambivalence that encompassed both control and chaos, social integration and the dissolution of relations. The "factory city of the 19th century"[1] developed from the dialectic of new workplaces and new settlements; one entailed the other.[2] The factory city formed "a spectrum that ranged architecturally from sprawling slums to the hallmarks of industrial prosperity, the new train stations and crystal palaces".[3]

From the early 19th century onwards, "industrialisation, urbanisation and eruptive modernisation"[4] had an enormous impact on the evolution of European cities, not just in terms of their urban structures but also their urban microclimates. Industrialization and urbanization altered both the indoor and outdoor atmospheres found in the cities. The anonymous architecture of industrialization created a new thermal geography of the city, with innumerable urban climatic focal points and consequent attempts at conscious human adaptation. Air pollution from the new factories, overheating in homes and unhealthy thermal conditions in workplaces provoked thorough investigations by architects, scientists and social reformers.[5] Decades before the advent of the comfort paradigm, large cities and new industrial landscapes stood for largely uncontrolled thermal conditions and experiences—a new diversity beyond moral judgment.

1.1 Schinkel in England

For the early historiographers of modern architecture, above all Sigfried Giedion, the anonymous functional buildings of industrialization in Europe were harbingers of a dawning new age of architecture. According to the argumentation of this historiography, the functionality and economy of industrial production promoted a new kind of structurally and constructively conceived architecture, which decisively contributed to the overcoming of the historicizing schools of style with their tendency towards decoration.[6] Within the framework of this historical argumentation, however, the microclimatic aspects that went hand in hand with industrial production and its architecture were omitted.

Karl Friedrich Schinkel's diary documenting the time he spent in Great Britain in 1826 not only includes a comprehensive analysis of the new building materials iron and glass; it also provides something generally overlooked—many keen observations concerning the associated microclimates both outside and inside of Britain's novel industrial buildings. The main fields of application of the building material iron, meticulously described by Schinkel, was in infrastructure (such as bridges) but also factory and warehouse buildings. These large buildings that characterized the new industrial landscapes also produced specific microclimates. Schinkel's journal registered the atmospheres of the emerging industrial landscapes and the associated capacity to influence them via building envelopes and the nascent technologies of building services. Schinkel's journal reveals how deeply he, as an architect, was impressed by new thermal conditions he experienced, and it provides an account of the new phenomena of urban climates and their associated social conditions.

As the art historian Gottfried Riemann emphasized, Schinkel's observations formed the "decisive starting point for the new tendencies that his late work exhibits"—first and foremost the Bauakademie in Berlin.[7] This late work, sublimating his observations of the new industrial landscapes of Great Britain, anticipated and pioneered the modern architectural culture of the 20th century.

1.1.1 Greenhouse Effects

Schinkel registered the enormous disparities of wealth that were also manifested in different urban microclimatic conditions. The extreme poverty of the new industrial workforce became visible in the hygienic conditions of their neighborhoods and dwellings. Whether in London, Birmingham, Edinburgh or Manchester, Schinkel made note of the thermal differences that would later so thoroughly preoccupy the European architects and urban designers of the 20th century. In one of his journal's entries, he describes walking to the Natural History Museum in Edinburgh: "We walked down one of the old streets to the Museum: no greater contrast can be imagined than between the filthy cramped conditions of the coarse black dwellings and the poverty of their occupants,

→ fig. 2

2
Karl Friedrich Schinkel, view of Edinburgh (UK), 1826.

3
Karl Friedrich Schinkel, view of Bath (UK), 1826.

4
William Atkinson, sun box, 1912.

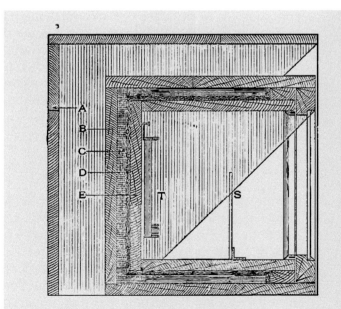

European Developments before 1945   18

5
Riding school, Brighton (UK). Schinkel's comment on June 10, 1826: "Too hot inside, really crazy: it was designed to be a greenhouse."

6
Joseph Paxton, Conservatory (84m long, 37m wide and 19m high) in Chatsworth (UK), 1840.

Thermal Geographies of the European City

and the magnificence, elegance and airiness of the new streets. Several fine wide streets have also been built through the old town, so that a visitor is usually unaware of such pockets of squalor."[8]

Technologies of Climate Control

As a guest of the British nobility and factory owners, Schinkel had the opportunity to become acquainted with various technical innovations in building services that concerned the microclimatic conditioning of the interiors of buildings. William Strutt, for example, owner of a cotton mill, is an example of such a technically developed imagination. As the operator of one of the largest spinning mills of his time, in 1806–1810 Strutt with the help of a local architect built a new type of hospital in Derby, which inspired Schinkel with its technical and architectural details.[9] In this hospital, the control of climate already has been put entirely at the service of a new kind of hygienic self-image. The comprehensive mechanical control of the interior microclimates and the mechanization of the washing of laundry single out this hospital as a significant pioneer of new climate-control technologies. In Schinkel's brief description, the program for the building services of the 20th century already appears; the building services, and not just the construction, are described here as a field of innovation in architecture: "Visited the famous Infirmary with Mr. Strutt, fine, pleasant building in every way. Magnificent staircase. The steps faced with lead plates. The famous hot-air heating, water-closet with shutters, movement of air in and out of the rooms, the stale air is drawn off by a rotating ventilator on the roof. Very practical cooking equipment. Magnificent baths, a whole room, the anteroom through a canvas curtain, warmed by air wafted in from the bath. The doors made of slate, so that the steam does no damage, everything thought out to the last detail. [...] Fairly large area in the Infirmary for drying clothes, steam washing-machines, hot and cold water is used (a continuously turning wheel with compartments.). Mangles, the washing is pressed after being placed in a square linen bag."[10]

In the case of the Lancaster School that Schinkel visited, again in Derby, a new kind of central heating, which circulated hot air in a closed system of pipes, was applied; Schinkel also encountered such closed systems in factories, for example to guide liquids in a system of tubes as in the case of a tannery. The uniformity of central heating was in clear contrast to the practice of heating zones of the room or parts of the body at that time. "Lancaster school, circulation heating, the floors on an inclined plane, lavatories visible from the teacher's chair through a glass door. Heating

with warm air to save wood, only moderately heated, but continuously. The flow of cold air always originates a long distance away underground from clean, healthy locations, gas light."[11]

It was not only in hospitals and schools but also in other institutions of bourgeois society that the new cultural technique of active air conditioning was used. The beginnings of central heating can be also found in courthouses, penal institutions and archives as well as the buildings of the new bourgeois public such as museums, theaters and cinemas, with heat generation outside the buildings. One advantage was the reduced risk of fire. These places were focal points of a new understanding of regulating the climate of urban societies.[12]

Heat Traps

Temporary buildings such as wide-span tents and the early glass architecture of southern England provided insights into novel thermal interior qualities. Schinkel visited, for instance, the round stable building in Brighton, which became later known as the "Dome". The building was built between 1804 and 1808 by the architect William Porden. It was one of the early glass-and-iron structures in England that Schinkel clearly recognized as a greenhouse: "The stables built around a large glass-domed building 85 ft in diameter. Each individual stable with 3–5 horses has its own ventilation. Too hot inside, really crazy: it was designed to be a greenhouse."[13] The description of this riding school made Schinkel aware of the thermal possibilities of such a building. In particular, however, its surprising function (as a riding school) prompted Schinkel to think beyond botanical uses that were common at the time. The size of the riding hall gave rise to the idea of an artificially created climate under a unifying cover (made of glass). The greenhouse effect represented, in a sense, a physical phenomenon with still open potential for social interpretation. Microclimatic experiences like those of the Dome have been conceived as experimental fields for new programs and functions that will change the character of architecture. The deliberate combination of solar heat and glass architecture equally founded and reinforced new architectural imaginations of the nobility of the rising bourgeoisie; imaginations that were based on the so-called "heat trap".[14]

As early as 1767, Horace de Saussure had simulated the greenhouse effect of glass buildings with the glass "hot box"[15] he had developed, thus initiating the systematic study of glass architecture in the field of building physics. In his empirical studies, the sun—in

→ figs. 4, 5, 6　　　　　European Developments before 1945　　　　**20**

combination with glass and the insulating layer—was recognized as a powerful source of heat. This could be desirable in winter, but could lead to overheating in summer. At the beginning of the 20th century, the American architect William Atkinson empirically deepened[16] this experimental approach in a specially manufactured arrangement, which he called "sun boxes", thus establishing a tradition that in the 20th century understood buildings as receivers and amplifiers of thermal conditions (see chapter 3). Atkinson summed up an insight that Schinkel might already have sensed during his visits to English greenhouses, that architecture itself can be seen as an experimental arrangement—a kind of *sun box*: "That the sun's rays are not of indifferent value in the heating of our houses in winter is shown by the last experiment (December 22), in which the air within the sun box reached a temperature of 115° F with the air outside at 25° F. Every dwelling may be converted into a sun box by properly insulating the outside walls."[17]

1.1.2 Black Fog

The new industrial landscapes were accompanied by new kinds of urban climatic phenomena. Schinkel made emblematic drawings of these that demonstrate his dual talent as an architect and a draftsman. The ability to penetrate industrial landscapes with technical expertise and aesthetic capacity gives these drawings a futurological dimension, in how they accurately anticipated the later atmospheres of European cities.

In his diary, Schinkel documented the "comprehensive change" from which the English countryside had suffered as a result of rapid industrialization, in which the "appearance of landscape and towns" underwent a profound transformation. "New industrial buildings, usually hastily and haphazardly erected, as in this case the ironworks with their smelting furnaces and workshops, form a strong contrast to the park-like landscape with a few villa buildings."[18] Schinkel captured one of these contrasting landscapes in his drawings of Dudley (*View of the industrial scene around Dudley*), which incidentally was also prominently depicted in a painting by William Turner in 1832 (*View of Dudley*). The dense atmosphere and peculiar colors in Turner's painting were due in equal measure to heavy air pollution and novel manufacturing techniques; the coal dust in the air made for colorful sunsets. In a diary entry, Schinkel speaks of the "overwhelming sight of thousands of smoking obe-

lisks. Coal, iron and lime are mostly brought up from the mines by winding engines."[19]

Buildings Without Architecture

In other architectural drawings, the novel industrial architecture of Dudley's industrial landscape was subjected to a drawing-based examination. One of the drawings shows the just completed Wednesbury Oaks ironworks near Dudley, depicting two parallel hall buildings with large chimneys as well as a prominent walled enclosure with arches. The smoke-billowing structures must have fascinated Schinkel in their articulation and sheer size, which he could only compare to major works of architecture built by the nobility. Schinkel acknowledges the funnels as new representative signs of his epoch. His perception of the chimneys as "smoking obelisks" connects the new man-made atmospheres to the picturesque taste of his time. The confrontation with a new type of architecture inspired the use of a metaphorical vocabulary. Referring simultaneously to the past and the future, Schinkel placed the new industrial architecture in a historical context, in which it was conceived as being part of ancient Egyptian, Roman and Classicist traditions of ruling. Further north, near Newcastle, in "a wider valley which contains as many potteries as Dudley has ironworks", Schinkel found "wonderfully Egyptian-oriental forms of the towns because of their factory buildings".[20] In Manchester, on the other hand, huge cotton mills are described in the context of social conditions and drawn as a kind of non-architecture due to the brute cubic appearance of the buildings. "Since the war 400 factories have been built in Lancastershire; one sees buildings standing where three years ago there were still fields, but these buildings appear as blackened with smoke as if they had been in use for a hundred years.—It makes a dreadful and dismal impression: monstrous shapeless buildings put up only by foreman without architecture, only the least that was necessary and out of red brick." Schinkel harshly judges these objects that so fascinated him as being "buildings without architecture".[21] But he would later take inspiration from the radicalism of the factory buildings he saw in Manchester and elsewhere in his magnum opus, the building for the Bauakademie in Berlin.

The smoke of factory chimneys is one of the recurring landscape-defining phenomena one finds in Schinkel's diary. More broadly, he saw the new signs of industrialization not only in the cities themselves but also in the open landscape. Urbanization was accompanied by a new kind of atmosphere, which gave the romantic view of the landscape a man-made

→ figs. 3, 7, 8, 9, 10   Thermal Geographies of the European City   **21**

touch. "Canal in Birmingham [...]. Pleasant location, in the distance you can see the smoke of ironworks, which stretch for miles."[22] Natural fog and man-made smoke were to be often found together and often mixed-up phenomena, which indicated the growing anthropogenic character of the climate of industrial landscapes.

London's Haze

When Schinkel visited England in 1826, its cities appeared as if shrouded in veils. London, capital of the world at that time, was subject to a man-made climate in a particularly dramatic way. Schinkel's notes on the city accordingly contain several remarks on the subject. The inversion of the weather caused by human activity led to severe air pollution.[23] At the beginning of June, Schinkel writes, "Although the weather was fine London was wrapped in fog and smoke; you could not see to the edge of the city, the towers were invisible in the haze."[24] Two days later he writes, "From everywhere one has a view over the Thames valley and London, which is however never visible owing to the smoke from the chimneys."[25]

Since the end of the 18th century, London had been infamous, not least among foreign visitors, for its "black rain", as the architectural historian Nikolaus Pevsner pointed out.[26] Schinkel anticipates Pevsner's observation in his diary by bringing *urbanization and climatization* to bear on their interactions. The atmospheric conditions in the outdoor spaces of the great cities become apparent as the result of industry and heating practices of the inhabitants. The atmospheres of London are *hybrids* or even *artifacts*, subject to human behavior and imagination. The industrial landscape of England literally enters the viewer's field of vision in Schinkel's descriptions. In London, Schinkel visited the Covent Garden Theatre where he was confronted with an "unbearable vapor and stench from the gas light".[27] The new urban microclimates had first and foremost an olfactory dimension, which was later also incorporated into literature by authors such as Emil Zola and Charles Dickens. In Dickens's story *Our Mutual Friend* (1864), "an atmosphere of mist, darkness, filth, and death spreads over the great city" of London.[28] An image of London emerges as a mixture of "amorphously growing urban masses, urban climatic disadvantages [and] the effects of urban working conditions on the inhabitants".[29]

In his essay *The Geography of Art,* Nikolaus Pevsner therefore questions the immutability of "climate" and, following on from this, the rigid cultural-theoretical argumentation that proclaims "the dependence of character and history on climate".[30] In times of industrialization the climate increasingly becomes a man-made phenomenon—the formula "black fog is moisture plus soot," launched by Pevsner, powerfully incorporates the anthropogenic character of this phenomenon, inversing the assumed deterministic relationship between mentality and climate. The climate was no longer a stable quantity. As Pevsner writes, "A moist climate may be the natural climate of England and as such be permanent. It will always be conducive to mists and fogs. [...] But black fog is moisture plus soot, and so what one complains of as climate is the combination of climate with such things as the exploitation of coal, a development of industry that calls for vast masses of coal, and, in the house, a system of heating evolved for wood fires and not yet universally adjusted to the use of coal. [...] Perhaps the early and ruthless development of mining and industry is English?"[31] The mixture of *fog* and *smoke* from the chimneys indicated a growing anthropogenization of the climate, which would still have the same problematic character in the middle of the 20th century.[32] In 1949, for instance, Charles Ernest Brooks addressed the necessary exchange between the meteorologist and the architect in order to reduce the environmental pollution in English cities. "Smoke is one of the most important climatological factors in this country. When you consider that nine-tenths of the dirt and dust and soot deposited in London comes from burning coal you realize what a very great problem that is. With a south-west wind, the smoke of London can be traced as far as Norwich. That is a problem which meteorologists can point to, but it is for the architects to solve it by getting rid of the smoke."[33]

The heavy air pollution[34] that industrialization created in European cities affected not just the lower classes but all strata of society, as Lewis Mumford pointed out in *The City in History* (1961). Accordingly, Gert Kähler speaks of the "megalopolis, which became the clearest sign of a dissolution of ties to the natural environment" and the "first large-scale 'destroyer of the environment'".[35] The technologies for burning; the shape and the sizes of the chimneys; and, above all, the siting of the factories remained of greatest concern throughout industrialization.[36]

→ fig. 11   European Developments before 1945

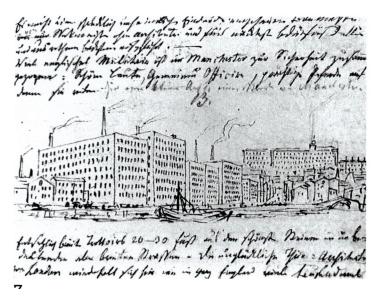

7
Karl Friedrich Schinkel, view of cotton mills with "smoking obelisks", Manchester (UK), 1826.

8
William Turner, view of Dudley, Worcestershire (UK), 1832.

9
Karl Friedrich Schinkel, view of the industrial scene around Dudley (UK), 1826.

10
Karl Friedrich Schinkel, Bauakademie (Building Academy), Berlin, 1832–36.

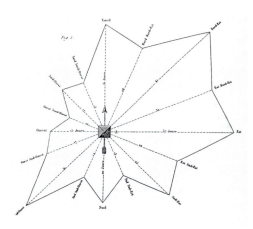

11
Jean-Pierre-Joseph d'Arcet: on the correlation of pollutant exposure and wind direction, 1843.

Thermal Geographies of the European City 23

1.2 Private Urban Development and the Heat Economy

The logic of manufacturing underwent a fundamental reorganization in the course of the Industrial Revolution. Home work was increasingly replaced by new factories, whereby two forms of development can be distinguished. On the one hand, workshops that were "still quite dispersedly embedded into neighborhood structures",[37] such as the Luisenstadt in Berlin (from the late 1840s), were located relatively close to the center of cities. There was a prevailing conviction that "small and medium-sized industries" should not have separate areas—i.e. that "special industrial and working-class neighborhoods [should be] avoided", which strongly exacerbated the problem of air pollution.[38] As late as 1927, Ludwig Hilberseimer stated, "Residential districts are interspersed with noisy and smoking factories."[39]

And on the other hand, new, "land-demanding large-scale enterprises"[40] emerged in the countryside and on the outskirts of the city, which depended on large numbers of both unskilled and skilled workers as well as transportation and energy infrastructures such as railroads and waterways. In Berlin, large-scale production plants were established along the "entire ¾ ring of workers' residential areas".[41] The shortest possible distance between the workplace and the home was of central importance for the working classes, due to the lack of means of transportation.[42]

With industrialization, the "heat economy" became a factor of both economic and hygienic dimensions.[43] On the one hand, the heat economy provided the energy resources for new industry and the housing sector; at the same time, however, as we have already seen, it caused the first serious man-made environmental problem in history—the result of air pollution caused by the insufficient combustion of heating fuels. It was not until the beginning of the 20th century that considerations about the efficient utilization of energy resources would transform into the discourse on energy efficiency. "Only the conditions of the war economy after 1914 as well as the coal emergency of the immediate post-war period created a broad awareness of the value of efficient coal utilization, so that heat management advanced from a marginal subject area to a key discipline within a very short time. [...] From a purely technical point of view, this development certainly accommodated the concerns of air pollution control. A movement dedicated to using fossil fuels as efficiently as possible was bound to be interested in reducing smoke and soot as the classic products of incomplete combustion."[44]

1.2.1 Factories: The Advantage of Location

Climate knowledge, as it matured in the context of the industrial production of a new a globalized economy, evolved as trade-off between inside and outside conditions. In this context, the aforementioned basic contradiction of economic rationality becomes apparent: it simultaneously promoted microclimatic conditions that were harmful to health and the planning of measures to overcome them. In many cases, the increase of industrial production was accompanied by both a neglect of urban climatic conditions and an ever more precise control of the indoor climate. Economic development also relied on the controllability of the indoor climate. In this sense, the factories of the 19th and early 20th centuries must be conceived in two ways: as part of an atmospheric plague of European cities on the one hand, and on the other as a scientific–technical field for innovation. In the *Dialectic of Enlightenment,* Horkheimer and Adorno speak of the "hygienic factory space" as a reform strategy of capitalism, which was based on a gradual reformation of living conditions—shown, for example, by the way in which the microclimates of factory spaces were dealt with.[45]

Outdoor and Indoor Climates

In addition to local coal deposits, humid climatic conditions were of decisive importance for the establishment of the cotton industry in areas of England from the beginning of the 19th century. The humidity in these regions had made possible a quality of textile products that was unparalleled anywhere in the world—something that even Schinkel noted.[46] The cotton industry required rather humid conditions with constant, moderate temperatures. Areas such as Lancashire, Yorkshire and the city of Liverpool particularly benefited from these specific climatic conditions.[47] The bioclimatic conditions favored the establishment of certain industries. Indoor conditions, which were highly dependent on external conditions, had to be considered with site-specific solutions. The location-dependency of industries meant that they were "often sited along riverbanks, lakes, or narrow valleys, and often equipped with thick exterior walls" to allow for uniform temperature and humidity in the work spaces.[48] This crucial location factor led to the migration of industry from Manchester to neighboring Oldham within a few decades, as Gerhard von Schulze-Gävernitz noted as early as 1892. "This humidity was later to make it possible to spin cotton to a fineness which, on the other hand, would be impossible to achieve elsewhere, or only at great extra cost. How much this climatic advantage comes into consid-

→ figs. 13, 15   European Developments before 1945   **24**

eration is shown by the fact that spinning increasingly was located in the shadow of those hills where the precipitation was strongest, so for instance, in Oldham rather than Manchester."[49]

This "natural climatic favor" discussed in the technical literature had a central influence on the choice of location for industries—and thus also for urbanization—in the 19th century. Their purpose was twofold: to provide "cost advantages for individual companies" and to protect "society from various kinds of damage".[50] Exemplary industries in which the advantage of location was taken into account were the textile industry (including yarn-making and bleaching processes); the food industry; and, later, the paper and film industries. In addition to the basic climatic advantages of a location, seasonal fluctuations also influenced the production and manufacturing processes. In the case of beer brewing, the tobacco industry and the confectionery industry, one can speak of truly *seasonal* work, as decisive steps in the production process were only carried out at specific times of the year. They might require entirely opposing conditions; very hot and humid in the case of the tobacco industry and cool and dry in the case of the confectionery industry. The importance of the season for certain technical processes in industrial production and their success was a huge consideration for certain industries until the middle of the 20th century.

The influence of climate on industrial production subsequently led to the study of microclimatic conditions in factories.[51] The "importance of climate for the industrial location", the "optimum climates of industrial production processes" and the "study of artificial climates and air-conditioning technology" were still closely linked.[52] In particular, the "air temperature, humidity, air movement and air purity"[53] formed key parameters that were incorporated into science. Proper climatic conditions in the factory and the warehouse were of great importance for industrial production in Europe and its colonies.[54] This involved the advantage of certain local microclimates for manufacturing processes of raw materials, semi-finished products and goods, as well as their persistence. In reference to Gabriel Guévrekian's "Batiments Industriels" of 1931, we can mention grain silos, sugar refineries, docks, dry fodder silos, cement silos, water storage facilities, wholesale market halls, benzene refineries and storage buildings for financial instruments (in paper form). Storage buildings in particular evolved at the intersection of microclimatic and structural requirements; in the center were the microclimatic strategies used to preserve things and to store goods. Storage buildings, where raw materials and semi-finished products were temporarily stored or kept for a longer period of time have to be distinguished from the production facilities themselves.

In contrast to the planning of human dwellings, as Hilberseimer notes, in the case of industrial factory farming, for example, careful attention was paid to the correct solar orientation (insolation) of the buildings. "It is only in our times that builders have flagrantly disregarded it in their construction for human dwelling, though, oddly enough, they seem to remember well the value of insolation when they build shelter for domestic animals. Poultry breeders, for instance, almost invariably take care to locate chicken houses toward the sun. It is good business for them to do so and they know it."[55]

Goods versus Workers

Goods and workers in factories have different needs. Even in the middle of the 20th century, the so-called field of *industrial climatology* still had to organize the adequate mediation of indoor and outdoor conditions. "Climates affects industry in two ways: by raising or lowering the general efficiency of the workers, and by the effect on the actual processes of manufacture."[56] The productive power of the workforce also appears as a function of the microclimates inside the factories. Charles Brooks points to the textile industry in England, where the basic microclimatic contradiction between labor and goods was particularly visible: "the high humidity required for good products is very unfavorable to the welfare of the workers. Strict control is necessary and the conditioned air must be evenly distributed through the factory by proper circulation."[57] Just as the raw materials required an adequate microclimatic environment, the workers—especially in the hot European colonies of the tropics—demanded an adequate physical rhythm. The imperative for comfort still appeared to be characterized by the real danger of heat stroke—a kind of total bodily exhaustion. The impairment of the "efficiency" of the workers was in contradiction with the things to be worked: "The theoretically best conditions for the product may not be conductive to sustained output by the workers. In such cases a compromise is necessary. Generally speaking, a suitable outdoor climate and surroundings, with their good effect on the energy and efficiency of the workers, are more important than the requirements of the process rooms, since conditions in the latter can, within limits, be adjusted by air conditioning."[58]

→ fig. 14   Thermal Geographies of the European City

12
Cités Ouvrières de Mulhouse, around 1855.

13
Siemensstadt, Berlin, 1907.

14
Adolph Menzel, *Eisenwalzwerk* (Iron Rolling Mill), 1875.

European Developments before 1945

15
Siemensstadt, Berlin, 1931.

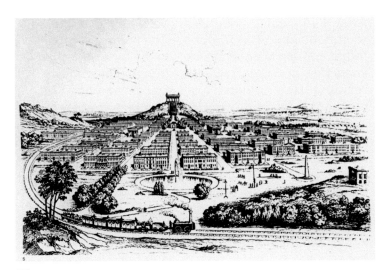

16
Bernhard Christoph Faust, view of the Sun City, 1829.

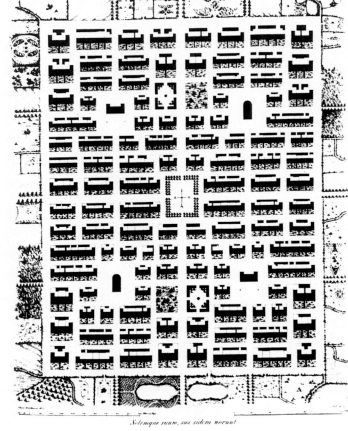

17
Bernhard Christoph Faust, layout of the Sun City, 1829.

Thermal Geographies of the European City

The microclimates of factories, which emerge as important components of urban development in European cities, provide a novel field of activity for engineers and physiologists. Statistical data, such as those collected on occupational injuries and mortality rates in the European and American industrial workforces, bear eloquent witness to microclimatic conditions in factories. Extreme forms of heat, prolonged humidity and rapid temperature changes were the cause of severe illness and premature death. The factories were not only built with new construction methods and building materials such as iron and reinforced concrete; they were also equipped with new microclimatic conditions, in which people often had to spend hours on end. The engineer André Missenard reports, "The disastrous consequences of the artificial, too warm and humid climates have since been confirmed by the English investigations on mortality to be due to simple pneumonia and tuberculosis among the weavers and the cotton spinners. Thus, mortality due to diseases of the respiratory tract has been compared among workers employed in the warm and humid environments required by cotton work and among woolen mill workers working under normal conditions. The mortality due to bronchitis is two or three times higher among cotton workers than among wool workers. The baleful influence of humidity is confirmed by the fact that in the same factory there is three times as high a mortality from bronchitis and pneumonia among the workers who work in the damp halls as among those who spend their days in the dry halls."[59] This problem of proper microclimate anticipated an issue that would later be given the name of "Sick Building Syndrome" in air-conditioned office buildings. This phenomenon, from the second half of the 20th century, was also an expression of a dilemma or "closure paradox". The growing need for airtight building envelopes—in the name of high employee productivity as well as energy efficiency—led to a gradual reduction of air exchange rates, resulting in health problems of the occupants of these buildings.[60]

Air-improvement Systems

The industrial production that accompanied the rapid industrialization of Europe led to the formation of two perspectives on the control of climates: one devoted to labor and the other devoted to things (raw materials and goods). Factories formed the venues of a new kind of climatic knowledge in which trade-offs were made between people and things. Which of the two aspects was prioritized depended on the priorities of the manufacturers. However, industrial production made necessary new thermal regimes, in which eco-nomic, biological and social factors were combined. Until the middle of the 20th century, natural ventilation and mechanical air conditioning were equally important means of combating sometimes extreme microclimatic conditions: "In offices and factories the effect of high temperature and humidity is accentuated by the heat and moisture produced by the workers themselves. Ventilation is only a partial remedy, for under extreme conditions the current of air required to give sufficient cooling would be so strong that it could raise dust and cause other inconveniences. The cooling power of the air increases only as the square root of the air speed, whereas the lifting of dust increases as the square of the speed. Air currents exceeding 500 feet per minute are impracticable for this reason. Under such conditions the only remedy is air conditioning."[61]

Since the beginning of the 20th century, different technical apparatuses and procedures that influenced microclimates in factories and warehouses began to appear, which variously comprised "heating systems, air improvement systems, ventilation, deaeration, dedusting and demisting".[62] In practice, such mechanical air-conditioning systems either compensated retroactively for buildings that were designed to be climatically inadequate or they provided a degree of control over indoor conditions that would be impossible to achieve by passive architectural means alone. "Engineering seeks to put the knowledge gained from climatology and medicine into practice. After all, we cannot demolish entire blocks of houses, factory buildings, etc. because the 'microclimate' found in them is unfavorable. On the other hand, we all know how uncomfortable we often feel in smoky restaurants or cramped work spaces, and how unpleasant it can be to stay in a cinema or theater in midsummer, despite all the ventilation. This is where technology steps in: it creates an artificial climate in place of the natural one."[63]

Industrialization, as a driver of urban development, contributed to the increasingly artificial character of urban indoor climates against the background of existing climatic conditions. The growing technical influence on the control of indoor conditions led to natural climatic advantage becoming less important. Locational advantage was overridden and eclipsed by other economic considerations. "Modern manufacturing processes are so complex that from start to finish a wide range of conditions is generally called for, and these can only be provided artificially."[64] The thermal continuum between inside and outside appeared increasingly as the enemy of productivity and economic efficiency. Colonial production geared to the global market demanded adequate microclimatic environments that were predictable and stable. Brooks speaks of the "stability of climate for special industries".[65]

Sun City and the Reform of Living

At the same time as Schinkel was astutely registering the effects of new construction technologies on architecture in 1826, the first thoughts were being made about a type of urban planning in which social reform, hygienic and thermal considerations intersected. The sun, with its symbolic significance for an enlightened understanding of health, was at the center of this new type of urban planning. The idea of a *Sun City* developed by the German physician Bernhard Christoph Faust in the 1820s represents a particularly comprehensive project in this regard. Faust operated and made proposals in the context of his immediate social environment—the city of Bückeburg (in today's Lower Saxony). His ideal of the *Sun City* must be understood in the context of considerations at the time regarding the reform of housing for the working class, and it was thanks to Ludwig Hilberseimer's coverage of Faust's ideas in *The New City* that his *Sonnenbaulehre* ("solar building doctrine") was not forgotten.[66]

The reform worker settlements built by European companies during the 19th century created orderly housing for their employees, but brought factories and housing into close proximity without further climatological considerations. Views of the now-famous Cité Ouvrière in Mulhouse, built in several stages between 1853 and 1897, is typical for a residential area that, while strikingly orderly, was laid out in close proximity to a factory complex. In addressing the issue of housing, the improvement of the micro-spatial living space of the workforce was a central concern, but little attention was paid to urban climatological aspects. Burtscheid, a neighboring town of Aachen, was, in an exemplary way, confronted by the "nexus of climatic factors and urban hygiene,"[67] as the historian Katja Esser attests. "As archival sources and historical maps document, the factory owners' residences, their cloth mills, and dye works with steam engines and the overcrowded working class quarters with poor sanitation, low air quality and characteristic diseases (cholera, abdominal typhus, variola) were situated closely in the city center throughout the 19th century." While the population between 1794 and 1900 "increased sixfold", "the construction of new factories was not forbidden until 1900", although there were "numerous complaints by residents about noxious effects caused by industrial plants and their steam engines".[68] The liberal, pro-business legislation of European countries in effect until the beginning of the 20th century prevented restrictions being made on the burning of coal and the siting of factories. It was only in the course of the 20th century that the urban-planning tool of zoning was to become an essential means of curbing anthropogenic air pollution. "The royal Prussian authorities had to deal in laws and regulation with the so-called 'smoke and soot plague' that endangered public health."[69]

The basic concern of Faust's ideal city was to orient life as comprehensively as possible towards the sun; in this respect, it can be seen as an early attempt at regulation under urban climatic auspices. Faust's "solar building doctrine" anticipated the methodological impetus that would also animate the heliomorphic designs of architects in the 20th century (see chapter 3). Already in his "Health Catechism", health was associated with sufficient sunlight in the rooms of a dwelling—"in the sense of the ancient concept of dietetics, rules for a healthy life were established".[70] Morality and empirical observation went hand in hand for Faust, and the dwelling appears as the pivotal site of physical and moral analyses. Faust, as a practicing physician, advocated a holistic view of life and was also an imaginative inventor who illustrated his ideas with his own drawings and sketches.[71]

The "solar building doctrine" was published "for the first time as an annotated text in short form" (eight pages) in 1829 in the *Monatsblatt für Bauwesen und Landesverschönerung* under the title "Andeutungen über das Bauen der Häuser und Städte zur Sonne" ("Suggestions for building houses and cities according to the sun").[72] He suggested that the long sides of buildings should be oriented "to the south" because, "uniquely and only those living rooms and sleeping chambers which have the sun half of the day can be perfectly healthy".[73] All living, working and children's rooms were to be arranged "on the south side, the north side was intended solely for utility rooms such as kitchen, pantries, workshops and servants' quarters". Faust made suggestions for "cross-ventilation via doors and windows, and for regulating light and air supply by means of shutters, curtains and screens. The house was to stand on a plinth to prevent wall rot, sponge formation, and wood rot."[74] Faust even experimented with mud building techniques, in order to promote his own vision of the sun city.

Two main streets divided the urban layout into two sections of the same size, with a large public square in the middle, the so-called "Sonnenplatz". While each of the four districts had its own park, the entire city appears embedded in a larger green landscape.

→ figs. 12, 16, 17   Thermal Geographies of the European City   **29**

368 buildings were envisaged, divided into 2–7 rows up to 150 meters long. Remarkably, Faust proposed a variety of building types (with flat roofs), taking social differences into account. "An equality of living quarters was not intended."[75] Following his own hygienic premises, he insisted on sufficient distance between the buildings in order to provide access to light, air and sun. The streets were to be broad and well supported. Each lot had a lawn on the south side, which he considered an extension of the habitable area, equipped with "planting, furnishing, irrigation and shading by linen tents".[76]

Faust's construction consultant, Gustav Vorherr, started to implement the ideas of the "solar building doctrine" from 1817 onwards at the so-called "Baugewerkschule" in Munich, which can be considered an early promoter of heliomorphic design practices in Germany and beyond. The teaching of sun-oriented building quickly developed into "a peculiarity" of the school. "In plans and construction drawings, all designs for residential buildings had to be oriented to the south. In 1824, for example, the master class drew a plan for the 'Sonnenvorstadt' in Munich. Since the school took in about 150 students a year from Germany and abroad, several thousand students were taught the ideas of 'solar building doctrine' during the 25 years of Vorherrs's activity here."[77] This development shaped an entire interdisciplinary and evidence-based line of education for architects and building professionals in Germany, which was to reach a new culmination with the Bauhaus. It is obvious that the modernist slogan of "air, light and sun" inherited or was connected to the vocabulary of Faust; "free light, free air, free life from pole to pole", was Faust's central motto. All theories of solar urban design, with their tendency towards symbolic culmination, are also implicit social theories, transcending the limited sphere of action of a single housing project. Heliomorphic design is invariably accompanied by a kind of morality, be it explicit or implicit. In the sun, the dream of the perfectly organized city found one of its strongest symbols of regulation. Geometrized light symbolism was paired in these urban visions with the beginnings of an urban climatologically thought-out city, in which equal consideration was given to interior and exterior spaces.

1.2.2 Tenement Houses: Poor Peoples' Air

Economic development in the second half of the 19th century led to the rapid expansion of European cities and the emergence of new types of dwellings on the outskirts of cities: speculative tenements. With them came a growing population of tenants, which in Berlin represented about 95 percent of the whole population by 1875.[78] According to Hilberseimer, the tenement house is "the type of urban, especially metropolitan dwelling house".[79] Accordingly, it has received a great deal of attention from hygienists, climatologists and architects.

With the rapid growth of cities, the regulation of climatic conditions inside urban dwellings also became an increasingly urgent problem, with biopolitical implications. Thus, there was talk of "harmful influences emanating from dwellings."[80] Lack of warmth (in winter) and overheating (in summer) turned out to be complex problems related to outdoor environmental conditions, construction methods, materials used, the arrangement of buildings and blocks, the still rudimentary building services of the time (stoves, toilets etc.), and the occupancy densities—in short, all those multi-scalar problems that urban climatology and modern architecture would later deal with. A comprehensive answer for "the use of the dwelling—residential density, division and treatment of the rooms"[81]— was only found gradually over time. Accordingly, important preliminary considerations for *an architectural knowledge of the urban climate* can be found in the writings and visual documents[82] on high-density tenement blocks. The severe thermal conditions in the apartments promoted a discourse on reformation that would lead to new floor plans and new hygienic facilities.

Epidemics in the City

Infectious diseases such as plague, cholera, typhoid, tuberculosis and malaria—only scientifically understood in the second half of the 19th century—were recurrent phenomena in European cities.[83] As can be read in the *Report on the Mortality of Cholera in England* from 1852, the causes of how diseases spread were by no means determined at the time. "The form of the cholera curve for all England is very remarkable", it reads; "The successive terraces and pinnacles of the Plate resemble sections of the primitive mountain formations, [...] or recall the lines of a strange Gothic architecture."[84] The metaphorical comparison between the shape of the curve and architectural style is some indication of how great the struggle was for a correct interpretation of the

European Developments before 1945

30

epidemiological data. It was only in the second half of the 19th century that an understanding of the connection between urban development, hygiene and epidemiology emerged through a scientific consideration of the city. For example, high outside temperatures in combination with poor hygienic conditions in cities were increasingly recognized as drivers of various types of illness. Investigations into the relationship between sewerage and drinking-water quality or between swamplands and neighborhood structures were equal drivers for innovations in medicine and urban planning.

The central subject of the studies was the city as a slum,[85] which together with considerations of hygiene represented, as the architectural historian Julius Posener termed it, a kind of "slumology".[86] At the *First International Hygiene Exhibition* in Dresden in 1911—pioneering for the discourse of modern urban planning—approaches were presented from an interdisciplinary perspective that aimed at the reformation of prevailing ways of living.[87] The exhibition included sections on infectious and tropical diseases, medical care and rescue, settlements and housing, professions and work, food and stimulants, games and sports, as well as clothing and personal care. It served to make a broader public aware of the pervasive relevance of "hygiene" as a domain of knowledge.

An important source for the assessment of the atmospheric conditions of large cities of the period is doctors' accounts "describing climate, topography, housing and working conditions."[88] The German physician and hygienist Carl Flügge, for instance, extensively reported on the hygienic conditions in the dwellings of German cities around 1900. He recognized the fundamental problem of the slum as the uncontrolled continuity between inside and outside: "We must come to the conclusion from these surveys that in the large German cities there is an exceedingly tight crowding of houses into the available space. Thus, masses of stone of immense volume are created, which in midsummer store the heat of the day, and in winter collect the smoke from thousands of meals between them; the traffic of the countless people flooding out of and into the houses takes place in the streets using the most varied means of transport, which bring noise, odors and dangers with them; in the courtyards, crafts and industry are established and fill them with disturbing noises and nasty fumes. The air in streets and courtyards has a breathless effect on people not accustomed to it, the restlessness of traffic and noise are irritating to the nerves. Only rarely do trees or lawns interrupt the sea of buildings; and it is only by negotiating several flights of stairs and long walks

that many families manage to get out into the open and enjoy nature. Within the individual buildings, numerous households are crowded together, coming into constant contact through stairwells, corridors, and lavatories. The individual cannot live in isolation; he is dependent on the customs and manners of the neighbor, he has difficulty in preserving his own character and is constantly exposed to collisions and temptations. An attachment to the home cannot be formed."[89]

At the beginning of the Industrial Revolution, there was still a close spatial connection between living and working. Many craftsmen still had their workroom in the immediate vicinity of their dwelling (as in the Middle Ages), or even in the same room. The consequence of this was the so-called "poor people's air" of home work.[90] While in summer, due to the lack of cross-ventilation, overheating was a recurring problem in the dwellings, in winter it was the absence of heating infrastructure and the affordability of the means of heating (coal, firewood) that were especially problematic. The hygienic reformers understood such "unhealthy" living conditions as being part of the so-called "pauperism",[91] which in Germany was to lead to the "law on the right to air".[92] These conditions were still to be found even at the beginning of the 20th century, for example by Carl Kassner: "Poorer people, who are already limited in living space, often have a real fear of fresh air in the rooms, and therefore the notorious 'poor people's air' develops there due to the exhalations and vapors of people living close together and due to the often damp walls. In addition, many adults are forced by their occupation (tailors, cobblers, washerwomen, etc.) to stay in it continuously, which further worsens the air."[93]

The Nature of High-Density Construction

The different microclimatic conditions of the housing typologies of European cities reflected different social conditions. In Vienna, for example, the blocks of houses in the "old town" with its "aristocratic and patrician palaces", the "new parts of the big city" in the area of the Ringstrasse and the "proletarian areas" had very different microclimates. The "possibility of cross-ventilation" was not given in the "rented barracks of the proletariat", while the "larger bourgeois apartments" had "rooms with different orientation, i.e. warmer and cooler".[94] "Due to differently intense effects of sun, humidity and wind, different interior climates develop initially depending on the way the dwellings are used, according to the floor and orientation."[95] In the microclimatic descriptions, the ideal floor appears to be located between great humidity

→ figs. 18, 19, 59   Thermal Geographies of the European City   31

18
Berlin-Charlottenburg: factories are mixed with housing estates
Source: Heiligenthal 1926.

19
Berlin housing conditions in the days of the German Empire.

European Developments before 1945   32

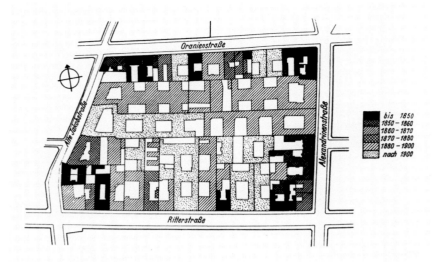

20
The intensification of land use of a block in Berlin between 1895 and 1914.

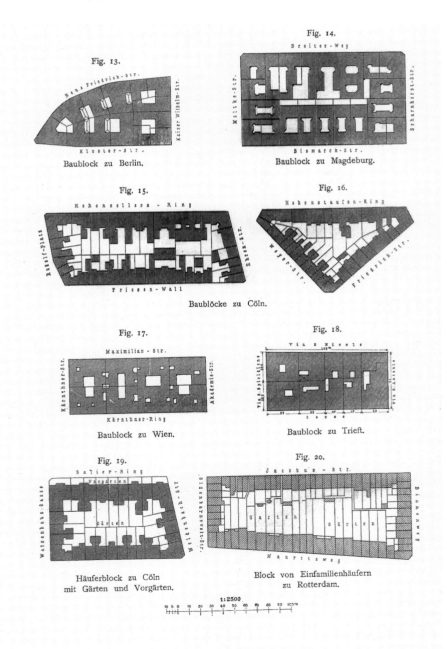

21
The urban block as the basic "climate unit" of the city.
Source: Stüben 1907.

Thermal Geographies of the European City

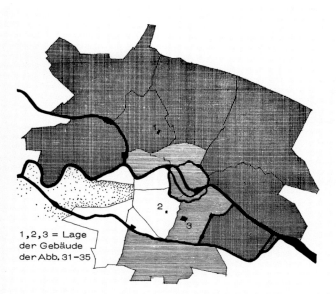

22
Housing conditions in the various districts of Berlin. From top to bottom: inhabitants per heatable room; inhabitants per room overall; proportion of inhabitants in dwellings with less than 1 heatable room (in %); proportion of servants in the city population (in %).

Im Stadtteildurchschnitt – Stadtteilbezeichnungen siehe Abb.23 – wurden im Dezember 1875 gezählt:

1,0-1,8	1,9-2,3	2,5-3,3	Einwohner je heizbares Zimmer
0,9-1,5	1,6-1,9	2,0-2,6	Einwohner je Zimmer überhaupt
15 - 32	33 - 47	50 - 74	Anteil der Einwohner in Whngen mit weniger als 1 heizbaren Zi. in %
4,8-2,6	2,4-2,0	1,9-1,3	heizbare Zimmer je Wohnung
26 - 12	11 - 7	5,6-2,7	Anteil der Dienstboten an der Stadtteilbevölkerung in %

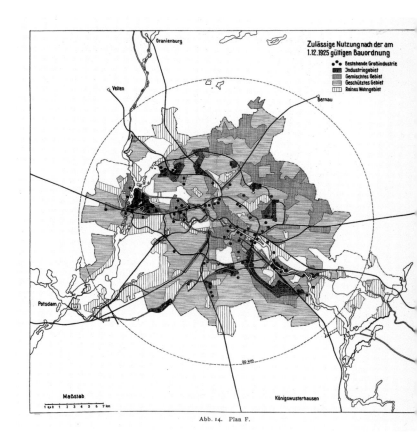

23
Permitted use according to the building code in effect on 1/12/1925. Hatchings (from top to bottom): existing large-scale industry; industrial area; mixed area; protected area; pure residential area.

European Developments before 1945

(in the basement and first floor) and risk of overheating or great cold (in the upper stories). Furthermore, "dust" (pollution) and "excessive heating in summer" are considered the main causes of health disadvantages of the urban climate. "The innumerable sources of heat, while reducing the consumption of heating material in winter, lead to higher degrees of heat in summer than in the countryside, which, moreover, are more difficult to bear because the movement of air is diminished. The nature of urban construction results in a great residential and settlement density, which itself becomes a damaging, and possibly sickening factor."[96]

Speculation in the housing market led in all European countries to the greatest possible building on available plots of land, which always included only as much unbuilt space as was required by law (access for fire trucks formed a key parameter for this). The basis of Berlin's tenements was provided by the 1853 building code, which permitted the building up of the entire plot (with only the 5.3 x 5.3 m² courtyards as exceptions), and the 1862 *Hobrecht Plan*, which combined deep building blocks with wide streets.[97] The private urban development of Berlin, for example, established a road network of 20–30 m "to the outermost periphery", which fueled land speculation and favored the creation of the largest possible building blocks on plots of land 70–80 m deep and with the highest possible utilization.[98] This resulted in blocks of buildings with high front buildings and side and rear buildings of almost the same height. In cities such as Berlin, Vienna and Trieste, blocks were built upon, leaving only "7 to 8 % [as] open space."

Courtyards formed a basic thermal infrastructure for the city; often reduced to non-accessible light wells. "The windows open on to canal-like courtyards",[99] say Brezina and Schmidt, who refer to layouts in Joseph Stübben's standard work *Der Städtebau*.[100] These analyses show how much, at the time, the interior was still considered as being dependent on the exterior and the surrounding built structure. The climatic interdependence within the block made it difficult, if not impossible, for building owners to find individual solutions, as the two authors point out. "The decisive factor for the internal climate that can develop in the buildings of the block, depending on the number of stories and the density of the development, is, apart from the width of the surrounding streets and the height of the opposite houses, the division into individual plots, their width and especially their depth."[101] The enormous densification that took place went hand in hand with the absence of publicly accessible green outdoor spaces. This led to a culture of

staying indoors, which was strongly criticized by physicians. Berlin tenements, equipped with three courtyards, are described by Julius Posener as "not intimate, but stuffy, smelly, oppressive".[102]

Winter: "The Heatable Room"

The central category used in the transition from the 19th to the 20th century to describe the quality of housing in the lower income classes was the *number of people per heatable room*. In the overcrowded apartments of the city of Berlin around 1905, for example, an average of "4-13 people" were crowded into "each heatable room".[103] Where only individual rooms could be actively heated, climatically supplementary measures had to be considered for the other rooms. This explains why architects made such concerted attempts to provide improved thermal conditions in winter by means of greater sunlight in the apartments. Flügge recommended the implementation of "well-sheathed and easily regulated iron stoves" as the primary source of heating in the dwellings of the lower classes.[104]

"Two forms of heating" can be distinguished for buildings of the period: individual room heating and central heating.[105] Beginning in the 1880s, the cast-iron radiator, "as we know it today", was developed in the United States; this represented a radiator made of cast-iron elements and connected by nipples. However, while centrally integrated steam-heating systems gained relevance "very quickly and early" (employing waste heat) in commercial and industrial buildings, central heating was delayed in the private sector—especially in the residential sector.[106] Because the boiler and system were designed for the entire building but only individual rooms in the apartments were heated, central heating in the residential sector remained uneconomical for a long time (due to the partial loading). This explains the dominance of individually heated rooms in European housing. "The last major turn toward almost exclusive central heat supply did not occur until the beginning of the 2nd half [of the 20th] century."[107]

Accordingly, collective approaches to thermal regulation of the built environment remained critical. In the case of cities, large wind loads necessitated greater sealing of the windows on the upper floors. "Since, as I said, the wind in cities increases from the bottom to the top, the wind bracing of these giant buildings must be very carefully calculated and designed. Moreover, the windows on the upper floors must be much more carefully sealed than those below, otherwise the wind will work too strongly against the heating and chase rain through."[108] Solar heat gains and reciprocal insulation of rooms by other rooms and buildings were aspects that received close attention from climatologists and hygienists. Flügge accordingly questioned the detached single house as a reasonable urban model, since its heating costs were considerably higher.[109]

Summer: "The Urban Sea of Houses"

Excessive solar radiation (overheating) into urban dwellings was seen as highly related to the spread of infectious diseases. Accordingly, solar radiation and with it "the midsummer climate of the apartments"[110] were seen as problems to be solved.[111] To some extent, city dwellers found their own strategies to counteract overheated apartments. Flügge mentions the "balcony-like projections on the houses" of American cities that allowed the city dweller to get access to fresh air and "to keep the upper body in free air as much as possible".[112] At the same time, Flügge recommends "externally mounted awnings, blinds, etc." and the use of "gas stoves" to "counteract excessive summer temperatures".[113]

However, the typological and constructive design of the buildings was recognized as having a much greater influence on interior thermal conditions in summer. Different construction methods were used for the multi-story tenements (massive construction) and for one- or two-story "small houses [...] for the less well-off population" (filigree construction).[114] "For the disastrous high temperatures in the apartments, the solar radiation of the building's walls and roof is to blame. The air temperature inside the building depends on the temperature of the walls, which are immense reservoirs of heat and can store about 2000 times more heat than the same volume of air. On the outer surface of the walls, the temperature reaches 40-50° on hot summer days, varying according to the exposure to the sky, the angle of incidence of the rays, the duration of irradiation, the color of the wall, and so on. The absorbed heat is then very gradually transferred through the walls to their inner surface, depending on their thickness, in 4-8 hours".[115]

Flügge saw that the thermal inertia and the insulating effect of the walls becomes ineffective in high summer, as massive buildings with walls "38–62 cm thick" have a large storage capacity and react slowly to temperature fluctuations, but are less sensitive to the cooling effect of winds. During hot periods, the outside temperatures are soon linearly reproduced inside, and even "opening the windows in these periods by day only brings air of much higher temperature into the room, and even at night the air heat inside the urban sea of houses does not tend to fall below the temperature of the closed room".[116] Filigree constructions "12½–25 cm thick," on the other hand, have a smaller thermal resistance, and they can be cooled down again more quickly, as Flügge notes.[117] Cross-ventilation is only useful as an uninterrupted flow of air that must be experienced by the bodies of the occupants themselves. "Only continuous ventilation can therefore have a cooling effect on the apartment itself", but "once the perceptible air movement has ceased, the air in the interior has become warmer than before".[118]

The aporia of providing thermal relief during heat waves is clearly stated by Flügge. Shading or ventilation form two basic and often conflicting strategies for cooling, which should influence the design blocks of buildings and their individual dwellings. In particular, tenement apartments, which faced narrow courtyards, provided little cross-ventilation. Flügge compares free-standing buildings with "buildings situated in rows" or in "closed building blocks" and emphasizes the different relevance of winds and shadows in each case.[119] He refers to "southern regions" to support the need for a differentiated approach: "Accordingly, in southern regions, one tends to try to keep houses cool merely by protecting them against insolation, but not by exposing them to the wind as much as possible."[120] Free-standing buildings are far more exposed to the sun and are correspondingly dependent on the cooling effect of the winds.

1.2.3 Public Arcades: Thermal Gradualism

Writers such as Walter Benjamin and Siegfried Kracauer provided evidence of the new thermal geographies of the modern metropolis in the form of atmospheric descriptions. Their writings reveal the peculiarities of urban atmospheres in and among the new metropolitan architecture. In addition to factories and apartments, it was the public spaces of the emerging bourgeois society that were characterized by new microclimates.

In Kurt Pinthus's anthology *Menschendämmerung* ("Dawn of Man") of 1919, Expressionist poets seismographically examine the metropolis of Berlin in light of its new urban atmospheres. They provide "evidence of the dominance of negative notions of the city", characterized by "smoke-blackened houses, factory chimneys, impersonal crowds of people", "human expressions such as sweating, groaning, crying out", and "swamp, mud, scum, fornication".[121] The new multi-sensory atmospheres of the metropolis appear in futuristic and Expressionist paintings such as Umberto Boccioni's *La città che sale* (1910) or Otto Möller's *Strassenlärm* (1920). These representations show the sensorial complexity that characterizes the large modern city. The different atmospheres—from smoky coffeehouses to the stuffy arcade—made the city a focal point for new kinds of thermal and olfactory experiences; air pollution, stench and heat went hand in hand.

Climate-based Public Spaces

Not unlike the cathedral building of the Middle Ages, 19th-century arcades represented a pan-European urban phenomenon. Prominent arcades were built not only in Paris but also in Berlin, London, Brussels, Budapest, Vienna, Moscow, Milan and many smaller cities such as Aarhus, Bath, Genoa and Leipzig.[122] Like the tenement, the arcade is "an invention of private building speculation",[123] which appeared after the French Revolution and gradually disappeared after the First World War.[124] Arcades formed glass-roofed corridors through the masses of buildings in which the products of industrial luxury were available to purchase; they were exhibition halls of the bourgeois material culture, as Walter Benjamin emphasized in his fragmentary notes in the *Passagenwerk* ("Arcades Project"). The sensations of prostitution and the colonially influenced world of goods were main characteristics, as were the thermal conditions that could be found inside these spaces. The arcades were accompanied by a new kind of thermal geography, which turned the urban climate into an object of architecture and urban development. They did not simply separate themselves thermally from their surrounding urban climate; rather, they became artifacts themselves where heat could be trapped and made into a setting. The greenhouse effect, as with the aristocratic greenhouse, was the decisive physical and sensory principle underlying the arcade.

From the late 18th century on, arcades were "narrow private streets that open up the interior of large building blocks".[125] "The passage is always an independent building with its own plot."[126] They could only exist as long as and only where they were integrated into the overall urban fabric. As a shortcut between two busy streets, the arcade simultaneously represented a public space on private land and, with its glass roof, a "shelter from the weather".[127] "Intermediate environments [...] represent a unique environmental answer to the densification imperative. Set in the heart of the urban block, it has nevertheless maintained a reference to the street and increased the permeability of the urban fabric."[128] An arcade transforms a street into a public interior of the city, by introducing buffer zones into it; it is an attempt to rethink the public space of the city by thermal means. The arcade was an expression of a new "emanated urban public sphere", which contributed to the "reorganization of public space" by elevating the pedestrian (or *flâneur*) to being the central actor of the city.[129] "The arcade space differs from the street only in that it is covered with a glass roof, has symmetrical facades, and serves exclusively the pedestrian."[130]

Stealthiness

The arcade oscillates between street, building and colonnade, making the distinction between inside and outside a complex process to be deciphered. Accordingly, the striking ambiguity of arcades has to be emphasized, which explains their almost erotic charm. The greenhouses of the English aristocracy described by Schinkel, conceived as galleries and riding halls, were further conceived in the European arcades into clandestine interiors of the awakening bourgeoisie. In its transitory character, the arcade forms an exemplary in-between space that mediates between inside and outside, or puts the interior (secretly) to work as an experience in the public realm. Johann Geist has rightly noted, "the illusionistic element of the arcade is the passage space: imagined exterior as interior—facade drawn into the interior with exterior architecture".[131] In Berlin, a total of only three arcades were built (Kaisergallerie, Lindenpassage, Friedrichstrassenpassage), the first of them in 1869. But a preoccupation with the arcade as an architectural idea took place decades earlier. The impetus for this, as Friedrich Geist has noted, can be traced back to Karl Friedrich Schinkel, who had seen arcades on his travels to France (1804) and England (1826) and reported on them in letters and diary entries. In a letter to his wife of April 29, 1826, en route to England, he writes, "On the way (from the hotel to Humboldt) we already viewed some of the glass-covered arcades, which are highly elegant and comfortably furnished, including the Palais-Royal."[132] In various designs, Schinkel attempted to establish the arcade as a new spatial idea in Berlin—for example, in his design for Neue Wilhelmstrasse (1819) or the town houses for Berlin designed in 1826, where the "arcade" is also already found as a spatial element of upscale living.

Arcades commonly combined massive construction and, in the form of glass roofs, filigree construction (cases where the arcade building was designed entirely as a filigree construction were rare). In addition to iron structures, timber structures were also used as supporting frameworks for the glass roofs. "It is only

→ figs. 24, 25, 26, 27, 28, 29   Thermal Geographies of the European City   **37**

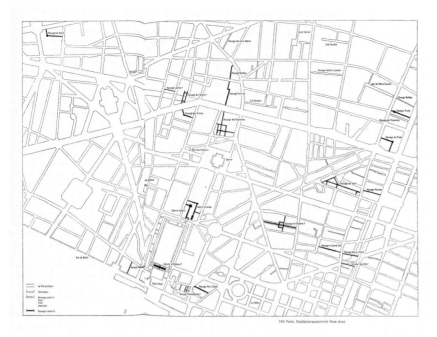

24
New arcades in the Paris area, *Rive Droite*.

25
View into an arcade.
New Trade Rows, Moscow,
1888–1893. Source: Geist
1969.

26
Kaisergalerie,
Berlin,
cross-section.

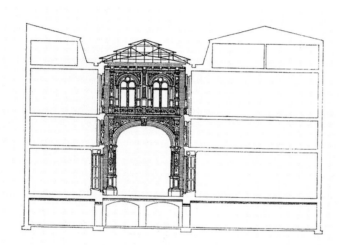

27
City of London "Arcade",
project around 1830.

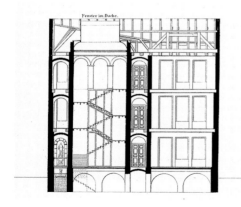

28
Karl Friedrich Schinkel, town
house for Berlin with skylight,
designed in 1826. The idea
of the "arcade" appears as a
spatial element of upscale
living, integrating skylights.

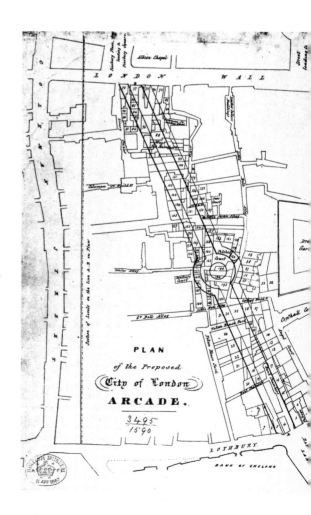

European Developments before 1945   38

29
View into an arcade.
Passage Coiseul, Paris,
1825–1827. Source:
Geist 1969.

30
View into an arcade. Kaisergalerie, Berlin,
1871–1873. Source: Geist 1969.

31
Entrance of an arcade.
Boulevard Montmartre, Paris.
Source: Geist 1969.

32
Glass roof: base and detail. Galerie de la
Reine, Brussels. Source: Geist 1969.

Thermal Geographies of the European City

in the 1920s that the full, continuous glass roof becomes common, feasible in terms of production technology, and the cost is affordable for the builder. The profile of the glass roof serves as another horizontal division. It may be saddled, arched, pointed-arched, or have the contour of a circular segment."[133] Arcades did provide protection from the rain; however, in reality, sealing the iron-and-glass roofs remained an unsolved challenge for decades. Leaks in the glass roof led to numerous puddles in the public space below. Dust settling on the glass surfaces soon led to the dimming of the spaces, which often were several floors deep. The twilight, according to Siegfried Kracauer, was integral to the furnishings of the passages, which allowed for all kinds of secrecy: "Above all, they satisfy the physical need and the greed for images as they appear in waking dreams. Both the very near and the very far escape the bourgeois public, which does not tolerate them, and gladly retreat into the secret twilight of the arcade, where they flourish as in a swamp."[134] That which has no place in the official self-image of bourgeois society had been outsourced to the arcade: Kracauer therefore speaks (instead of the unconscious) of the "inner Siberia of the arcade".[135]

Over time, the glass plates were sometimes "laid against each other with an air gap" to prevent soiling of the edges and to allow improved ventilation of the arcade. It was not until the second half of the 19th century that "with the help of plastically deformed sealing compounds" sufficient protection against water penetration was found.[136] With the gradual transformation of the arcade, from an "object of private speculation" to a clearly "public building", the "decay of the idea of space" was also prefigured.[137] For this reason, it is not so much the actual representative buildings—such as the Galleria Vittorio Emanuele II in Milan, built by Giuseppe Mengoni between 1865 and 1877—but the more typical, often anonymous arcades that have to be highlighted.

1.2.4 Attic Stories: Solar Access

As early as the 19th century, experiments were carried out to improve sunlight on streets and building facades, while at the same time achieving the highest possible economic utilization. Newly elaborated building laws were the result of the balance of these two concerns. Paris became a model for the European city, which under Louis XVI introduced the *Déclaration du Roi sur les alignements et ouvertures des Rues de Paris* ("Declaration of the King on the align-

ments and openings of the streets of Paris"), which "relates the heights of buildings to the widths of the streets" and introduced the "much admired unifying feature into the streetscape of Paris". The urban planning approach of "classifying streets and their buildability"[138] would be further elaborated into actual building codes under Haussmann and would attain its valid form in the building laws of 1902. This legislation regulated not only the street width and building height but also the building form in the transition area between facade and roof.[139] While "solar access to an individual building" can be described by four parameters—latitude, slope, building shape and orientation—"solar access to a city" includes (besides the four mentioned parameters) the height of the buildings, the width and the orientation of the streets.[140]

Mortality Rates

In 1908, Carl Kassner pointed out the changing qualities of different floors in London mansion blocks; from a climatic perspective, the best locations were on the middle floors, as they were less affected by the rampant air pollution and dust from the streets. In fact, a bourgeois desire for representation inherited from the aristocracy had meant that the price structure did not follow the actual thermal-comfort levels. "In London the comparatively purest air is to be found at 9–12 m. above the ground, i.e., about the level of the second to fourth floors; the most expensive apartments, namely, those on the first floor and on the first and second floors, must be said to be the worst from this point of view. Towards the top it is polluted by smoke, towards the bottom by dust."[141] In one of the statistics cited by Brezina and Schmidt, such considerations are made thermally and the "mortality in Berlin" (between 1875 and 1886) is correlated with the "altitude of the apartments".[142] One sees the increased mortality of those living in basement and ground-floor apartments as well as those on the top floors with the greatest exposure to intense solar radiation. Accordingly, for a long time, servants and the city's proverbially destitute artists had to live in the Parisian attic floors.[143]

The higher up and the lower down the dwelling, the higher the probability of dying comparatively prematurely; physicians recognized a clear connection between the increased mortality of infants and the elderly during the summer months and the construction of the speculative tenements. Flügge speaks of "two opposing influences, one cooling from the ground, the other heating from the roof", which have different effects depending on the number of stories; the "differences in the temperature of the individual stories" proving to be "quite considerable". According to

33
Mortality in Berlin and elevation of housing. Out of every 1,000 inhabitants to die: see the 3 columns on the right.

Tabelle 28
Sterblichkeit in Berlin und Höhenlage der Wohnungen¹)

Höhenlage der Wohnungen	Von je 1000 Bewohnern starben 1875—76	1880—81	1885—86
Keller	35,6	23,6	21,1
Erdgeschoß	29,4	21,8	20,4
1. Stock	28,6	20,6	18,4
2. Stock	29,2	22,3	18,8
3. Stock	32,9	22,0	19,0
4. und 5. Stock	36,5	25,8	21,4

¹) Aus Hdb. d. Hyg. herausg. v. Gruber, Rubner, Ficker, II. Bd., S. 133. Leipzig, S. Hirzel, 1911.

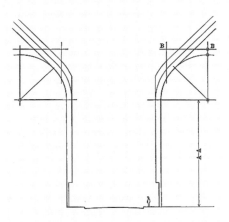

34
Diagram illustrating the building regulation of Paris applying to the height of buildings. The radius of the circular arc varies with the width of the street.

35
Courtyards in Barcelona (*Eixample*), Spain.

36
Aerial view of Barcelona (*Eixample*), Spain.

Thermal Geographies of the European City

measurements by Rietschel and Flügge, temperatures of 30–35°C can be measured on the "highest floors of tenement buildings". According to Flügge, this was also related to the high occupancy density and the poorly insulated stoves used for cooking purposes.[144]

Eixample (Barcelona)

In general, the attic became a mediating area between the horizontal and the vertical by introducing a 45° angle into its regulation. The basic rule was that the attic story must be set back from the building facade on all sides; however, numerous deviations from this rule could also be found, in that, for example, one third could be built flush with the facade or one or two facades could even be built on completely with an attic story. In interaction with the topography, further differentiations of such regulations arose. The quasi-heliomorphic proposals of the Catalan civil engineer and urban planner Ildefons Cerdà i Sunyers (1815–1876) were exemplary in this regard.

With Cerdà's development concept for a new quarter of Eixample in Barcelona, there was an explicit regulation committed to the control of the sun and winds. Already here the tension between "solar access" on the one hand and a development vision on the other hand appears. In 1855, Cerdà proposed (on an area of 7.46 km²) a street grid with streets 20 m wide (interrupted by boulevards 50 m wide) and large urban blocks of 113 m square. He also regulated the solar access and the ventilation of the individual apartments in "four ways": 1. for the 20 m-wide streets the building heights were limited to 16 m; 2. the blocks were to be built on only two sides, which would create parallel or L-shaped block figures with huge courtyards that guaranteed sunlight and ventilation from two sides; 3. All urban blocks had truncated corners, which further increased solar access; and, finally, 4. with a north-east/south-west orientation of the blocks, an insolation regime was also implemented that ensured strong shading of the street spaces while providing good daylighting for the apartments.[145] However, due to subsequently relaxed regulations, the terracing of the parapet areas tended to result in a high utilization of the neighborhood (in some cases buildings of up to 30 m were allowed) and over time the inner courtyards were filled with lower buildings.

Although the attic story emerged from the speculation and exploitation pressures of the 19th century, it has contributed more than any other typological development in European architecture to sensitizing

architecture to urban climatic issues. Design research and empirical research come together in the attic for the first time, as the "warmth of individual floors"[146] becomes a comprehensive problem of design (see chapter 3).

Conclusion

Private urbanism, with its slums of interwoven living and working zones, and with its arcades and attic floors, produced a new kind of thermal-sensory geography in European cities. Blending and gradualism characterized the spatial and thermal order in the 19th-century city. The lack of separation of industrial and residential areas led to unbearable hygienic conditions. Only the separation of functions and, consequently, zoning according to building law would later make it possible to improve the air conditions in residential areas by the appropriate siting of industries. As techniques of disinfection were not yet known, the (lack of) "hygiene", the great concern around 1900, was therefore concerned with a precise analysis of the transitions, the entanglements and interactions between inside and outside. Hygienic descriptions provided the first comprehensive understanding of the context-bound nature of housing. The thermal character of the "dwelling" could only be explained from its embeddedness in large-scale tenements and high-density residential quarters. High mortality rates were due to the enormous occupancy density of rented apartments—an ultimately social problem that was not comprehensively remedied until the second half of the 20th century. As Julius Posener rightly pointed out, "Whether a slum develops basically depends on only one factor: occupancy rates."[147]

The history of the attic floor shows in an exemplary way that urban climate phenomena were not discussed solely in a climate-deterministic manner, in the sense of a cause–effect scheme. Rather, it opened up room for architectural research under ambiguous atmospheric conditions, which transcended the framework of scientific thought and in turn opened up new scope for it. On the one hand, the attic floor enabled a higher than hitherto economic utilization; at the same time, it guaranteed that sunlight reached the deeper layers of the streets. This speculative development of the 19th century brought about an architecture in which the potentials of urban climates through design were explored.

→ figs. 35, 36   European Developments before 1945   42

2 Man-made Climate by Design

Urban Climatology and Modern Architecture

Historically, modern architecture and urban climatology (as a science) emerged at the same time. One can speak of a co-evolution of the modern architectural movement and the new field of urban climatology: in the same period as the Dammerstock estate (1929) was erected in Karlsruhe, empirical studies on urban climates were undertaken in the city. Although the pan-European character of modern urban climatology must be acknowledged, the scientific foundations were developed in Germany and Austria in the interwar period.[1] The city of Vienna for instance—with research carried out by Schmidt, Tollner and Steinhauser—was the setting for the the first comprehensive understanding of the microclimatic differences within the city induced by high "building densities." From 1927 onwards, systematic measurements suggested not only that the temperature pattern between city and countryside showed differences but that there were also large thermal differences within the cities themselves, due to the structure of the built environment. The "width and narrowness of streets" were examined with regard to the "temperature pattern" and, in the case of Vienna, "wide squares" (Schwarzenbergplatz), "narrow alleys" (Habsburgerstrasse) and "avenues" (Parkring) were compared with one another.[2] "Wide streets and squares without trees are very hot in the middle of the day, but cool down more dramatically at night (Schwarzenbergplatz). Avenues and tree-covered squares are much cooler and have less diurnal variation (Parkring). Narrow alleys, on the other hand, are characterized by particularly pronounced cooling during the midday and afternoon hours, amounting to temperatures of 5 to 6° C lower than their surroundings. At night, the difference disappears or even reverses (Wollzeile, Habsburgergasse). It is the strong impediment of solar irradiation and thermal radiance that causes this major dampening of the daily variation."[3]

→ figs. 54, 55

2.1 Visions of the Modern City and Climate

With the notion of the "man-made climate" (*künstliches Klima*), the numerous observations and individual investigations on the microclimates of the city were combined into a coherent approach—whereby the interdisciplinary hygiene research of that period further differentiated itself as *a discipline* of urban climatology and its application.[4] Among meteorologists, physicians, architects and urban planners, the notion of man-made climate became common in the interwar period for describing quite different—today one would say anthropogenic—climatic phenomena and their architectural–technical influence. There was talk of "climatic conditions" that "came about with the intervention of creative man, as a man-made climate".[5] As we will see, an urban understanding of climate control was at stake.

As discussed in the previous chapter, the modern research on man-made climates was strongly motivated and influenced by the air-polluting effects of factories on the one hand and the poorly equipped and ventilated residential districts of the working class on the other hand. In Europe, the notion of man-made climate became the central concept of long-smoldering discussions on issues of climate in urban development. Urban climatology produced a body of knowledge that was directed towards application and aimed to support the scientification of urban development. The fundamental goal of early urban climatology was not mere description but rather an application that could contribute to a new modern city. The notion of the man-made climate formed the decisive interface along which the mutual appropriation of urban climatology and modern architecture and planning occurred. The scientific exploitation of different climatic scales was accompanied by analogous developments in the field of modern architecture.

2.1.1 Interscalarity between the Apartment and the City

The year 1937 saw the publication of two significant works that were seminal for the scientific discourse on man-made climate: on the one hand, the fundamental climatological treatise *Das Stadtklima* ("The Urban Climate") by the geographer Albert Kratzer and, on the other hand, the more hygiene-oriented *Das künstliche Klima in der Umgebung des Menschen* ("The Artificial Climate in the Human Environment") by the meteorologists Ernst Brezina and Wilhelm Schmidt. The publications were united by a

43

fundamental interest in the man-made character of the climates in cities; in both, the urban planning and architectural implications were prominently showcased, leaving no doubt on the actual correspondences between meteorology and buildings.

These two publications had been preceded ten years earlier by the 1927 publication *Das Klima der bodennahen Luftschicht* ("The Climate of the Air Layer Near the Ground") by the meteorologist Rudolf Geiger. In particular, it was Geiger who had introduced the key category of urban climatic thinking: the microclimate. In the second edition, he stated that "there is probably no microclimate whose impact on human beings is as far-reaching as the city climate".[6] The notion of "man-made climate" was also introduced in 1937 by French engineer André Missenard–influential advisor of Le Corbusier–in *L'homme et le climat* ("Man and his Climate").[7] Missenard conceived "the elaboration of the doctrine of man-made climates" as one of his "essential aims". He was concerned with "improving the unreasonable, if not inhumane, conditions of industrial living conditions that are only too numerous".[8]

Three Urban Deficits

The idea of the artificiality of the climate was literally in the air among European research circles in the interwar period. The industrial city and its social inequality was the political background of this research; the shortcomings of urban atmospheres were the starting point of an affirmative understanding of man-made climate, consciously created by architects and urban planners. The interdisciplinary endeavor that brought these authors together was "hygiene", which since the middle of the 19th century had shaped bourgeois social reforms (see chapter 1). The question of hygiene occupied architecture and urban climatology between 1850 and 1950 as obsessively as the question of energy would from the mid-20th century onwards. Until the middle of the 20th century, meteorological considerations in urban planning remained part of the general discourse on hygiene and health.[9] The hygienic examination of the city was a driver for the elaboration of urban climate concerns (and vice versa). In 1914, August Schmauss succinctly stated that "the meteorological aspects to be taken into account in the planning of buildings and towns form part of the hygienic requirements".[10] In 1919, Bruno Taut spoke of a new type of "urbanization according to social, economic and sanitary aspects".[11]

Interdisciplinary hygiene was a science of the proper distances between people and things. Walter Gropius's Bauhaus exercises of 1930 were exemplary for this approach, controlling the distances between buildings by means of *solar diagrams*. In writings on hygiene, health-related and thermal aspects were hardly separated. Accordingly, there is a close conceptual connection between health and the microclimatology of the metropolis in the interwar period; hygiene depended both on climatological and medical expertise. According to Albert Peppler, empirical meteorological "investigations", as they emerged in the late 1920s onwards, had "considerable urban hygienic significance".[12] In addition to the microclimatic investigations on urban form, foundational research in building physics was undertaken, which understood the notion of man-made climate from the perspective of building construction. In 1936, the German engineer Joseph Sebastian Cammerer, in the preface to *Die konstruktiven Grundlagen des Wärme- und Kälteschutzes im Wohn- und Industriebau* ("The Constructive Principles of Heat and Cold Protection in Residential and Industrial Building"), referred to the "serious structural, hygienic and thermal flaws [...] with which countless large housing developments of recent years are afflicted".[13]

According to Ludwig Hilberseimer in *The New City*, the industrial city is characterized by "three deficits"; all three are closely linked to urban climatic phenomena. In particular, he criticized the lack of planning in the development of the contemporary metropolis: "The great cities of our time are the product of empirical growth rather than of planning principle. The inadequacy of these cities can be traced to three principal deficiencies brought about by the rapid and random growth of urban settlements. In the first place, no effort was made to locate industry in proper relation to residences. No thought was given to prevailing winds. Therefore, the smoke, soot, and fumes of our industrial cities constitute an evil with grave consequences to the health of the people who live there. In the second place, houses in the residential districts were built without the slightest thought to the need for sunlight on the part of the people who would live in those houses. The population density is highest in the worst and unhealthiest parts of the city. Recreational areas in those sections are direly needed. As they exist at present such residential districts are a danger, not only to the people who live there, but to the whole community. Crime and health statistics are witness to this fact. In the third place, the disorder within the city area, the indiscriminate conglomeration of industrial, commercial, and residential districts, gives rise to almost insoluble traffic problems. Far more traffic con-

→ fig. 43   European Developments before 1945   **44**

37
Tony Garnier, the *Industrial City*, 1904.

38
Otto Wagner, the *Big City*, 1911.

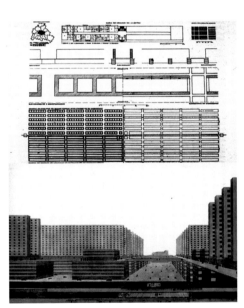

39
Ludwig Hilberseimer, the *High-rise City*, 1927.

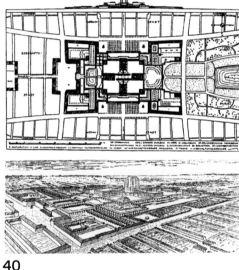

40
Bruno Taut, the *City Crown*, 1919.

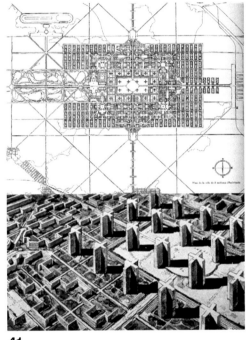

41
Le Corbusier, the *City of the Present*, 1922.

42
Nikolay Alexandrovich Milyutin, the *Linear City*, 1930.

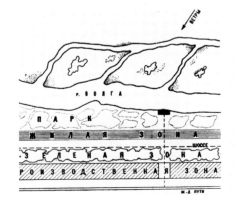

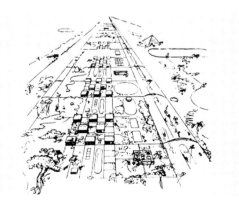

Man-made Climate by Design

45

43
Map indicating correlations between the distribution of hygiene and crime. Source: Hilberseimer 1944.

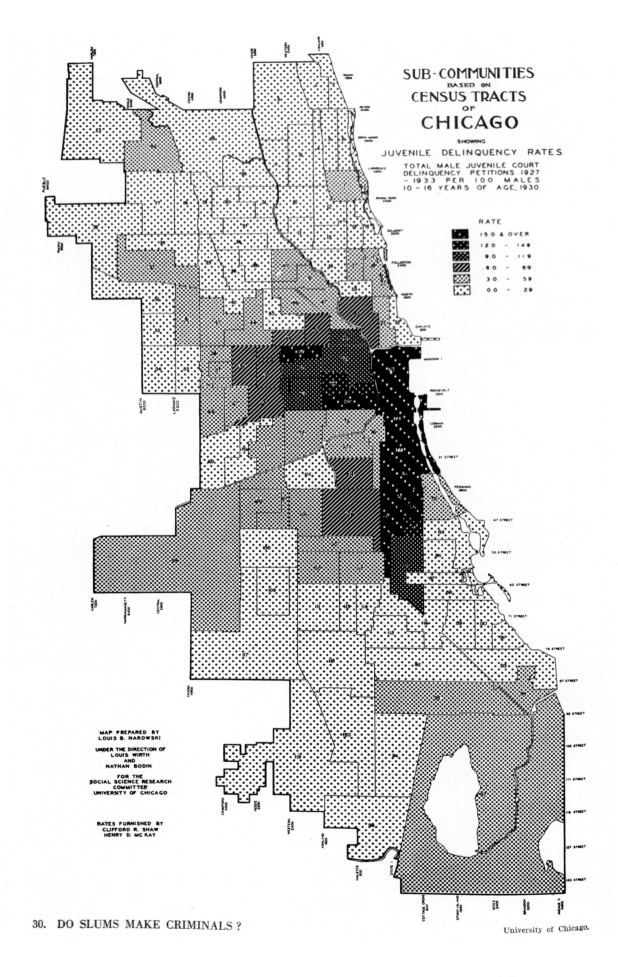

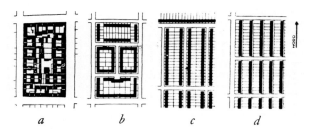

44
Evolution of block planning in Germany from 1925 to 1930. Source: Wright 1935.

45
Le Corbusier, facade of a high-rise building, 1925.

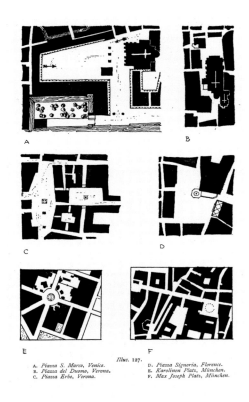

47
Le Corbusier, Pavillon de l'Esprit Nouveau, simulating the vista of future high-rise buildings, Paris 1925.

46
Camillo Sitte, the medieval city as a model for urban renewal.

48
Tony Garnier, rendering of the residential area of the *Cité industrielle*, 1904.

Man-made Climate by Design

veyances than should be needed must be used. And even then, traffic facilities continue to be inadequate. The antiquated street system, faithfully followed, adds danger for pedestrians and motorists alike and this danger mounts as traffic increases."[14]

What emerges in this list of shortcomings of the contemporary metropolis is the various scales of the urban climate, which, as Hilberseimer notes in his second point, affects the population unevenly. The residential areas with the least sunlight are often the most densely populated and, accordingly, the most unhealthy. Thus, as he points out with reference to studies by the Chicago sociologists Louis Wirth and Nathan Bodin, there is a correlation between hygiene and crime rates, between the climatic and social structuring of cities.[15] This correlation was recorded in many places of the world in the course of the 20th century (see chapter 6); here, they are both addressed in terms of urban climate issues and as an underlying conflict of the industrial city. It would be incumbent on the "municipalities" as "public authorities" to enact "building codes", as Brezina and Schmidt emphasized, that would bring about thermal improvements in new development areas. The unrestricted primacy of economy and private property determines urban development against better knowledge: "The municipalities are almost powerless against the disadvantages of the man-made climate in built-up urban areas; they can only use palliative means, and even these only under favorable economic conditions and in a struggle with the interests of those circles whose economic prosperity depends on an inexpedient way of building that causes damage to the climate."[16] Only a "new city", as it is programmatically outlined by Hilberseimer, promised solutions for the thermal shortcomings of the districts and buildings.

The Entangled Formation of Two Disciplines

The "desolate development of the metropolis resulting from industrialization"[17] led to an actual crisis in urban-development thinking, but it also released the energies that were to lead to the new discipline of urban planning at the beginning of the 20th century. The "private urbanism" that had characterized urban development in the 19th century was increasingly questioned, accompanied with "lively debates" about a new, socially negotiated form of urban design.[18] Julius Posener spoke of the problem of "how to channel this growth of the city brought about by private initiatives",[19] leading to the two disciplines of urban design and urban climatology.[20] Even before the First World War, an understanding of "urban planning as a comprehensive and integral task" had emerged,

"which was condensed into a method of designing the city under the direction of architects in collaboration with artists, landscape architects, traffic and service engineers, economists, municipal scientists, and politicians". One can speak of a "multidisciplinary theoretical approach and, ultimately, of a multi-scale design practice" that shaped urban design.[21] According to Joseph Brix and Felix Genzmer, who launched the *Seminar für Städtebau* ("Seminar for Urban Design") at the TU Berlin in 1907, urban design was "art, science and engineering achievement at the same time".[22] Only three years after the seminar was launched, explicit urban climatological considerations–*Die meteorologische Grundlagen des Städtebaues* ("The Meteorological Foundations of Urban Design")–were incorporated into the new urban planning thinking.[23]

However, the topos of urban climatology as an auxiliary science of emerging urban design is too one-sided. It overlooks the fact that, at least in the interwar period, the nascent field of urban design was equally an auxiliary science of urban climatology;[24] the fertilization was mutual. In *Das Stadtklima*, Albert Kratzer refers to Martin Wagner's 1915 dissertation on *Das sanitäre Grün der Städte* ("The Sanitary Green of Cities") and, by way of example, to Le Corbusier's depiction of New York and his new design for a high-rise city in *Urbanism* of 1925. In one of the illustrations in *Urbanism*, Manhattan and its skyscrapers are compared with Le Corbusier's 60-story and 6-story apartment buildings. The "City for Three Million Inhabitants"–also shown in 1925 in the exhibition pavilion of *L'Esprit Nouveau* in Paris–must be considered an important reference used by Kratzer as a figure of thought in urban climatology. The well thought-out multi-scale nature of this proposal, which included facade elevations in addition to the urban scale, played into the hands of an urban climatic argument. Le Corbusier was striving "to formulate the fundamental principles of modern urbanism". Whether it was Paris, London, Berlin or New York–with these principles, any city could be "urbanized".[25] The pavilion comprised two parts: in one part architectural models and in the other urban models were displayed. It was surrounded by a garden, which provided backdrops for imagined views from a high-rise building. Le Corbusier's multimediality contributed to the success of the *Plan Voisin* by combining exhibition material with physically tangible views at a 1:1 scale. The model and reality were blended in a kind of augmented reality (avant la lettre). One can speak of a scenographic approach to the urban climate.

→ figs. 41, 45, 47   European Developments before 1945   **48**

As the most significant urban representations in Kratzer's work, Manhattan and Le Corbusier's high-rise city of Paris served as a conceptual pivot of the young urban climatology, thus adopting the progressive premises of the Modern Movement. Beyond this, however, one should emphasize in particular the importance of German and Austrian urban planning of the interwar period for the formulation of new climatological principles of the city. Indeed, Germany played a leading role in architectural research and practice at the beginning of the 20th century, as the American urban planner Henry Wright emphasized. "The evolution of modern German housing and community planning" also contributed to the foundation of modern urban climatology. Hamburg, Cologne, Frankfurt am Main, Stuttgart and Berlin were the cities where the modern approaches of urban design were comprehensively applied for the first time. Wright speaks of "the remarkable evolutionary advancement made in the short period of five years following the resumption of building in Germany in 1925. During this period Germany built, for the most part in organized communities, low-cost housing equal in extent to almost 25 per cent of its existing housing equipment at the beginning of the period."[26]

New Urban-planning Paradigms

In the early 20th century, "theoretical urban projects" emerged in rapid succession, which conceived of architecture as the central element of the new city.[27] "A new idea directs all these minds and hands, it is the idea of the new city", Bruno Taut wrote.[28] He speaks of how "today we have an idea of how best to organize a modern city. The distribution of residential, industrial, and commercial districts in the urban fabric, the accommodation of public buildings, schools, administrations, etc., all finally found a definite form, at least in theory."[29] In addition to Taut's own visions elaborated during the First World War, the *City Crown* of 1919, one can mention Ebenezer Howard's *Garden City* of 1902; Tony Garnier's *Industrial City* of 1904; Otto Wagner's *Big City* of 1911; Antonio Sant'Elia's design for a *central traffic station* of 1914; Le Corbusier's *City of the Present* of 1922; Ludwig Hilberseimer's *High-rise City* of 1927; and, later, Frank Lloyd Wright's *Broadacre City* of 1932. These urban models formed architectural surrogates of a much broader discourse on hygiene and urbanism; all of them had eminent urban-climatic implications, which subsequently found their way into the designs and buildings based on them.

What all these models had in common was the assumption that architecture should no longer be understood as a problem of the individual building but as a configuration of groups of buildings (see chapter 6). Such an understanding of architecture as interconnected urban buildings also shaped the theoretical conception of climate-conscious urban design. However, the different spatial patterns and the associated material orders invariably also had differing urban-climatic implications. The debate on "block versus row, deck-access versus staircase access, inner-city reform of the metropolis versus its dissolution into satellites"[30] was also a debate about the climate in the city. Garden city or high-rise city, housing with a common inner courtyard or row housing—the theoretical urban projects were also discussed in terms of urban microclimates. Along two axes of thought, from *horizontal* to *vertical* and from *organic* to *geometrical*, references to the man-made climate of the city were established and its influence and control conceptualized. References to the *horizontal versus vertical* character or the *organic versus geometrical* qualities were accompanied by preferences of the architects for buildings, green spaces or topographies.

a) Horizontal versus vertical development

The two models of the garden city and the high-rise city shaped the discussions on urban green spaces in the evolving modern city. Within the framework of CIAM 3 (1930) with the programmatic title "flat, middle or high-rise construction", these two models of modern city planning were prominently discussed. A primarily horizontal development by aggregation was counterbalanced by a vertical perspective: that of spatial development upwards and not just across the existing territory. They were merged most prominently by Le Corbusier. The fusion of high-rise city and garden city anticipated many aspects of an architecture adapted to the urban climate. From urban gardens to garden cities and high-rise cities—this was, among other things, the main vision of a deliberate planning of urban microclimates in 20th-century urbanization.[31]

Horizontal and vertical spatial development imply different references to nature in the city and, more broadly, to the physicochemical conditions of the urban environment. High-rise housing developments establish references to the sky, allowing for different experiences of the weather—be it the sun, the wind, or the rain—through the great number of stories. Horizontally oriented settlements, on the other hand, are more likely to relate to the ecological patterns of a site; buildings are found embedded in patches of trees or surrounded by water. Hilberseimer's *High-rise City* ties in quite naturally, at lofty heights, with climatic factors of influence, while Garnier's *Cité industrielle*, close to the ground, seeks the interlocking of green spaces and buildings in residential neighborhoods. In both cases, the question of an artificial territoriality was posed, in which architecture, topography and green spaces contribute to a new man-made climate. This relationality between architecture and urban climate formed the starting point for the urban-climatic thinking of the design disciplines, as well as their conceptions of new infrastructures of man-made climate (see chapter 3).

The discussion on horizontal or vertical spatial development also brought into play the idea of the dissolution or delimitation of major cities. Three schools of thought emerged: those that called for the *dissolution* of cities (Howard, Taut), those that sought their *delimitation* (Unwin, May), and those that aimed at *renewal* (Eberstadt and Gurlitt, Schumacher).[32] Soon after it first emerged as an idea, the garden city was increasingly seen in conjunction with, and not merely as an alternative to, the metropolis. Raymond Unwin had brought this idea into play, and it was his former assistant Ernst May who would later realize this in the form of green belts around Frankfurt am Main. Manageable urban islands–so-called satellites–were considered an extension of the metropolis.

b) Organic versus formalized geometries

Alongside the dichotomy of horizontal and vertical spatial development, it was another dichotomy, of organic and formal geometries, that shaped the correspondences between urban climatology and architecture. In 1927, Hilberseimer pointed out two forms of urban formation that would come to shape architectural approaches to urban climate. "Over the centuries, two types of cities have developed, which stand against each other as two opposed worldviews, but in practice have often intermingled: the natural, grown and the artificial, geometric city layout."[33] In 1909, Raymond Unwin already distinguished between "formal" and "informal" urban development, focusing strongly on the geometric character of "formal and informal beauty".[34] In contrast to Camillo Sitte's *Der Städte-Bau nach seinen künstlerischen Grundsätzen* ("City Planning According to Artistic Principles"), which was dedicated to the medieval city as a model, Hilberseimer advocated a geometricized form of city planning, which addressed new social and economic challenges.[35]

The question of the specific geometry of the city was always also a question of the integration of climate in city planning. Strict geometric street grids, for example, might have favored the use of winds to improve cooling in the city, while informal geometries tended to allow for reciprocal shading, emphasizing the influence of sun and shade. The playfully informal and the rigidly formal were examined for their microclimatic implications and controlling potentials in the context of new urban climatological thinking. Urban design with a sense for vegetation would always emphasize its importance for shading, while a rigidly designed city, such as Hilberseimer's *High-Rise City*, would emphasize its importance for channeling the winds.

A romantic line of thought posited that the city should be planned to appear as if it had grown organically. In this respect, the debate on organic versus formalized geometries reiterated problems of art theory that had already been discussed as the Picturesque in the context of the emerging English landscape garden in the 18th century (see chapter 5). Like vegetation and topography, architecture could also be elaborated organically—although, as history shows, strict geometric solutions often prevailed in the case of modern architecture. However, both developmental approaches of *organic* and *formal* geometry served as models for modern urban design that addressed issues of urban climate.

Interconnecting Indoor and Outdoor Environments

The emergence of modern urban design in Europe was inextricably linked with the question of housing; urban design meant first and foremost finding solutions to the urgent question of housing.[36] Modern housing typologies evolved in close interaction with the industrial city. In an article in 1924, Hilberseimer addressed the individual building in a dialectical relationship with the superordinate scale of the city: "Metropolitan architecture is essentially dependent on the solution of two factors: the individual cell of the room and the entire urban organism. The space as a component of the house grouped in street blocks will determine this in its appearance, thus becomes the design factor of the urban organism, the real goal of architecture. Conversely, the constructive design of the urban plan will gain essential influence on the formation of the space and the house."[37]

While rapid industrialization intensified the search for a conscious planning of urban areas, the architectural study of urban climates promoted a new modern sensitivity for the interdependence of the building and the

→ figs. 37, 39, 46, 48   European Developments before 1945   50

city. "Housing and urban development"[38] form a unit, as August Schmauss formulated in his 1914 article *Meteorologische Grundsätze im Haus- und Städtebau* ("Climatological Principles in Architectural and Urban Design"). The notion of man-made climate included dwellings, buildings, districts and whole territories. Accordingly, "clothing–building–city" were described self-evidently as climatic factors of influence to be treated in a unified manner, as "within the entities created by human hands." Meteorological parameters (such as temperature, air movement, moisture, carbon dioxide, dust, radiation, as well as electric conditions in the air) are all applied to "individuals" (metabolism, clothing); the "individual house"; and the superordinate "urban climate".[39]

Rather than referring to specific means of climate control, the notion of man-made climate emphasizes its *multi-scalar* nature. The notion of man-made climate did not primarily refer to specific technical means; the control of climate was not thought of yet as a mechanical mechanism of machines. The distinction between *passive* and *active* thermal strategies, which became paradigmatic from the 1960s onwards, did not yet exist in the first half of the century; rather, the order of the day was to gain a sufficient multi-scalar understanding of the structural–physical phenomena and the thermal dynamics between indoor and outdoor areas. For example, Brezina and Schmidt determined that particular exposure to winds and proximity to green spaces resulted in lower infant mortality during the summer. "The high infant mortality in hot summers in large cities [...] is well known. According to Rietschel, strong winds immediately lower the number of deaths. Prausnitz found that the summer peak of infant mortality in Graz was absent near the city park."[40] In general, winds and green spaces were fundamental elements of actively promoted man-made climate of the modern city. They were recognized as important parameters that influenced people's subjective experience when spending time outdoors during periods of extreme heat. The control of climate promoted a new modern understanding of the interdependence of scales between the building and the city; urban management of the climate meant accounting for this interdependence of scales. The cognizance launched by urban climatologists was particularly well adapted by Ludwig Hilberseimer: the linking of macro- and microclimatological perspectives was mirrored architecturally in the conjunction of city, house and individual rooms. "A satisfactory solution of the problems of city planning can be achieved only when plans for the whole city and plans for the houses in it are both taken into consideration. Only then will it be possible to meet the social, economic, psychological, and hygienic requirements of good human living."[41]

The research on man-made climates, developing in parallel with the genesis of the modern city, established a dialectical understanding between inside and outside. Spaces in the city, as well as buildings and their individual dwellings, were subject to analysis in terms of their reciprocal thermal implications. Rather than relying on infrastructures, the notion of man-made climate conceived the entire city as the critical zone for thermo-regulating interventions; it was simultaneously multi-scalar and interactive, connecting indoors and outdoors. Accordingly, *interactions* and *transitions* were at the center of the investigations by scientists at the time. The formation of knowledge was based on the fundamental assumption of thermal continuities and mutual reactions—and not yet on an understanding of buildings as sealed-off thermal interiors. It was about a new heat economy that relied on appropriate urban-planning measures. Rather than the establishment of a thermal infrastructure, climate control in the interwar period meant taking the interdependence of scales into account. The main means for the design of the man-made climates were the new elements of modern urban design.

2.1.2 Empirical Studies on Urban Microclimates

"Microclimatology" as a new field of research was established in the 1920s by meteorologist Rudolph Geiger and others by systematically examining the relevance and autonomous behavior of air layers two meters above the ground.[42] In this novel area of research, the global perspective of meteorology was expanded by including microclimatic and mesoclimatic factors. The phenomena that came to be known as "microclimates" are formed in the "boundary layer" where the higher-level global climate and local conditions (partly man-made) are superimposed on and mutually impact one another. Or, in the words of the climate historians James Fleming and Vladimir Jankovic, where the "abstract three-dimensional geophysical system" and the "intimate ground-level experience" overlap.[43] In the preface to the second edition of his book, Geiger mentions the fruitful "progress of microclimatic research methods in allied sciences, in the teaching of habitat in botany, in forestry and gardening, in zoology, biology and medicine, in agriculture, regional planning and even in the technical aspects of traffic and construction work".[44] This wide

European Developments before 1945

49
Aerial view of Munich, 1934.

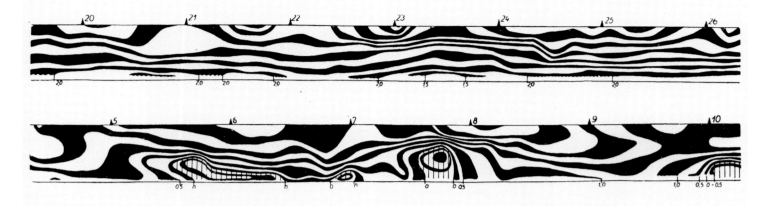

50
The different wind structure over different types of fields. Recordings by Wilhelm Schmidt.

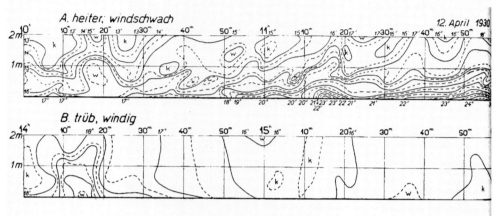

51
Thermal fluctuation as a function of the weather, 2 m above the ground. Source: Geiger 1927.

52
Temperatures above an asphalt road (at different heights).

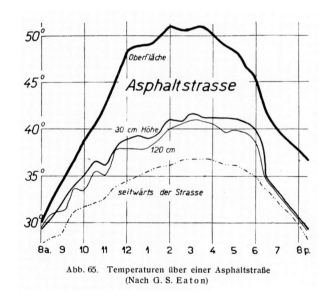

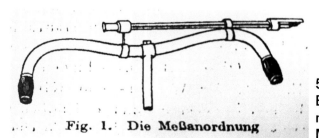

53
Bicycles were used for the measurement of microclimatic differences across the city of Munich (1933). Source: Büdel and Wolf 1933.

European Developments before 1945

54
"The Urban and Rural Sky".
Source: Hilberseimer 1944.

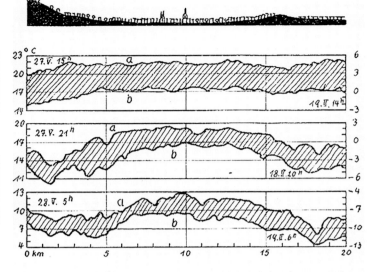

55
Section through Vienna and its surrounding rural areas indicating the urban heat-island effect. Feb. and May 1932. Source: Kratzer 1937.

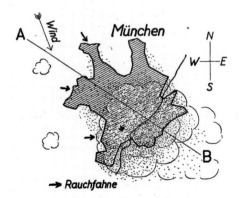

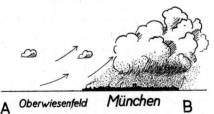

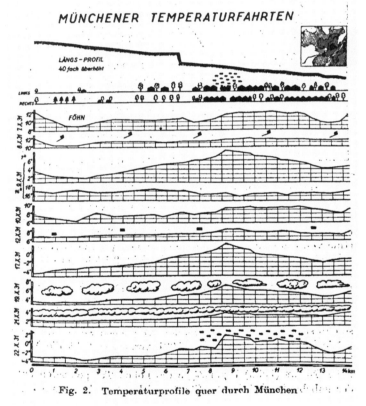

57
Temperature profiles across Munich. Source: Büdel and Wolf 1933.

56
Cloud formation over an iron-works (top). Cloud formation and rainfall over Munich. Source: Kratzer 1937.

Man-made Climate by Design

range of fields shows the interdisciplinary character of the research on microclimates, as well its practical applications for architecture and urban development.

The Climate Near the Ground

Although the scientific foundations of the concept of microclimate lie in forestry and plant biology, its conceptualization was brought to fruition particularly in relation to urban environments. The urban microclimate thus came to be viewed as a man-made artifact. From Geiger's perspective, to build means to change the microclimate: "man like the other animals avoids unfavorable habitat and seeks the favorable";[45] "unmolested nature", he says, has "an enormous number of microclimates", which humans, however, are destroying.[46] On the other hand, architecture and urban development open up the chance to deliberately influence microclimates; according to Geiger, a "rational search for the best climate" only takes place "with increasing civilization".[47] In this respect, city planning is also a kind of microclimatology. Geiger summarizes the anthropological rationale of his theory as follows: "The labor of man does not always lead to destruction of the microclimate. He also establishes new microclimates, especially through his building activities. Every newly built dwelling makes a number of separate climates out of the climate that was already present, just above the ground, on the building site. On south-facing walls the microclimate will be so favorable that good fruit, perhaps even grapevines, can be grown. This gain is at the expense of the north side, which is dark, cold, damp and rough. Still different are the east and west sides. The climates of the various rooms are modifications of these four outdoor climates. Beyond these, there is also the cellar climate and the attic climate."[48]

The distinction between the various microclimates of a building was further elaborated at the level of urban microclimates, as the detailed concept of the microclimate was also closely tied to research on the thermal conditions of cities. Urban microclimates are the result of the interaction between meteorological and architectural factors. They exhibit almost infinitesimal differences: microclimates in cities show thermal differences even within and between the individual rooms of a building. The notion of microclimate addresses both inside and outside conditions; they form a complex mesh of superimposed scales.[49] "One may now speak not only of a specific city climate, but also of a specific climate of broad streets, avenues and squares, and narrow alleyways" and of the individual rooms of the buildings.[50] As Albert Pepper noted, the dynamic character of temperature conditions must be also emphasized: "In our rapidly growing major cities, special climatic types emerge that deviate (more or less) from the ambient environment, sometimes lasting only a few hours."[51]

The man-made climate of the city represents a superposition of "natural" (bioclimatic) and man-made components. With Bruno Latour, we can speak of the man-made climate as a *hybrid object* or *quasi-object*[52] which integrates characteristics that are equally meteorological and architectural, empirical and designed, natural and cultural. By setting up a reciprocal relationship between climate and architecture, between weather and human activity, urban climatology both undermined and transformed the mono-causal approach of bio-climatic architecture. Urban climate phenomena such as heat islands or air pollution are strongly related to, and often intensified by the built environment and its technologies.[53] Patterns of urbanization and density affect the man-made character of urban microclimates: "The urban climate is more pronounced, the more extensive the city is and the fewer the areas of semi-rural construction it has in it".[54] Such remarks came from the observation of industrial cities such as Berlin or Vienna, which were characterized by many incorporated rural enclaves and urban development on green (rural) meadows. Already in 1910, the urban planner Carl Kassner summarized the outcome of these phenomena from a design point of view: "The dryness in more extensive cities is mainly caused by the higher temperature, and this again is a consequence of the stone masses of the houses, because when these are exposed to the sun in the course of the summer, they heat up considerably and collect such a quantity of heat that at night there is no significant cooling as occurs outside in the countryside."[55]

Empirical studies on urban microclimates were conducted, for instance, in Vienna, Linz, Berlin, Munich, Stuttgart and Karlsruhe. In addition to stationary measuring stations, dynamic measurements were also made in the city, using cars and bicycles to establish the temperature profiles of the big city. Büdel and Wolf, for example, used bicycles for their measurements in Munich (1933),[56] while Schmidt in Vienna (1927) and Peppler in Karlsruhe (1929) resorted to cars.[57] In each case, the aim was to "establish temperature cross-sections" in order to elicit the microclimatic differences.[58] "In order to determine the meteorological conditions of the air layers near the ground quickly and conveniently over large areas, the motor vehicle is the ideal instrument for research. Without greater expenditure of personnel, it is possible to establish cross-sections of the meteorological

→ figs. 50, 51, 52   European Developments before 1945

elements through plains and mountains, depressions, valleys and hills, metropolitan areas and forests, and to investigate the peculiarities of the atmospheric conditions that are influenced by the ground structure and cover. In order to obtain an equally detailed picture of the atmospheric condition with fixed observation stations, an extremely dense network of measurements would have to be established, which for financial reasons is not possible. The car, on the other hand, is a low-cost research tool that should be used extensively."[59]

In the case of bicycles, the station thermometers were simply attached to the handlebars in an almost horizontal position. A minimum riding speed and a brass cover ensured the necessary radiation protection and ventilation of the thermometer, respectively. In addition, most of the driving was done during the early morning hours to further mitigate the distortion due to irradiation. Humidity was also measured. To generate co-presence of urban data, measurements were taken by several observers "simultaneously". On the one hand, "within half an hour a large number" of measurements were made over the entire urban area; on the other hand, in the autumn of 1931, temperature profiles were made daily "along an approximately 14-km-long route across the city".[60]

Microclimatic Measurements in Karlsruhe, Munich and Berlin

The microclimates of a city are marked by four fundamental, mostly man-made factors: 1) the hermetic sealing of the urban surface (the materialization); 2) the vertical profile of the urban fabric and the arrangement of the buildings;[61] 3) the "artificial input of heat"; and 4) the "pollution of the air".[62] However, the simple equating of inner cities with heat islands is not supported by these original measurement series, which provide a more complex climatic picture. Dense "cloud cover" may largely cancel out significant thermal differences between the city and the countryside, but also within the city.[63] Weather conditions, in other words, can "enable" or hinder certain urban microclimates.[64] In addition to the notion of a "heat island", these studies provided an idea of the multi-factorial genesis of urban climates. In the center of the empirical research was the question of how the main influencing factors affected one another—in other words, how bioclimatic conditions such as wind or topography altered or balanced certain man-made temperature conditions. The "influence of air pollution"[65] as well as "building density" on "urban temperature" was discussed in particular.[66]

a) City Profiles: The Superimposition of Influencing Factors

In the case of Karlsruhe, topography played a key role in supplying cold air to downtown areas. "Since the vertical thickness of the surrounding cold air reservoir is only a few meters, cold air can only flow into the urban area if the latter is at the same altitude or slightly lower than the surrounding area."[67] In the case of Munich however, the influence of topography could largely be excluded from the outset. In a series of measurements—for example, on November 5, 1931—the "influence of air pollution on the city temperature" was particularly evident. "After a cloudless, almost windless night of radiation (Munich airfield: on the ground -2.3°, at 300 m altitude +6.7°), a dense layer of haze lay up to 100 m above the city. Under its protection, islands of warmth of up to +4° developed in the old town and the station area, compared with -3° at the northern edge of the sea of buildings." [68] In many cases, it was determined that air pollution created a kind of inverted weather condition, which led to the additional warming of inner cities. Accordingly, cooling alone could occur from the outskirts of the city. "The cooling of the large city on still summer nights is apparently effected not so much by radiation from the city air itself, which is too filled with masses of smoke and dust, as by the radiation-cooled air masses of the surrounding open country slowly advancing on all sides toward the interior of the city, gradually breaking up the urban heat island."[69]

Besides air pollution, "the influence of building density on the city temperature" was also investigated.[70] On November 14, 1931—a day with typical late-autumn weather, with closed cloud cover and corresponding "lack of radiation"—the thermal significance of Munich's building density was unequivocal: "Again, the thermal core lies around the train station; with astonishing exactness, the 5° isotherm circumscribes the part of the city densely built in dense construction; the park in the northeast of the old town has a cooling effect far into the city core; likewise, the influence of the course of the river is unmistakable. The looser development of the outer districts also has a noticeable influence on the temperature, as illustrated by the isothermal curve in the east of the city."[71] In Karlsruhe, too, the thermal difference between built-up and unbuilt zones became evident during summer "evening drives from 7pm to 10pm with almost absolutely still air" (July 23, 1929). "Now hot, stagnant air masses lay over the city area, while a cold air lake had formed around the city by radiation. The temperature difference between the interior of the city and the surrounding area was as much as 7°."[72]

→ figs. 49, 53, 56, 57   Man-made Climate by Design

"Clear" and "mild" nights with foehn winds (as on October 7, 1931)[73] as well as "cloudless, windless" nights (as on October 9, 1931),[74] on the other hand, led to a far-reaching climatic alignment of city and countryside. "The interior of the city is no warmer than the open countryside", was the finding for the first case; for the second, "the more openly built-up or open outer districts are usually even warmer than the interior of the city under the influence of the more favorable radiation conditions".[75] Existing winds significantly influence the thermal conditions of cities, as was also largely confirmed in the territorial measurement series in Karlsruhe. "Even with slight ventilation (wind force 2–3 m/sec), according to this, the temperature differences between the large city and the open country are not very considerable and in our case hardly exceed 1½ degrees, on average they are considerably lower still."[76]

b) Street Profiles: Interactions of Wind and Radiation

Rather than one influencing factor alone (such as the great mass of houses), it is only a combination of such factors that gives rise to heat islands. It was in fact the systematic meteorological comparison between city and countryside that first showed the stunning autonomy of the city climate; however, the main theoretical and practical interest was focused on the city-wide temperature differences. Urban climatology, as it emerged in the interwar period in Germany and Austria, was primarily a theory of the man-made climate *in and between* the buildings of large cities. In contrast to the focus of the second half of the 20th century, it was not only the climatopes of larger scales that were investigated but also, through the measurements made of streets and courtyards, more detailed microclimates. This approach was fostered by the knowledge that the thermal conditions of a building depend on the built context, and that the control of climate cannot be meaningfully developed without reference to it. As Brezina and Schmidt determined, in the countryside, unlike in the city, "a climate effect of neighboring buildings" was "virtually unknown".[77]

Accordingly, the temperature profiles of individual streets and courtyards were measured in order to be able to precisely reproduce the thermal interdependency and the effective causalities. Examples of this were measurements "in some streets of the west of Berlin", which were carried out over three days (Feb. 18, March 1, June 29, 1933) on the advice of Karl Stodieck, professor for architecture at the Technische Hochschule Charlottenburg (see chapter 3). The authors, Albrecht and Grunow, speak of an actual

"observation program under improved experimental conditions".[78] The "air temperature" and the "horizontal and vertical wind direction and velocity components" were measured "at various heights" in the "adjacent street sections", "in the courtyards" and "on the roofs". These were supplemented by the measurement of "flow profiles over the entire cross-section of the streets and courtyards".[79] Geisbergstrasse, with an east–west course, and Passauerstrasse, with a north–south course, were chosen for the study. The former, in contrast to the latter, was lined with trees.

The microclimatic profiles were created "by stretching a wire across the road from chimney to chimney, on which a small suspended trolley could roll on two wheels. This wagon was moved from the roof on one side by strings that ran over a pulley at the chimney on the other side of the street. At the bottom of the cart hung a pulley over which ran a string. At the end of this was suspended the measuring apparatus; it could be raised to any desired height by the pulley elevator". The movable measuring apparatus "consisted of a resistance air thermometer (Th), an electric anemometer for horizontal air velocity (W) and for vertical air displacement (V)".[80] Four people were used to measure the vertical, in accordance with Rudolph Geiger's standard work, at 2, 5, 10, 15 and 20 m above the ground, and also the street cross-sections by taking "readings at the house wall, at the curb, in the middle of the street and at the curb and the house wall on the other side of the street".[81] In addition to temperature response, wind conditions were also measured throughout. Observations were made of "vertical air displacement near the house walls (street and yard) at various heights and distances, and on the roofs".[82] This revealed, as the actual "main result" of the investigation, "that thermal circulation is almost completely suppressed by dynamic circulation". At Geisbergstrasse on March 1, 1934, between 14:30 and 15:00, it was found "that the temperature conditions were not decisive for the formation of the air circulation, but that under the influence of the prevailing S-wind on the warm N-side a descending movement occurs, while on the cold S-side ascending movement takes place. So the S-wind was forced to descend from the opposite side of the house and sucked up the air on the other side of the house."[83]

In the courtyards, on the other hand, the air has no "possibility to freely flow", which explains the "predominantly thermal circulation" during the "transition from day to night"[84] and thus also the delayed cooling of the courtyards. Nonetheless, the study on the

→ figs. 51, 62, 66   European Developments before 1945   58

"interaction of wind and radiation" showed "that the dynamic effect of wind at low wind speeds is sufficient to completely eliminate the thermal effect".[85] The consideration of wind and thermal conditions finally revealed the different scales that are at work in the genesis of microclimates in cities. For this, the layout of the streets plays a decisive role—as does the arrangement of the building volumes. This empirical thermal knowledge was to be made usable for the design of buildings and districts.

Empiriocriticism

The "logical empiricism" of the *Wiener Kreis* ("Vienna Circle")—founded in the Austrian capital at the beginning of the 20th century by Bertrand Russell, Ludwig Wittgenstein and Ernst Mach, among others—served as the philosophical model for the scientific rationalism that was applied by meteorologists and architects in examining the climate as a cultural artifact of the city. The anti-metaphysical stance, as it was elaborated among intellectuals in Vienna and Berlin, encouraged a specific and new type of empirical assessment of the city. One can speak of the "problem of an interdisciplinary scientific view of the world under the influence of a need for practicality".[86] The so-called "empiriocriticism" represented the leading epistemology of urban climatology and its application as climate control on urban scales through architecture and urban design from 1900 onwards. Empiriocriticism was the starting point of a new scientific understanding of the world.

One should particularly emphasize the importance of Ernst Mach, who as a physicist and philosopher renewed the epistemic foundations and methods in equal measure. Mach was concerned with the relationship between human sensory data and the material conditions of the world.[87] His primary question was "why sensations are also able to represent objective reality, whether sense data also have truth values".[88] The psycho-physical methodology, as it was developed towards the end of the 19th century by Mach but also by Gustav Theodor Fechner and Wilhelm Wundt in Leipzig, further deepened the empiriocritical approaches under sensory premises. According to Mach, the "outer life" and the "inner life" should be integrated in such a way that "'sensations' are grasped as common 'elements' of all possible physical and psychic experiences".[89] The psychophysics of sensory perceptions were to be subjected to objective measurability. According to Mach, science (e.g. teaching of mechanics) emerges from pre-scientific everyday experiences (handicraft). The "linking of physics, physiology and psychophysics"[90]

as a peculiarity of Mach's scientific methodology also had great relevance for the way urban climatology linked the city's exterior with apartments' interiors.

Epistemologically, urban climatology's interscalarity relied on a systematic linking of the objective external world and the subjectively experienced internal world.[91] Accordingly, urban microclimates are to be understood at the interface of the thermal environment of the city and the thermal experiences of the citizens. In an article on metropolitan architecture written by Hilberseimer for *Der Sturm* in 1924, it was stated: Metropolitan architecture "seeks to free itself from everything that is not immediate, from emotional content and emphasis on the soul. It strives for reduction to the most essential, for the greatest expression of energy, for the utmost possibility of tension, for ultimate exactitude. It corresponds to the way of life of people today. Is an expression of a new attitude towards life, which is not subjective-individual, but objective-collective in nature."[92] Through the notion of man-made climate, the ambivalence of the metropolis, which is also a thermal-climatic one, was transformed into a subject of design. The architectural appropriation of the insights of urban climatology must be understood within the framework of the great ambiguity of the man-made climate. The urban microclimate is characterized by the ambivalence of the uncontrollable, which one seeks to control. Only in light of this ambivalence does the richness of the climatic approaches come into view, as they emerged in the interwar period.

2.2 Creating Urban Microclimates in Space and Time

From the outset, the practical implications of urban climatology for architecture, landscape architecture and urban design were recognized and emphasized. "The density of built-up areas, the heights of the houses, their distance from one another, the width of the streets and squares, their orientation and their plant life—all of these have their effect on the temperature picture of a city", was how Albert Kratzer formulated his research hypothesis in 1937.[93] In a nutshell, these aspects were the ingredients for an urban theory of climate control. From the perspective of such a theory, urban microclimates appear consciously designed and constructed; they are man-made artifacts, not only empirically but also as the result of deliberately designed architecture and landscape architecture.

Man-made Climateby Design

58
Density distribution of the city of Zurich as a function of social and topographical conditions. Illustration shown at the CIAM 4 exhibition on "Housing, Working, Traffic, Recreation in the Contemporary City" at the Stedelijk Museum, Amsterdam, 1935.

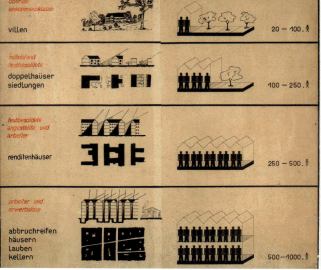

59
The residential areas of the total population by income. Illustration shown at the CIAM 4 exhibition on "Housing, Working, Traffic, Recreation in the Contemporary City" at the Stedelijk Museum, Amsterdam, 1935.

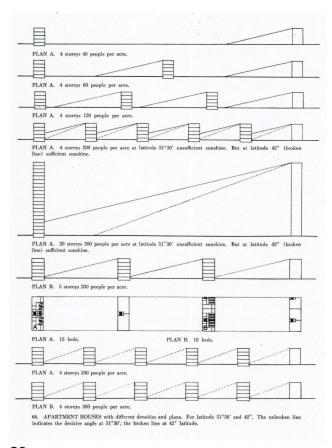

60
Apartment houses with different densities and plans. Source: Hilberseimer 1944.

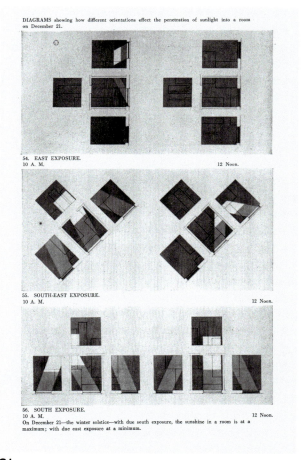

61
Investigation of solar radiation in residential interiors. Source: Hilberseimer 1935 and 1944.

European Developments before 1945

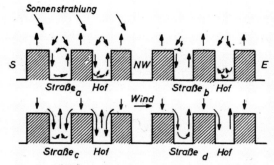

62
Schematic representation of some characteristic air circulations in streets and courtyards of a city during the day (a and b) and at night (c and d). Source: Kratzer 1956.

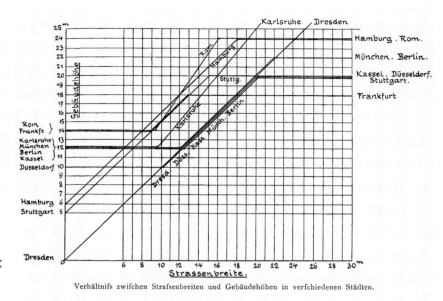

63
Relationship between street widths and building heights in different cities.

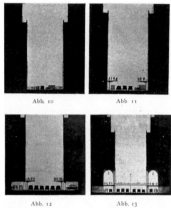

64
Possible improvement of the road system. Source: Hilberseimer 1927.

65
Street with skyscrapers in New York. Source: Hilberseimer 1927.

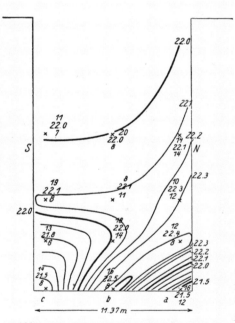

66
Temperature distribution in a courtyard in Berlin-Charlottenburg on July 27, 1931, at 6 p.m. One can see the influence of the outer wall heated by the sun. Source: Kratzer 1956.

Man-made Climate by Design 61

Two aspects of climate control at the urban scale need to be emphasized: the focus on *groups of buildings* and the emphasis on *thermal variation*. By focusing on these two aspects, the notion of man-made climate opens two fundamental lines in thinking about climate control in cities. It relates to both the spatial and the temporal aspects of urban climate. The notion of man-made climate promoted a *systemic* understanding of thermal regulation throughout the city; in this respect, it can be described as proto-ecological.

2.2.1 Groups of Buildings: Climate and Density

The high density of the major European cities was the main driver for the new urban climatological thinking in architecture in the interwar period. The thermal interdependency of building structures was thought of as being decisive for the genesis of microclimates in the city and their conscious control. "In contrast to the countryside with its individual houses, in the city the buildings are mostly built against one another, so that the houses are adjacent to the open air on two sides at most. [...] The mutual proximity of the individual houses has the consequence that each also strongly influences the climate of the neighboring house and that of the houses opposite. Thus, within each house and each living space there is a much more pronounced internal climate than in the countryside, and when leaving the house, the city dweller enters the street, that is, an area that has just as much of an internal climate as the house."[94] According to Brezina and Schmidt, the "block of houses" constitutes the basic "climatic unit" of the city.[95] "If in rural buildings the single house is the unit from which one has to start when considering the interior climate, in the case of block-forming construction in the interior of cities the more appropriate unit is the block of houses."[96] With this, Brezina and Schmidt succinctly summed up the difference between general knowledge on climate and the more specific insights of urban climatology. The block of houses must be analyzed in detail "because the mutual adjacency of the building walls, the many commonalities of the courtyards, both in terms of heat conduction and the supply and radiation of heat and light, give rise to very important climatic relationships, far more so than, for example, between two houses facing each other on a street".[97]

Courtyards

Even more than the street width, inner courtyards of the urban block appear (with their overlapping microclimates) as the main field of thermal analysis. Empirical climatic studies went in parallel with design proposals that introduced linear buildings (*Zeilenbauten*) or large courtyards. The "receding of the building facade, and the insertion of small park and play areas immediately lower the temperature. Likewise, in buildings around large courtyards, the cold air from the courtyard flows onto the street and creates a difference of up to 2°C."[98] The reorganization of the courtyard can be considered as the central task of the architecture of the 1910s to 1930s which led to a new sense of urban interiority (see chapter 3). The harnessing of the sun, for example, to dehumidify overcrowded apartments, represented the symbolic center of this discourse, although the construction-related problems were far broader and also included problems of overheating. "The climate of each urban building, and the possibility of improving it, is dependent on the location and composition of the houses in the whole neighborhood."[99] Accordingly, the distances between buildings, the street and courtyard widths were of central concern of the Modern Movement. The demand h=b (building height h=distance between buildings b) must be seen in the context of the high densities such as those found in Hamburg (h=3b) or Paris (h=5b).[100]

In Ludwig Hilberseimer's work, the sun appears as the city's main regulator of density. As an architect with sophisticated methodological standards, Hilberseimer developed a comprehensive approach of how to use the sun as a design factor of cities. *Heliomorphic design* appears as the decisive lever to regulate the population density in cities with all their hygienic and physical requirements (see chapter 3). From the early 1930s, Hilberseimer had been concerned with the solar penetration of spaces and the control of the distances between buildings. In 1930, he presented an investigation on building distances, the so-called *Bauhaus solar diagrams*, created with Sigfried Giedion and Walter Gropius.[101] In 1935, he published the study *Raumdurchsonnung* ("Solar Penetration of Spaces") followed in 1936 by *Raumdurchsonnung und Siedlungsdichtigkeit* ("Solar Penetration of Spaces and Settlement Density").[102] In it, the insolation of a bedroom was studied in the context of the different orientations of the building and the times of day and year. Hilberseimer emphasized the sun as a source of heat for the buildings, especially during the winter period. "It is of primary importance that the rooms should get the sun."[103] The overheating of external walls however must be countered by adequate shading and by means of "cross ventilation".[104] Hilberseimer emphasized the *reciprocity of inside and outside*, architectural and urban perspectives; what had been introduced with regard to the

→ figs. 21, 58, 61, 62, 66   European Developments before 1945

VORSCHLAG ZUR CITY-BEBAUUNG
LUDWIG HILBERSEIMER

Eines der wichtigsten und aktuellsten Probleme des heutigen Städtebaus ist die Umorganisierung und der Umbau der City. Heute stellt sie eine Mischform von Wohnstadt und Geschäftsstadt dar, mit dem Resultat, daß sie weder als Wohnstadt noch als Geschäftsstadt zweckmäßig ist. Es müssen daher alle Wohnungen aus der City entfernt werden, damit sie systematisch für ihren Zweck umgebaut werden kann. Wie die City selbst stellt auch ihre Bebauung eine Mischung von Wohn- und Geschäftshäusern. Das heutige Geschäftshaus ist aus dem Miethaus hervorgegangen, aus dem von Etage zu Etage die Zwischenwände entfernt und größere Fenster ausgebrochen wurden. Reichte ein Haus nicht mehr aus, wurde ein zweites und drittes hinzugefügt, bis eines Tages die Unübersichtlichkeit dieses Zufallsprodukts im Interesse einer rationellen Betriebsführung zur Errichtung eines neuen Gebäudes zwang. Aber auch dieses neu erbaute Haus reichte bei der fortschreitenden Entwicklung und Vergrößerung des Betriebes, besonders bei Warenhäusern, bald nicht mehr aus und machte daher weitere Neubauten erforderlich, die, so gut sie auch dem alten Gebäude angefügt wurden, im Grunde genommen auf verbesserter Grundlage dieselben Nachteile hatten wie die ursprünglich aneinandergefügten und für Geschäftszwecke ausgebauten Miethäuser.

Ein Beispiel, wie ein großes Warenhaus aus einzelnen Teilen nacheinander zusammengefügt wurde, ist das Warenhaus Wertheim in Berlin, am Leipziger Platz, das heute fast einen ganzen Straßenblock von sehr erheblichen Ausmaßen umfaßt. Als man vor mehr als dreißig Jahren den ersten Bauteil plante und ausführte, konnte noch niemand ahnen, daß der Raumbedarf des Hauses Wertheim einstmals den Umfang, den es heute hat, annehmen würde.

Eine weitere Phase dieser Entwicklung stellt das Warenhaus Tietz am Alexanderplatz in Berlin dar. Hier ist ein ungefähr gleich großer Gebäudekomplex wie das Warenhaus Wertheim nicht stückweise durch Vergrößerung, nacheinander, sondern planmäßig auf einmal entstanden. Dabei konnten natürlich alle die Vorteile, die eine übersichtliche Planung ermöglicht, für den Betrieb nutzbar gemacht werden. Nicht nur das Warenhaus, auch das Bürohaus hat diese Entwicklungsphase durchlaufen. So besteht der dem Scherlschen Zeitungskonzern in Berlin gehörige Gebäudekomplex heute noch aus vielen einzelnen ehemaligen Miethäusern, die untereinander verbunden ein phantastisches Durcheinander von Räumen und Gängen in verschiedenen Höhenlagen darstellen. Infolge der damit verknüpften Schwierigkeiten hat auch der Scherlsche Konzern bereits mit einem Neubau begonnen. Für große Konzerne ist es verhältnismäßig einfach, ihre Raumbedürfnisse durch Neubauten zu befriedigen. Anders ist es hingegen für kleinere Geschäfte, die gezwungen sind, sich in irgendeinem Bürohaus einzumieten, dessen Räume oft sehr unzweckmäßig sind. Soll ein solches Bürohaus seinen Zweck erfüllen, so muß es dem einzelnen Mieter, abgesehen von den unmittel-

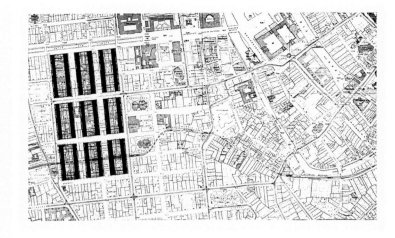

67–69
"Proposal for City Development" (Berlin).
Source: Hilberseimer 1930.

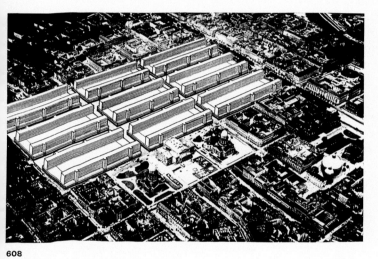

608

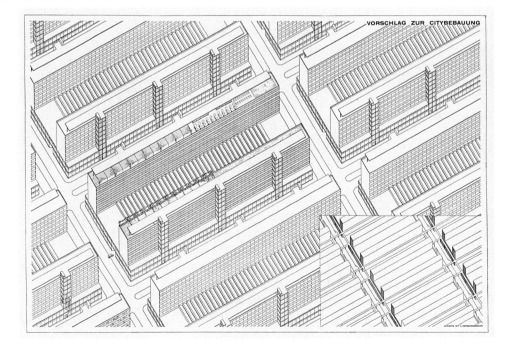

Man-made Climate by Design 63

individual room must be extended to the settlement as a whole. "The solutions we seek for the settlements as a whole must rest upon the same understanding of the importance of insolation which we sought in considering the plan of a single room in a single dwelling."[105]

Hilberseimer carefully examined a three-part approach for harnessing the sun and determining a city's density: first, one must consider geographic circumstances such as the latitude or topography; second, the distances between buildings as a function of shading must be examined; and third, one must examine the proportional relationships of buildings (length, depth, and height) as well as their use of materials. "If we are to arrange our buildings according to good insolation principles, the distance between the houses must be determined by the shadows cast by these houses. The length of such shadows will be governed not only by the height of the shadow-casting structure, but also by the latitude in which we build, the time of the year and the day, and the altitude and azimuth angles of the sun."[106] If one takes insolation as a benchmark, then in different latitudes only different densities could be achieved. Comparing Moscow, Paris and Chicago, Hilberseimer emphasized that in Moscow only half the density of that of Paris or even only one third of that of Chicago could be achieved. Like the latitude of a city, the topography was seen as a given natural condition that significantly influences the amount of sunlight and thus the possible density. On slopes with a northerly orientation, a much lower density should be foreseen than on those with a southerly orientation. "If the angle of the slope is identical with the altitude angle of the sun, permissible density reaches the zero point."[107] Hilberseimer's approach appears as a principle of zoning and of the regulation of urban densities. "Orientation and duration of insolation must be considered together in relation to density."[108]

Streets

With the emergence of the modern city, traffic became a determining factor of urban form. The search for a new city went along with the search for new ways of organizing traffic. This invariably involved rethinking the street spaces and occupying the city with new types of volumes. In the case of Brezina and Schmidt, the ratio of street width to building height is put into a systematic relationship as a means of determining the quality of microclimates. Either the "building height only slightly exceeds the street width" (type 1) or the width of the street accounts for "only a fraction of the building height" (type 2).[109] A third category of

streets, that are much wider than the height of the buildings, has different "climate characteristics" because those streets are usually "occupied by lines of trees". In general, urban structures of type 2 (e.g. the historic city centers) are replaced by those of type 1. "Modernization of the old centers involves virtually nothing more than widening narrow streets and cutting through blocks of houses that are inappropriate from a traffic point of view."[110] However, the microclimates of a street are significantly influenced by the "kind of construction" of the buildings, and are by no means only dependent on the width of the street. In terms of urban microclimates, the matter is thus much more complex. "The question now is whether the climatic advantage of wide streets as a whole really outweighs their disadvantages. The answer to this question cannot be given in general, it must be different, first of all depending on the natural climate to which the city belongs: latitude, average and extreme temperatures of the summer months, wind strength, the way the buildings are made are all important. The influence of the construction method is probably also noticeable differently in the individual floors. So, for the time being, we have to rely on an emotional judgment. Investigations into which of the two types of building is more beneficial overall, i.e. in which the advantages outweigh the disadvantages, would have to take into account the climate and the state of health of the inhabitants, while at the same time eliminating differences in income and occupation".[111]

The architect Alfred Schmidt emphasized the differences between the Classical street formed by buildings of the turn of the century and the new type of free-standing linear blocks (*Zeilenbauten*). The latter received relatively evenly distributed sunlight on all floors for about five hours. In contrast, streets built on both sides always had uneven sunlight in the apartments. "In the case of streets with buildings on both sides, significant differences in sunlight compared to the free-standing blocks are now apparent, which are particularly noticeable on the lower floors".[112] Schmidt identifies street directions "which form angles between 22 and 45 degrees with the meridian" as being the most favorable. However, he also emphasizes the importance of site-specific architectural solutions, since ideal conditions are usually not given. Suboptimal lighting conditions can be improved by means of the arrangement of windows and walls. The climatological concern is thus limited to a small-scale, construction-related task: "The illumination effect for the interiors could be increased in case of unfavorable conditions by means of technical aids (oriel-like window projections and sloping reveals). Consider the Romanesque window, which, with its inside and out-

→ figs. 60, 63, 64, 65   European Developments before 1945   **64**

side sloping, is exemplary in terms of illumination and heat retention."[113]

The idea of the correct orientation of the buildings—a central concern of the early Modern Movement—is put into perspective by Schmidt. The absolute north–south orientation with "evenly sunlit sides" has numerous disadvantages, such as "the daily loss of the warm midday sun, the almost complete lack of sunlight in the winter months, and the fact that the streets themselves are without shade at midday and the buildings expose their entire broadside to the weather generally approaching from the west. This makes the making of north-south aligned streets very questionable".[114] Schmidt mentions several disadvantages of the new modern typology that affect the thermal conditions of both the buildings and the streets. Thus, he follows an urban climatological argumentation that goes beyond the focus on the alignment question. A consideration of different seasons also reveals the contradiction between desirable heat gains (in winter) and prevention of overheating (in summer). In Schmidt's view, the wind also plays a central thermal role, which also has to be taken into account.

Land Reform

For Ludwig Hilberseimer, a new city was only conceivable on the basis of comprehensive land reform. Accordingly, he had made his architectural proposals dependent on political reforms. In his writings, the urban climate proves to be the driver for land reform and appropriate settlement density; designing with the urban climate in mind always means taking these two aspects into account. In his *Vorschlag zur City-Bebauung* ("Proposal for City-Building") of 1930, the architectural reconstruction of Berlin's city center was linked to the revision of the prevailing privatist prerequisites of urban planning. "The reorganization and reconstruction of the City" (from a "mixed form of residential city and business city"[115] to a pure business city) was to be accompanied by the overcoming of the "system of individual buildings".[116] Based on his theoretical "proposal" for a *High-Rise City*, Hilberseimer proposed slabs of high-rise buildings that could be designed integrally and used collectively. The integral approach would make things possible "that the individual cannot afford. The office building can best meet these demands when it encompasses entire blocks, freed from the individual building, and thus also offers a completely different possibility of use."[117] In contrast to the already more advanced high-rise city in the US, the goal was to develop an "urban layout appropriate to the high-rise" that emancipates itself from its feudal heritage. "It is

not based, as are the high-rise cities of America, on the system of the individual house that originated in the Middle Ages."[118]

Hilberseimer's proposal represented literally an over-writing of existing city layouts by the new high-rise slabs. His proposal is not about higher "building utilization" and "building density" but rather a "different distribution of building mass" and "concentration" on the land. Improved climatic conditions emanate from the new primacy of superordinate planning perspectives and a new scale relying on groups of buildings. The new urban layout allows "the necessary supply of light and air by maintaining sufficient distances, avoiding courtyards and providing appropriate orientation towards the sun".[119] Even in his American exile, Hilberseimer would advocate for the integration of multiple lots into new entities; this was intended to overcome the disadvantage of the "division of land into unsuitable lots".[120] In *The New City*, a proposed new zoning law would promote "large land units"[121] taking the adequate solar exposure of the buildings and the desired densities of occupancy and the "minimum dwelling area" into account.[122]

2.2.2 Thermal Variations: Climate and the Rhythm of Life

While mechanical air conditioning is predicated on the elimination of the time dimension for users and occupants, passive forms of climate control are always characterized by temporal cycles and interferences of varying character. In his À *la recherche du temps et du rythme* of 1940, André Missenard ambiguously linked the term of "time", as it also appears in Marcel Proust's À *la recherche du temps perdu*, with the term "weather", thus playing with their congruency in the French language. *Weather* and *time* appear as two sides of the same coin, which are placed in the center of the notion of man-made climate. Making control of the urban climate dependent on the (everyday) "rhythms of life" was one of the central epistemic prerequisites of the interwar period's thermal thinking. The (indoor) climate could not yet be stabilized in a broadly effective way; *comfort* was not yet a hegemonic idea. Urban microclimates were perceived as dynamic rather than static entities. In this respect, winds or parks, if they are consciously brought into play, are to be understood as part of an architectural repertoire of coping with the urban climate.

→ figs. 67, 68, 69   Man-made Climate by Design   65

The Biology of the Big City

One of the central considerations of the promoters of man-made climate was the *variation of climate* throughout the day and year. City dwellers are hardly exposed to the natural climate anymore; Missenard describes the conditions accordingly as contradictory: "Heating does indeed produce comfortable temperatures in homes and factories when the outside temperature becomes too low, and the difference from outside plays the role of stimulating variation. [...] But this is accessible only to a part of the population. Besides, these man-made climates are dangerous because they produce a monotonous temperature that is [...] sleep-inducing".[123] As heating warms and puts people to sleep, the mechanization of climate control and the thermal homogenization of interiors is a questionable development, while the notion of man-made climate is ultimately committed to the productivity of the individual and the population as a whole, contributing to a biopolitical agenda. Missenard's main concern was about *labor* rather than *housing*. Climatic foundations were sought for the emerging labor physiology under industrial capitalism. The notion of "heat island", which later became the paradigm of urban climatology, (still) reflects this biopolitical concern of a lack of productivity found in early urban climatology. Missenard's central hypothesis was that "Talented workers cannot do much work unless they are stimulated. In climates that are too monotonous, people flag".[124] The productivity of workers is largely dependent on their environmental conditions. The threat of living and working under inappropriate climatic conditions relates to the possible reduction of economic productivity. This was also the great disadvantage of the hot tropics, which were only suited to serve as regions of raw-material production but not for their processing. "In the cold zone, temperatures are probably low enough to permit intense physical exertion. But since vitality is tied to maximum metabolic rate, which itself depends on the minimum value of temperature, it is desirable that this minimum can reach 0 degrees."[125]

Hence, thermally induced stimulation (via variation)—and not the stabilization of climate as such—is at the center of Missenard's reflections; the "ideal climate" consists of a great thermal variation: "Thus, the average temperature of the ideal climate will vary [...] between 0 and 22 degrees during the day."[126] The variation of temperatures must be set up and controlled in equal measure. Missenard speaks of the "application of balancing climates", in order to get the best out of the individual—namely, "making latent systems effective".[127] The appropriate "man-made climates"

must be ensured by means of adequate architectural and technical solutions.[128] In this respect, climatic intervention in the ambiguous space of the metropolis always represents a form of "climatotherapy",[129] which makes man-made climates relevant not solely "for the healing" of the individual.[130] Urban climatology is always also a form of climatic therapy for the metropolitan society, which continuously subjects the biopolitical demands of work to adjustments by means of architecture and technology.

Thermal variation forms the blueprint with which the climate of the ambivalent metropolis is to be analyzed. This requires a regulation of the ranges of variation of urban microclimates. The man-made climate of the modern city consists of *thermal differences*, which have to be created and regulated in practice. In this respect, it is not the opposition of active and passive, technological and architectural means that is decisive but a climatic interventionism that systematically allows *microclimatic time regimes* to unfold. To control the climate, as outlined by urban climatology of the interwar period, means correlating the different rhythms of modern life with the thermal geographies of the city. "Average temperature" as well as "frequency and scope of variations" must be "brought into a correct relationship"[131]—this is one of the incantations of the man-made climate.

Such an understanding, as developed as a hallmark of the interwar period, did not end with the National Socialists' coming into power. The "clarification of the metropolitan question" remained a critical concern of German-speaking scientists at the congress "Biology of the Big City" that took place in 1940 in Frankfurt. Several contributions dealt with urban climate and microclimatic issues.[132] The focus on *natural processes* in the metropolis anticipated—although with a pronounced political agenda—considerations that resurfaced in American "ecological design" 30 years later (see chapter 6). The metropolitan question became apparent also as a question of urban climate.[133] Treating the "housing question", as raised by Friedrich Engels,[134] not as a class issue but as a "metropolitan question" was the reformist path taken by interdisciplinary research on hygiene. While communist authors such as Engels understood the "housing question" strictly as part of a much broader social agenda, hygienic considerations aimed at the control of industrial production and residential areas.

→ figs. 70, 71, 72   European Developments before 1945

70
"80–90% of the population has no time for recreation". Illustration shown at the CIAM 4 exhibition on "Housing, Working, Traffic, Recreation in the Contemporary City" at the Stedelijk Museum, Amsterdam, 1935.

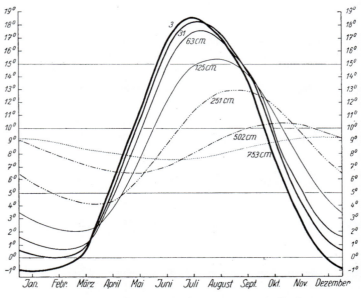

71
Annual course of soil temperatures in Königsberg (today's Kaliningrad, Russia). Source: Geiger 1942.

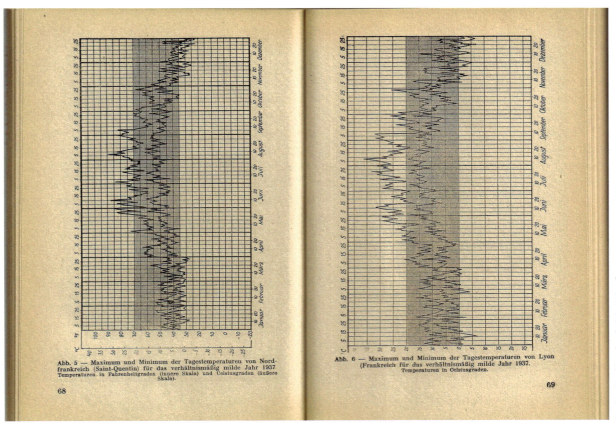

72
Maximum and minimum daily temperatures of the French cities of Saint-Quentin and Lyon in 1937. Source: Missenard 1949 (1940).

Man-made Climate by Design   67

The Eugenic Legacy of Man-made Climate

Missenard's reflections on productivity in light of the urban climate go far beyond purely economic considerations. Both Brenzina and Schmidt, as well as Missenard, display a penchant for eugenics that links climate to culture and heredity to the environment.[135] However, in sharp contrast to traditional *climatic determinism*, urban climatology promoted climate as an artifact of man. The man-made climate, as it emerged with the modern metropolis, is full of ambivalences; it is situated in a field of tension between natural and artificial influencing factors. Missenard's research had a dual focus: on "the impact" of the climate on human beings, and on the creation of "artificial climates" by human beings.[136] According to Alexis Carrel (a famous representative of eugenics), Missenard was concerned "to distinguish the influence of the environment from that of heredity".[137] There was a belief in the feasibility of combining thermal optimization with a eugenic optimization of the individual. Under the auspices of eugenics, heaters–wrongly used–promote the effeminacy and weakening of man; this begins already with the "fire" of early man, which is also understood as a "man-made climate".[138]

The inhabitant of the modern metropolis is particularly subject to the danger of turning soft. In addition to the prevailing "way of building", the "way of life"[139] of the city dwellers is considered to be detrimental to their general health. Brezina and Schmidt speak of the "unfavorable effect of the urban climate on health." They explain that this is due not only to the "extreme interior climate of the rooms" of many city dwellers but also to their "way of life" in general, which leads to a "greater sensitivity to temperature differences" and thus to the "need for a high, constant indoor temperature."[140] Compared with the countryside, working and living conditions negatively affect health. Missenard speaks of the "deceptive pretense of comfort" that "is more conducive to natural inertia than to trying to keep physical and mental qualities intact."[141] Hence, he demanded that resilience must be promoted through appropriate, artificially induced, quasi-natural climates indoors. The notion of man-made climate meant the anthropogenic character of the urban climate could be recognized and accordingly modified in architectural–technical terms.

The ambivalent line of reasoning developed by the European theorists of the man-made climate accounted for the eugenicist flavor that the debate on climate took on in the interwar period. European climatologists and architects speculated about the connections between climate and society, as well as climate and the constitution of individuals. The study of man-made climate of the period appears as part of a science on man in which physiological and psychological, scientific and (quasi-)anthropological findings converged. While the question of housing formed the social dimension of urban climatology, eugenics was the broader biopolitical agenda from which theorists of the man-made climate (such as Missenard) drew. Only after the Second World War would American-influenced research on comfort expel the eugenic legacy from the discourse on climate.

2.3   The Pre-Industrial City as Model

The pre-industrial and the industrial city form a charged pair. In the intellectual history of the climate-adapted city, the reference to traditional, pre-industrial urban environments within and outside of Europe is a consistent feature reflecting the future of the climate-adapted city. Central to this is the trope of climatic knowledge that has been forgotten and must now be restored in a scientific manner. The technical focus on traffic and sewage systems had downgraded the relevance of climatic factors in modern urban planning; according to Alfred Schmidt in 1926, "we must now laboriously recover, through scientific investigation, what our ancestors possessed in the way of rules of thumb".[142]

However, the architectural adaptation of traditional knowledge on urban climates relies on the recognition of the inseparability of imagination and science, visual culture and physics. Historically developed cities form an inexhaustible reservoir of concepts for the design of climate-adapted modern urban environments precisely because in them physical principles have been transposed into a designed urban space, thus bringing the exceptional and the principle-derived together. There was a consensus among climatologists and urban designers of the interwar period that, prior to industrialization, a unity of climatic knowledge and architectural translation prevailed. This conjectured unity of climatic and architectural solution-finding is ultimately what allows architects and climatologists to point to the exemplary nature of pre-industrial and vernacular urban formations.

2.3.1 "Ancient Town Planners"

Urban climatologists assumed that "the double aspect" of the influence "of the city on the climate on the one hand and of the climate on the city on the other hand had been observed by separate authors over the centuries".[143] Thus, if one follows the observations of these climatologists, urban climatic phenomena as man-made artifacts were by no means an issue of industrialization alone. Rather, already in the Neolithic period, dense accumulations of buildings were accompanied by microclimatic transformations that reshaped naturally given climates and produced man-made climates within such settlements.

The medieval old town of Korčula (Croatia) is an example of a street grid oriented according to the complex wind conditions; it represents how modern architects conceived pre-industrial cities as models for consciously designed urban microclimates. Ludwig Hilberseimer discussed the case in his book *The New City*. Korčula "faces a mountain on the mainland, the Monte Vipere. The position of this mountain influenced the location of the city; winds from its summit influenced the city's structure. The streets are arranged in a herringbone pattern so that the cold mountain winds cannot penetrate them. These streets laid out to an angle to the direction of the prevailing winds." Not surprisingly, Hilberseimer continues, "To protect residential areas from the atmospheric discharge from industrial areas, the layout of both areas must be determined by the prevailing winds."[144]

Climate and Urbanization

It is precisely the *lower* complexity of pre-industrial urban environments that gives rise to studying their operative climatological principles in detail. As such, pre-industrial cities appear as a kind of epistemological testing ground (living labs) to explore and to illustrate the scientific character of the modern approaches. The pre-industrial city represents a model for the assessment of the effectiveness and relevance of urban climatological parameters; Alfred Schmidt speaks of the climate-relevant "rules of the early town planners".[145] Carl Kassner emphasized in 1910 the need for a critical evaluation of previous climate knowledge and the scientific questioning of common sense. On the one hand, historically grown urban layouts, such as those in northern Italy from the medieval and early modern periods, show the presence of urban climatic considerations in their design. "Where such winds blow–the Bora in Trieste, the Mistral in Marseille–no long, straight streets may be laid out in the direction of the wind, say toward the harbor, but

only obliquely to it and broken. The direction of the prevailing wind in a place can easily be seen by the inclination of the trees."[146] On the other hand, there is a danger that our perspective is steered in certain directions by contemporary (scientific) readings, without taking into account the truly original solutions of the period. At issue is how one can judge "the action of the weather as a cause" for urban planning arrangements. "If one deals with the question whether meteorological influences on the location and layout of cities can be proven, then it is not difficult to find such influences almost everywhere on superficial examination. However, as soon as one goes back to the history of the foundation of the towns, one experiences many disappointments, because often, where one thought to have found the effect of the weather as a cause, completely different reasons for the establishment of the place emerged. One must consider that only in very few cases were the weather conditions there were already sufficiently known at the time of the first settlement or the foundation of a larger place."[147] Kassner thus emphasizes the multiple reasons for the location and arrangement of cities and thus a basic indeterminacy of what caused them to be that way. An urban arrangement can have more than one cause and in many cases it is impossible to trace what ultimately was the decisive factor. Caution is advised when viewing historic city structures through the lens of contemporary perspectives.[148]

As a whole, the scientific knowledge of urban climatology appears also as *a heuristic of planetary urbanization*. Climatic knowledge represents an interpretation aid for assessing city formations. In the context of the colonial exploration of the world, the global scale of climate, as systematized by Köppen and Geiger, provides a better understanding of regional territories. The appearance of a pre-industrial city might relate to its climatic zone but not to its multifactorial influences, as Kassner points out. "As the climate influences the building in its structure, so does the whole city, and this has been repeatedly pointed out. Even aside from vegetation, looking at the majority of city views one can at least say in which climatic zone the city lies; but one cannot necessarily be more precise, since, for example, if only a limited amount of space is available for the city, as is the case on steep coastlines, this necessitates such narrow streets as one would otherwise only expect to find in the subtropics for shade."[149] In other words, the climatic-zone approach depends on micro-spatial investigations. Urban vegetation, for example, appears as an indicator of the prevailing thermal conditions, which architects should use to evaluate the prevailing microclimates, combining architectural and

→ figs. 73, 74, 124 Man-made Climate by Design

landscape-architectural elements. "In addition to the climate of the individual zones, equally important for urban design and its particularities is the specific climate of the site and its surroundings, which is dependent on its location. At the sea or inland, on the plains or in the mountains."[150]

With the climatic-zone approach, a comparison of different countries becomes possible; the consideration of climatic zones provides numerous insights into what is to be considered a reasonable building method in a particular place. Rules of thumb for contemporary building can be derived from this. "In general, the rule is: the airier the construction of the buildings and the larger the doors and windows, the more southerly the location; cities with predominantly horizontal roofs cannot be located in areas with noticeable snowfall, because they would be too heavily loaded in winter."[151] The following 1941 statement by Helmut Landsberg is exemplary in characterizing the interdisciplinary relationship between meteorology and architecture: "In Europe, old buildings show a curious relationship to continentality. The further inland, the thicker are the walls. This is presumably an early architectural attempt to compensate for the wider temperature swings, both diurnal and annual, inland. [...] The wall thickness [...][is] a function of distance from the ocean."[152] This statement expresses particularly clearly the merging of macro and micro perspectives in applied urban climatology. The continental view, which is created by observations of climate zones, is made directly applicable to the interpretation of urban and architectural phenomena. "The climatic effect of the narrow alleyways helped bring about their widespread use in the cities of the Orient as well as the Mediterranean area."[153] The macro-scale has direct implications for the microclimatology of buildings.

Building types and climatization appear as indexes of national idiosyncrasies. Kassner quotes, for example, the Russian climatologist Aleksandr Ivanovič Voejkov (1842–1916): "It is a general observation that the colder the winter and the longer the cold season, the warmer the dwellings: Italian or Japanese are colder than French and English, the latter colder than German, the latter colder than Russian and Scandinavian, and the dwellings of the Eskimo and Chukchi are so warm that the people there are naked, while the Japanese in winter put on one padded kimono over another."[154] What results from the consideration of large climatic scales, according to the historian of science Deborah Coen, is "the appreciation of cultural variation";[155] but also, as she has pointed out with regard to the Austro-Hungarian Empire, "unity in

diversity".[156] A (cross-cultural) comparative methodology provides quasi-ethnographic insights into the ways of living and building in given countries, and into overarching principles of the rationale of buildings. The arrangement of buildings in a city has a direct impact on heating needs, depending on the climatic zone. "While the open building method has already its disadvantages in our country, it is even less appropriate in the far north; [...] that is why in former times in our country and now, for example, in Norwegian cities, the houses were only placed with the narrow gable end facing the street, whereas the long sides abutted one another. Every city dweller who moves into a detached house in the suburbs suddenly realizes that he needs much more heating material outside of the city than in the city."[157]

Vines along the facades are a proven way to protect the facade from solar radiation. Hilberseimer refers to an "old Chinese philosopher" who recommends greening or shading facades. "The shade of these trees would protect the house in the summer [...] and in the winter, when the trees were bare, the sun could find uninterrupted access into the rooms. There is much wisdom in his suggestion."[158] Other proven means include arcades, eaves and cantilevering the floor to protect the floors below from too much sunlight (see chapter 1). Buildings are subject to constant erosion due to the weather conditions, and require adequate protection. The constant exposure to sunlight, rain, snow, wind, as well as urban air pollution all contribute to heavy wear and tear on urban buildings, but also provide an unmistakable weather-related patina. Weather, in the form of driving rain for example, is seen as a force threatening architecture that must accordingly be taken into account. The ageing of buildings is in no small part a function of their exposure to the weather.[159]

2.3.2 Terrace Construction (on Steep Hillsides)

Historically, the urban reasoning on the thermal use of sun and winds was reflected in written and unwritten form in recommendations, regulations and laws concerning urban development, and in visual-graphic representations of street maps, zoning plans and sets of standards. Ancient cities such as Olynthus (Greece) and Priene (Turkey), as well as the pueblos in the southwestern United States (such as the Acoma Pueblo), have been repeatedly cited since the interwar period as examples of regulated "solar access". Richard Döcker dealt with pueblo-building (*Pueblobau*) as part of his study of modern terrace

→ figs. 75, 76, 77, 217   European Developments before 1945   **70**

construction (*Terrassentyp*).[160] Döcker, who was the construction manager of the Weissenhofsiedlung in 1927 and built two buildings on the site himself, published *Terrassentyp* in 1929, the first systematic work on this pre-industrial construction method and its relevance for modern housing development.[161] In particular, Döcker criticized the "architecturally poor hillside architecture" of Stuttgart with its scattered buildings, which to a certain extent transferred the pattern of planning on the flat terrain to the hillside.[162] At almost the same time, Alfred Schmidt was working in a similar direction, calling for parallel consideration of different climatic factors such as wind and sun: "It should be emphasized that it would be one-sided to try to orient the streets only according to the sun. For very many places the consideration of the winds is as necessary as that of the sun. One thinks, for example, of the coastal cities under the influence of the sea and land winds, or of mountain cities which have to suffer from the ferocity of the mountain winds (Föhn, Bora, Mistral) that descend into the valley."[163]

Ludwig Hilberseimer dealt with questions of solar accessibility in the context of his investigations on the climate-adapted new city and thus created a whole modern tradition, which was deepened some 30 years later by Ralph Knowles and others (see chapter 6). Hilberseimer speaks of "the Greek spirit" that shaped the heliomorphic planning of cities. In light of such a spirit, modern heliomorphic design seeks to overcome social differences by creating more uniform, one could say more democratic, forms of *solar access* through architecture (see chapter 3). For example, he mentions the Hellenistic city of Priene, which was built on a steep hillside in 350 BC taking the orientation of buildings into account. Residential streets had an east–west orientation, while links of these streets ran from north to south.[164] The important rooms of the houses and their porches were oriented to the south. In Xenophon's *Memorabilia*, written about 371 BC and also mentioned by Hilberseimer,[165] Socrates emphasizes the value of the sun in urban planning.[166] Hilberseimer also addresses the "Pueblo dwellers of San Ildefonso",[167] of which he had acquired knowledge through the writings of Edgar Lee Hewett. Similarly to the case of the Hellenic city of Priene, the pueblo dwellings were oriented to the south.

Sara Bronin has pointed out the trans-cultural significance of solar-access regulations: "Ancient Romans protected the right to solar heat and light through prescriptive easements, government allocations, and court decrees. Ancient Greeks protected solar rights through rigid land planning schemes that oriented streets and buildings to take advantage of light and passive solar heat. More recent rules–such as the so-called 'ancient lights' rule established in medieval England or the permit system currently used by Japan–have continued to refine the concept of solar rights. Each regime has recognized that sunlight, in reaching any one parcel, may travel across multiple parcels, and its route may vary throughout the day and from day to day. By necessity, then, the creation of solar rights implicates the rights of neighbors, both immediate and further afield."[168] Historically, access to sunlight and solar heat had to be regulated within the framework of a neighborly balancing of interests. According to the historian Katja Esser, conflicts arising from the "privation of air and light by neighbouring houses" have been known since the Middle Ages in Aachen, for example. Aachen, as a spa town, introduced regulations following the great fire of 1656 that allowed the town to be well ventilated. "A guidebook for spa guests from 1762 describes remarkable features: a building density not high enough to obstruct ventilation plus numerous gardens and meadows within the city walls, spaces for cultivating vegetables and fruits."[169]

The pre-industrial city appears equally as a model of a comprehensive understanding of applied climatology and as a symbol of a pre-scientific, i.e. less exact, approach to the climate of the city. In 1930, Paul Schmitt emphasized the still insufficient base of knowledge concerning climate control in/of/through cities. He mentioned, for example, the still largely unexplored interaction of sun and wind as heating and cooling resources for urban housing. Likewise, he criticized common principles of hygiene that were the result of a lack of scientific expertise on the one hand and a lack of knowledge of urban history on the other. Schmitt questioned the paradigm of the southern orientation of the houses (rows), which should be approached with accurate urban climatological knowledge. "In the Urals, in windy and snowy areas, all houses are built east-west with the entrance facing south and the barn facing west. Embankments are heaped up as protection against north winds. The building arrangement is adapted to the particular climatic conditions on the basis of long-term experience. Such experience must be used by the town planner."[170]

In the urban climatological literature of the interwar period, pre-industrial cities are marked by this ambivalence: on the one hand, they form recurring models for the new climate-adapted city; on the other hand, however, they are characterized by innumerable deficits, which form the starting point for the new city. In

→ fig. 119   Man-made Climate by Design   **71**

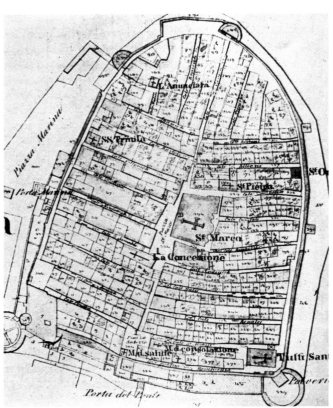

73
Layout of the old town of Korčula (Croatia). Source: Hilberseimer 1944.

74
Old town of Korčula (Croatia). Source: Hilberseimer 1944.

75
Rendering of the city of Priene, Asia Minor (today's Turkey). Source: Hilberseimer 1944.

European Developments before 1945

76
Four-story buildings of the Pueblo Taos (USA). Source: Döcker 1929.

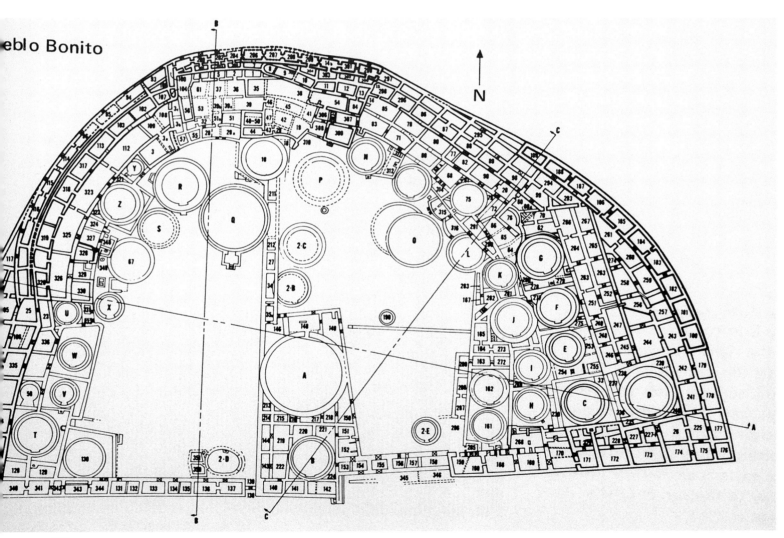

77
Layout of Pueblo Bonito (USA). Source: Hilberseimer 1944.

Man-made Climate by Design

particular, diseases such as plague, cholera and malaria—which were related, among other things, to climatic conditions—formed fundamental problems in pre-industrial cities. Therefore, as much as the pre-industrial city is recurrently cited for its exemplary climatic qualities, the deficiencies of the traditional (grown) city are also pointed out. In this respect, the pre-industrial city is comparable with the speculative factory town of the late 19th century. The industrially shaped city has numerous aspects that make it itself an unplanned, grown city and thus it almost resembles the pre-industrial city—it was precisely the nature-like character of industrialization that was registered and criticized at the time. Hilberseimer speaks in 1927 of the problems of the "natural, grown [...] city layout" and was referring, in particular, to the city of his time.[171]

3 Democratizing Urban Nature

Infrastructures of Climate Control at Large Scales

In the interwar period, climate control was still seen as part of a collective urban endeavor rather than as a private matter. The fact that the majority of new housing did not yet have central heating sheds light on the importance of climate control at the urban scale at the time. As I will show, it was *urban nature*, *the arrangement of buildings* and *zoning* that appeared as the three central levers for a conscious design of the microclimates in the city. Thus, it was not yet primarily technical infrastructures that shaped the logic of climate control but rather the elements of the emerging modern urbanism itself: transport routes, green spaces and entire urban blocks that, considered together, formed a "large technological system" of urban climate control.[1] These components, which also shaped the new discipline of urban planning, formed the infrastructures of the man-made climate. Its climatological principles emerged in parallel with modern urban development.

3.1 Access to Urban Nature: Public Green

Discussions about man-made climate in the interwar period were shaped by a concern for the accessibility of urban nature (such as green spaces and bodies of water). The Social Democrats who came to power in cities such as Berlin and Vienna after the First World War understood "residential building as a means of proving the superiority of their policies".[2] With the social-democratic urban development under Martin Wagner, urban nature came to be viewed as representing important infrastructure for broad sections of the population, and access to it was thus systematically enhanced. The democratization of access to urban nature thereby anticipated concerns that were later also propagated in the United States under the ecological auspices of "solar access" (see chapter 6).

As the geographer Matthew Gandy emphasized, "the place of urban nature in architecture, planning, and urban design" played a significant role in interwar Berlin. In particular, green spaces and bodies of water were at the center of Berlin's regional and urban planning efforts, which were developed in close connection with the new housing developments. Gandy speaks of "a new set of interactions between water, the human body, and urban space", which was conceptualized by architects in that period.[3] There was "growing popularity of bathing at the urban fringe" and new forms of "semicultural landscapes".[4]

3.1.1 Decorative or Sanitary Green

Berlin was the first ever metropolis to develop and test the quite contrary aspirations of the *stony* and the *green* city.[5] Negotiating between the conscious planning and dissolution of cities, new concepts for the integration of nature into the city and the transformation of the city into urban nature were elaborated. Matthew Gandy speaks of "a series of modernist architectural interventions that explored the possibility of creating a new synthesis between nature and culture".[6] In particular, *networks of green and blue infrastructures* were connected with the new housing estates and industrial sites.[7] Greater Berlin was the setting for numerous contrasts that suggested experimentation with such infrastructures. The regional climate was characterized, as the geographer Friedrich Leyden put it, "by its volatility, not to say unpredictability",[8] while the urban landscape alternated between highly dense urban areas and still largely undeveloped wooded regions, lakes and semi-natural landscapes. In addition, there were gardens, such as in Westend and Finkenkrug near Spandau, which were habitats for Berlin-specific flora and fauna, including animals such as salamanders and numerous bird species.

The Architecture of Urban Green

The omnipresent nature in Berlin opened a new "space of urban imagination",[9] which gave the architectural reflection on the city a specific direction and linked, as I would like to show, the design of the man-made climate to the future of cities in general. The "connection between the city and nature [appears] in the existence of natural spaces within Berlin [...] that inscribe themselves in the discourse of the modern and seminal city".[10] The omnipresent natural spaces of Berlin were seen in Siegfried Kracauer's writings as "signs of an urban potential for the future".[11] In his texts *Strassen in Berlin und anderswo* ("Streets in Berlin and Elsewhere")*,* written between 1925 and 1933, Kracauer describes a roof garden where experiences of nature and the city are inseparably combined: "One flies in an elevator through ten floors, where business life, if not flourishing, is vegetating, and then reaches a platform that can claim the status of a high-altitude health resort. For it is not simply a rectangular area of asphalt, but a kind of artificial alpine pasture. Lush green meadows stretch out directly above the stuffy offices, and lush flora sprouts in countless tubs from the soil of the cash books and files. Here the sun shines more brightly than down in the depths, here the wind blows as if around peaks. But the special magic of this skyscape is that it contains a lot of deck chairs, which are free to use."[12] The *green terrace* and the *green facade* became architectural means for mitigating the city climate. In 1931, Goldmerstein and Stodieck published a booklet relevant to urban climatology, which is characterized by its focus on feasibility. Entitled *Grossstadtsanierung* "Metropolitan Redevelopment", the authors put forward green proposals for Berlin as a *city of stones*, which include intensive greening of vertical and horizontal surfaces.[13]

However, modern urban development increasingly followed the requirements of traffic, alongside those of industrial production. Motorized traffic demanded new and wider roads, competing with urban green. In 1937, Brezina and Schmidt complained about the inadequate greening of large cities. "Thus the street ground grows at the expense of the yard area, but a reduction of the built-up area or even a gaining of area for greenery hardly occurs at all".[14] In addition to the materials used, and the shape and alignment of the buildings, green areas have a crucial impact on the microclimates of the city. Albert Kratzer emphasizes the "climatic impact" of green areas: "They mitigate the strong warming effect on the bordering city districts and collect the dust of the streets."[15] Le Corbusier's city is mentioned as an example of urban territory being made available for the development of green areas through the construction of high-rises; according to Le Corbusier, Kratzer maintains, "85 to 95% of city land" is to "remain available" for green areas.[16] The central (hygienic) concept in the context of an urban-planning integration of housing estate and green space was aeration (*Durchlüftung*), which was applied at different scales. Green infrastructures were much more than mere vegetation; they were an agency to enable the ventilation of the city. In this respect, they appeared as the counterpart to the overheated and air-polluted streetscapes in inner cities; transportation systems and green infrastructures were understood as complementary networks of the city.

In fact, from the end of the 19th century a discourse was developing that assigned a comprehensive role to greenery in cities that went beyond mere

78
Hermann Jansen, general map in the frame of the competition for "Greater Berlin", 1910.

→ figs. 79, 80, 81, 82   European Developments before 1945   76

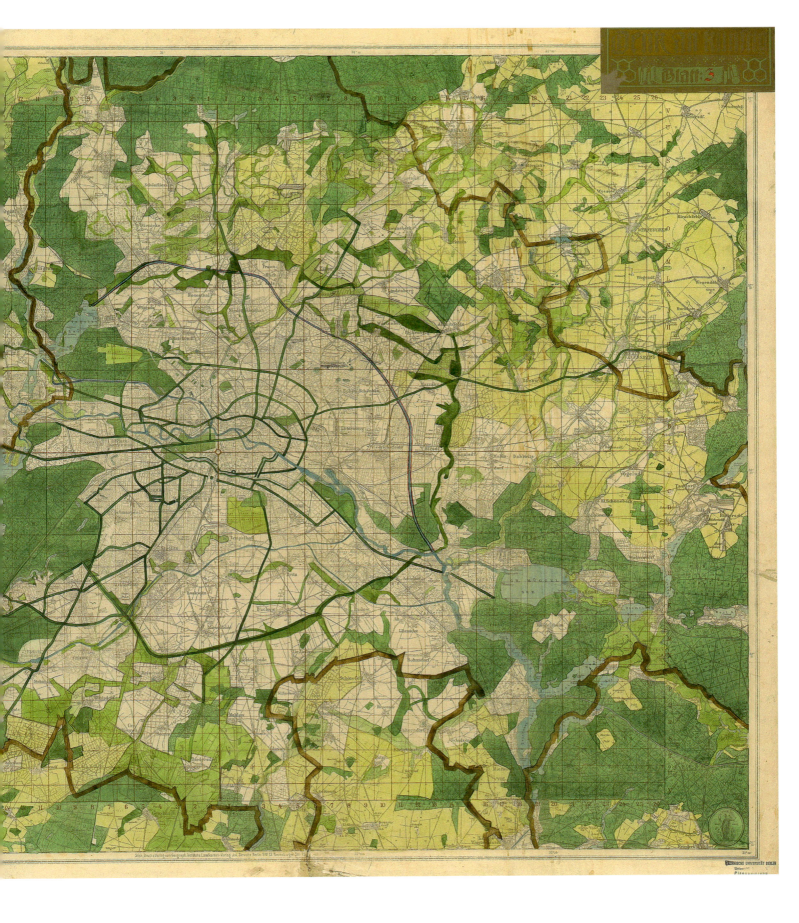

beautification or recreation. In *Der Städtebau* ("Town Planning"), published from 1890 onwards, Josef Stübben had already interpreted the advantages of open development as manifested in neighborhoods of villas in terms of urban climate. The greater permeability and greening have an impact on the "health" in the respective neighborhood and, moreover, in the city as a whole; compared with inner cities, the negative thermal effects of mutual shading of dense urban structures are insignificant. Stübben launched the idea of a general importance of green spaces (as green lungs), from which left-leaning architects such as Martin Wagner were later to distance themselves. The unequal accessibility to green spaces for different social classes in the industrializing city was not yet an issue. "The advantages of open building are not limited to a graceful, handsome appearance, to the better effect of architecture, and to greater convenience for the inhabitants; the freer building of individual districts is at the same time an important measure for the health of these districts and of the city as a whole. Due to their supply of unpolluted air and their abundance of plant life, they also benefit the neighboring parts of the city. Their effect on health is similar to that of public gardens, and their importance is therefore all the greater the poorer the city is in plantations and parks. The greatest advantages, of course, are enjoyed by the inhabitants of the villas themselves, since they are provided with abundant light and air, can direct the rooms of their houses to the rays of the sun, and can make themselves almost independent of the influences of the neighbors."[17]

Camillo Sitte, in particular, proved to be an important provider of key terminology for the planning of green urban infrastructures. Among other things, he introduced the notion of "sanitary green" into the urban-planning discussion in distinction to "decorative green." The systematic integration of green spaces in housing estates went hand in hand with Sitte's call for "artistic" urban planning, which was characterized by a medieval-like arrangement of the city.[18] The diversity of urban situations propagated in this context also had implications for an accompanying microclimatic diversity, characterized by a variation of shaded and sunlit, warmer and cooler zones. The freer geometry of vegetation was also a guarantor of this approach, which was directed more generally against the excessive geometrization and formalization of cities. In his 1903 zoning plan for Marienberg (Ostrava-Marianské Hory) in what is now the Czech Republic, Sitte deliberately worked with the concept of "sanitary green" as part of a new city plan. This plan aimed to develop a town from a small housing estate.[19]

Two architects and urban planners took up and continued Sitte's conceptual impulses in a particularly profitable way: Hermann Jansen (1869–1945),[20] one of the well-known representatives of an approach to urban design based on medieval architecture as advocated by Sitte, and Martin Wagner (1885–1957), who took up Sitte's significant distinction between "sanitary" and "decorative" greenery in the context of his 1915 dissertation in order to orient existing greenery to the requirements of the emerging modern metropolis. In Berlin immediately before and after the First World War, the insight that the urban climate was also shaped by its green infrastructures led to novel ideas towards planning, through which disparate hygienic concepts came together to form a new coherent planning approach.

Ventilation Corridors

An urban-planning milestone for this debate was the *Gross-Berlin Wettbewerb* ("Greater Berlin Competition"), which was conducted in 1910.[21] Towards the end of the 19th century, Berlin had become the largest industrial city in Europe.[22] Industrialization took off especially with the founding of the Zollverein after 1835, resulting in huge factory complexes that shaped the city's development as well as necessitating a huge amount of working-class housing to be built. In the late 1920s, Berlin was the fourth largest city in the world–after London, New York and Tokyo– with more than four million inhabitants. Large differences were apparent between high-density inner-city districts such as Kreuzberg, Wedding and Friedrichshain and suburban areas such as Steglitz and Zehlendorf.[23] It was not until 1920 that the nearby towns and suburbs were incorporated into the city.[24]

The competition for a new master plan was launched jointly by Berlin's two architectural chambers in 1907. It focused on a considerably larger area than Berlin's 1920 city limits would have encompassed. It included the region from Zossen to Oranienburg and from Königswusterhausen to Potsdam—an area of over 2000 km².[25] The competition has to be considered "a key transformation of international urban thought", in which the flexible interlocking of different scales from S to XL, from the consideration of the region to that of the building, is included, so that one can speak of "a new understanding of the territory" and "the inter-scalar relationships of the city".[26] "Instead of the bounded, compact city of the nineteenth century, 'Big Plans' announce the rise of the city region as a network of urban components drawn across the geography of the region."[27] Several competition entries made the greening of the city their central theme.

→ fig. 78   European Developments before 1945

Hermann Jansen worked out "a colossal, branching green corridor for the extended urban territory of Berlin", which would have been characterized by "numerous urban islands".[28] The jury awarded one of the first two prizes to Jansen. While urban islands were to provide space for housing and industry, the several 100-m-wide branches of the green network proposed were intended for public buildings (schools, kindergartens, courthouses), playgrounds and sports fields. He writes of a "large-scale belt of woods and meadows interspersed with hundreds of green tributary arteries",[29] which thus resembled, in a contemporary vocabulary, a "green network".[30] In a semicircular movement, a particularly "large forest and meadow belt" encompassed the metropolis from the north to the south. Green belts were supposed to run from the inner city to the outer districts, connecting Berlin's large forested areas with the core city. They formed the widest of all the newly created "forest, park and meadow areas".[31] In addition, there were radially laid out green zones, with numerous cross-connections, which in a sense penetrate the city from all points of the compass. It was the declared goal to bring the "existing woods and existing larger trees" "into relationship with each other" through reforestation and the creation of new meadow and park areas.[32] Thus, one can speak of a deliberate blending of natural and residential areas. The proposed breakthroughs through the existing urban structure were based on the approach of Georges-Eugène Haussmann in Paris, but now aimed more strongly at a network-like connection of the previously disparate urban districts and green spaces. On the plan *Durchbrüche* ("breakthroughs") this new network is indicated in its methodical intention. It was not enough merely to conceive of the new city as a network of apertures; beyond that, new apertures must be cut into the existing, speculatively generated city. An important working tool was the street cross-section, which showed the new structuring of the street lines and the deliberate upgrading of the tree stock. On the outskirts of the city, housing estates were to be built—as Jansen's numerous renderings show.

Although distancing himself from Vienna as model, Jansen referred in his competition paper to the Austrian metropolis and the *Volksring* ("Peoples' Ring") propagated by Eugen Fassbender in 1893—a "green ring" that was to enclose and limit the city and provide both better air for the largely unregulated industrial zones and working-class recreational spaces. "The main goal was not to create a geometrically closed green belt, but to make a sufficient area of open land available for each housing estate, i.e. to provide it with large sports and playground areas in addition to for-

est areas."[33] The guiding criteria for the planned afforestation should therefore not be mere quantity but the quality of the "tree cover" and the "proximity" to the "most densely populated residential areas". This housing-policy concern anticipated central features of the Weimar period. The urban-planning discourse on the role of green spaces was combined here with democratic theoretical claims that—in the long run—also affected thinking about the urban climate. The "natural beauties" of Berlin were to be "preserved in the main for the general public".[34] "So the children in the east and north of Berlin have the same right to forests, open spaces and [...] the banks of rivers and lakes, as those in the west. To want to make a distinction here is not quite understandable."[35]

The "strong working-class population" in the east of Berlin and the comparatively weak greening suggested the creation of "larger areas for parks". The division of the green spaces also followed urban climatological considerations: "These are laid out in such a way that they separate the industrial and residential quarters from each other and make it possible to walk to the train stations on park paths."[36] In an exemplary and early articulated manner, the design of urban greenery appears as a key element of urban planning. Along with industrial and residential development and transportation, green spaces were introduced by Jansen as a missing link that separates or brings together the functional zones of the city in a desirable way. Jansen speaks of "cross-connections" between the so-called "radials" of his plan, connecting the different urban zones. What was previously mixed in an uncontrolled manner was now to be kept clearly apart or cultivated as a "connecting link between these important, seminal industrial areas".[37] Thus, it was necessary to weigh up between the demands for the closest possible proximity of work and home and the arguments of hygienists and climatologists: "The workers' estates and other small housing estates" were to be built "as close as possible to the industrial sites"; "industrial sites, especially those with disruptive and unhealthy operations, were to be separated from residential areas by a green zone".[38] New rapid-transit lines were to provide relief from the "inadequate, time-consuming connections" of the past.[39]

Local Air Centers

Martin Wagner, using Camilo Sitte's vocabulary, took up Hermann Jansen's vision for Greater Berlin in his dissertation *Das sanitäre Grün der Städte. Ein Beitrag zur Freiflächentheorie* ("The Sanitary Green of Cities. A Contribution to the Theory of Open Spaces")

→ figs. 23, 83, 84   Democratizing Urban Nature

79
Adapted old town of Bern (Switzerland). Rendering shown at the CIAM 4 exhibition on "Housing, Working, Traffic, Recreation in the Contemporary City" at the Stedelijk Museum, Amsterdam, 1935.

80
Old town of Bern (Switzerland).

European Developments before 1945

81
View into the transformed Schillerstrasse in Berlin-Charlottenburg. Source: Goldmerstein and Stodieck 1931.

82
Example of the play, sand and green areas gained. Source: Goldmerstein and Stodieck 1931.

83
Different types of distribution of green areas in the city. Source: Heiligenthal 1921.

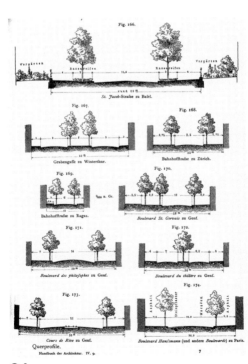

84
Greening of streets in different cities. Source: Stübben 1980 (1890).

Democratizing Urban Nature

from 1915. The study was prompted by the unhygienic living conditions of broad social strata, which led to numerous diseases of the "respiratory organs" (such as "pulmonary tuberculosis"). Wagner was primarily concerned with the deliberate use of green spaces in planning and construction to improve the quality of life of urban workers and employees, and called for a greening of the densely inhabited workers' quarters. Far more than representation, Wagner aimed at the everyday practical accessibility of green spaces for the general population. The accessibility of green spaces–the "physical possession of open spaces" by broad sections of the population—was his central concern.[40] "While decorative greenery was responsible for the beautification of the big city, sanitary greenery was able, on the one hand, to ensure urban ventilation and, on the other, to compensate for the unsanitary conditions in backyard and basement apartments."[41] Future greened outdoor areas were to provide a new kind of (large-scale) interior space for the urban population.

The function of green spaces as "air reservoirs and air improvers", described by Wagner as their *Daseinswert* ("existence value"), was to be subordinated to their *Nutzwert* ("utility value"). "What is 'sanitary green'? All green spaces and green areas that have a positive influence on human health will be able to be called sanitary green. The influence of green spaces on human health is particularly characteristic in two ways. The green areas can become indispensable as the great air reservoirs and air improvers for the metropolitan population, without the population needing to come into closer contact with the green areas; in this case, their sanitary value lies in their existence per se; in future, let us call this its existence value. On the other hand, the green spaces gain sanitary value to the extent that man uses them, whether in the form of playgrounds or sports fields, promenade roads or parks. This value of the green spaces can be called utility value. The existence value of open spaces has until now been greatly overestimated at the expense of use value."[42] This opened a line of thought by Wagner that would shape the modern treatment of green spaces as "air reservoirs".[43] Wagner opposed Jansen's vision of a territorially spanning network of green strips that would house, among other things, public buildings. Rather, Wagner aimed at a microclimatic redesign of residential neighborhoods themselves, linking urban climate issues with social concerns. His argumentation promoted the everyday accessibility of the green spaces, which were not conceived as mere recreational areas (at the edge of the city), as well as the microclimatic interdependence of outdoor spaces and housing.

Martin Wagner had thus taken up considerations that had already been made a few years earlier by Otto Wagner with regard to Vienna. In *Die Grossstadt* ("Large City") of 1911, Otto Wagner propagated the integration of green spaces within the individual quarters instead of a green belt around the city. These were to be understood as part of a comprehensive infrastructure and transport provision, which would be combined with the evenly expanding quarters. According to Wagner, "it seems more correct to give each individual district its own sufficient air centers in the form of parks, gardens and playgrounds than to project the adoption of a forest and meadow belt; after all, the creation of a belt stretching around the city is again merely a predetermined site development".[44] In contrast to Hermann Jansen's proposal for a "forest and meadow belt" and the radial city (with uniform distribution of green space) by Eberstadt, Möhring, Petersen, which propagated large-scale "ventilation", Otto Wagner sought to strengthen the local "air centers" within the dense city. The landscape architect Leberecht Migge was to take up this understanding of the two Wagners and developed it further. The "city-ordering function" of green space,[45] as Martin Wagner had in mind, must, however, be distinguished from the rigid block grid that Otto Wagner was simultaneously striving for with his "Large City" proposal. In parallel with modern mass housing, new political regulations were intended to restrict speculation and private power of disposal over green spaces.[46] With Martin Wagner, who had been appointed to the office of city planning councilor in 1918, new urban-planning principles were introduced aimed at putting a stop to the privatization of urban green spaces.

3.1.2 The Interiority of Settlements

Around 1900, a new awareness emerged in Europe of the importance of the planned urban quarter as part of a city whole. The architectural historian Katharina Borsi speaks of "the process of evolution from the continuous, undifferentiated fabric drawn in the 1860s to the emergence of spatially and programmatically distinct urban segments around 1900".[47] Borsi introduces the term "interiorization"[48] in order to describe the specific sense of formal unity of the new urban quarters. Such an interiorization must also be understood in terms of microclimatic conditions. The integration of green open spaces in the interior of the housing developments was an important driver of urban innovations. Bruno Taut called these open spaces *Aussenwohnraum* ("outdoor living space"). "Getting out into the open" was, as the hygienist Carl

→ figs. 38, 87, 88   European Developments before 1945   **82**

Flügge had already expressed, a central problem for architectural design (see chapter 1).[49] The adaptation to the changing seasons still played out in the field of tension between inside and outside. According to Flügge, there should be a clear preference for low buildings, which represented a counter-program to the rampant tenements of the period. The suburban housing estate becomes the hygienic ideal: "for the accommodation of the largest part of the population, it is absolutely necessary from the hygienic point of view to adhere to the small house of no more than two stories".[50]

Outdoor Living Space

In his competition entry of 1910, Hermann Jansen distinguished between "existing developments", "new closed developments", "narrow open developments", "wide open developments", "new industrial areas" and "waterways". Jansen aimed his design considerations at the approximately 90 percent of Berlin's population that counted as working class. A clear criticism of overly wide streets, which was due to the bourgeoisie's desire for a sense of status, he contrasted with "large recreational areas" and "plain residential districts".[51] Jansen proposed "small housing estates", especially for the outskirts of the city, which were to be built as terraced houses (and not as miniature villas) with generous gardens. An example of this is the small housing estate in Buckow-Rudow.[52] Here, large- and small-scale climatological considerations for new housing estates overlap: "Ventilation must be ensured to all dwellings, and this is most easily achieved by determining the distance of the rear side from the front side, which is based on the respective building block depths or the depth of the rear courtyard and garden. In no case should the distance of the back fronts of two streets be narrower than the width of the usual residential street with its 18–30 m or the height of the enclosing houses; the minimum length of the building block should also be determined for the purpose of strong passage, i.e. at least 100–200 m. The unfortunate principle of side and cross wings should be definitively abandoned in the case of residential quarters; except for speculation, no one has any interest in them. If the building block is opened on one side, if possible in the south by a small building sandwich, ventilation is facilitated–besides, a welcome motive for an architectural formation arises."[53]

The development plan for the Tempelhofer Feld, which had a 180-m-wide park belt giving the perimeter block structures their overall form, is exemplary. The long blocks, which are given openings by their "gate buildings", open onto these park belts "for the purpose of ventilation".[54] The new building blocks Jansen has in mind were based on the omission of the inner courtyards typical of Berlin's tenements; this created a more balanced relationship of "solids and voids".[55] He contrasts speculative examples on Joachim Friedrichstrasse (p. 54) and examples of self-designed small estates (p. 59) to make the differences visible. Opening up the blocks also allowed for ventilation of the interior of the housing estates. Jansen linked climatic demand with architectural aspiration by declaring ventilation a new motif of architecture.[56] In order to achieve the desired wind pressure, he recommended building blocks at least 100 m long. Especially in the planned urban extensions, he introduced perimeter-block development, which allowed for extensive green spaces inside the housing estates. He called them the "inner parks" [57] of the housing estates. Some 70 years later, architectural historian Julius Posener spoke of the "interior of the building block as a garden for walking".[58] Walkways with rows of trees inside the estates, combined with the tree-lined avenues of the streets, were to form a coherent vision of urban green spaces. On at least two sides of the blocks, large openings make the crossing of the individual estates possible. Hence, climatological considerations find their counterpart in the green space of the housing estates and streets. The microclimatic regime of greening the metropolis was combined with the control of winds. In this respect, with his competition entry Jansen was aiming for something that was arguably a precursor to *thermal governance*. He emphasized the importance of the "interaction of development plan and building regulations"[59] for the future feasibility of his proposals.

With the Lindenhof housing estate in Tempelhof-Schöneberg about ten years later (1918–1921), Jansen's ideas were taken up and put into practice.[60] The project was exemplary for a new type of settlement based on the integration of housing and green infrastructure. The estate offered low-cost housing, and had communal facilities as well as individually supervised planting gardens. This mix, novel for Berlin, owes much to the collaboration of Heinrich Lassen; Martin Wagner, then Schöneberg's city building councilor; and Leberecht Migge. In addition, Bruno Taut was responsible for the planning of the house for single people that was part of the estate. With its old tree populations, the Lindenhof complex was characterized by a park-like character, at the center of which was a pond that had emerged from a small lake. The importance of open spaces for modern urban development is exemplified by the *Helle Hölle Volkspark* ("Bright Hell People's Park") designed by Migge, which united various community facilities. It is here

→ figs. 85, 89, 90   Democratizing Urban Nature

that the idea of green infrastructure becomes exemplarily visible. One can speak of the creation of a micro-climatically rich outdoor space for poorer strata of society.

However, the interpenetration of urban nature and architecture was also explored in the case of villas and single-family houses. The Foerster-Mattern-Hammerbacher consortium was particularly prominent in the design of modern gardens and public spaces.[61] The philosophy that emerged from this working community was based on an integration of house, garden and surrounding forest. The natural space penetrates the house and the house continues sculpturally in the natural space of the city. Accordingly, it was not simply a matter of visual references but of a literal interpenetration of urban nature and architecture. An example of this is the Poelzig House in Berlin-Grunewald, whose garden was designed and realized by Herta Hammerbacher. A remarkable group of four houses was built in 1923–24 in Berlin-Zehlendorf by Erich Mendelsohn and Richard Neutra. These single-family houses are among the earliest modern residential buildings in Berlin.[62] Particularly noteworthy was a rotating, furniture-like device inside of two of the houses that exposed different segments of space as needed. It made seasonal and daily climatic adaptation and variation the basic principle of the floor plan. "Like the backdrop of the room in a theater", the floor plans of the apartments in the two houses could be "varied with the change of season and time of day".[63]

The Green Industrial Campus

It was not until the end of the 19th century that large industrial firms emerged with "innovative factory planning," as a result of which industrial architecture was to become the explicit forerunner of a new that means Modern kind of architecture.[64] Exemplary for this is the Brunnenstrasse complex of AEG in Berlin, which from 1907 onwards relied on Peter Behrens, an architect and artistic advisor, to create an architecture of corporate identity. At the beginning of the 20th century, AEG had numerous factories in and around Berlin. The complex adjoins the Humboldthain; it was the first example of a a usine verte (Le Corbusier), a "factory in the green."

The new type of industrial campus also contributed to housing reforms, as can be seen, for example, in the workers' housing estates of AEG, Siemens and Krupp. "They were on the way to real, comprehensive social reforms."[65] To put it in general terms, the trend here was a shift from the speculative perimeter-block development common at the turn of the century towards the direction of consistently oriented row building. Uniformly aligned row buildings allowed for particularly easy control of building exposure to sunlight by adjusting the spacing between the rows depending on the number of stories. "The block layout is shown to have gradually evolved from the original hollow square with four street frontages, b, first to an arrangement in which all building units faced the same direction, but faced the opposite sides of the lateral streets, c, to a final organization according to which all buildings faced in one direction, arranged without direct street frontage, but on footpaths running through from one street to the next, d. Examples of all three phases are to be noted in various cities, and especially in Cologne, Frankfurt, and Stuttgart. These changes were accompanied by a concurrent evolution in the exterior aspect of buildings, from pitched-roof to flat-roof structures, and toward a gradual severity of architectural expression, even to the omission of spandrels in the latest examples."[66]

The new districts were often built in relation to nearby workplaces, such as was the case for the large housing estates of Haselhorst in Berlin. In the Haselhorst housing estate in Spandau (by the Reichsforschungsgesellschaft für Bau- und Siedlungswesen), different construction methods and materials were studied in detail, including their physical and microclimatic effects. The focus was on the systematic study of all steps of planning and construction. Within the framework of a competition in 1928, 3,000 housing units for low-income earners were planned; however, it was not the prominent forces of modern architecture that prevailed.[67] In the case of Haselhorst, for example, Alexander Klein had considered how the winds should be taken into account in the urban design of the housing estate.[68] The estate was built between 1930 and 1935, with its most important designs by Fred Forbát, Paul Mebes and Paul Emmerich. With 12,000 inhabitants, Haselhorst became the largest housing estate in Berlin of the Weimar Republic period. In view of the climatic sensibility of industrial campuses, however, four housing estates of the interwar period can be highlighted as critical in the case of Berlin: the Britz housing estate (with the Hufeisensiedlung), the forest housing estate of Zehlendorf (Onkel Toms Hütte), the Siemensstadt and the Weisse Stadt.[69] In his buildings for Siemensstadt, Hugo Häring, for instance, made the different intensities of sunlight the subject of his design. While the south side with its semi-elliptical balconies has an organic and complex structured facade, the north side of the apartment buildings was designed as a huge flat surface. Box windows are

→ figs. 15, 44   European Developments before 1945   **84** embedded in it, which took the increased insulation requirements into account.

The large-scale forest housing estate Zehlendorf was built 1926–32 in seven construction phases by the architects Bruno Taut, Otto Rudolf Salvisberg and Hugo Häring for the housing company GEHAG. Taut was responsible for the overall urban planning; at the same time, he was also the architect of most of the 1,900 residential units (1,100 in multi-family houses and 800 in single-family row houses) for a total of 15,000 residents. The architects built the housing estate into the existing Grunewald. Although it borrowed from the Garden City idea, its metropolitan character was to be explicitly preserved. The gardens, front gardens, and public street and square spaces of the housing estate, which were designed for a primarily middle-class population, were subordinated to the loosely represented pine trees of the Grunewald, while references to allotment garden colonies (with fruit trees, for example) were deliberately avoided. The visions of the landscape architects involved contributed significantly to this. Leberecht Migge and Martha Willings-Göhre were in charge for the design of "the outdoor living space" (Bruno Taut).[70] Additionally, Georg Béla Pniower, who himself lived in the estate (Hochsitzweg 105), also contributed to the design of the gardens. In particular Pniower had suggested that the pine trees with their light canopy should be preserved.

The Zehlendorf forest estate was exemplary for its consideration of natural infrastructures in urban design. It possessed "a differentiated building concept, which includes landscape and nature as a means of design", and a color concept that further underlined references to nature.[71] By penetrating the housing estate space with a forest-like tree population, a microclimatic peculiarity was created that cannot be achieved by conventional gardens.[72] The existing pines and the old birch trees arranged as avenues were included in the arrangement of the protected building area. "In addition to the pine, the original forest area was characterized by the following species of wood: sand-birch (Betula verrucosa), sessile oak (Quercus petraea), common oak (Quercus pedunculata) and rowan (Sorbus aucuparia). Fir and spruce species such as Douglas spruce, larch, black pine and Weymouth pine are detrimental to the appearance."[73] The color scheme, in turn, had introduced a kind of abstract mimesis of nature, adding visual stimuli to the thermal effects of the tree populations. Not only were the exterior walls of the individual sections of the housing estate painted in different colors; in addition, specific details, such as the window and door frames, were also subtly colored in an unconventional manner. Going beyond the communicative dimensions and orienting sense of the colors, the differently colored building sections also represent a complexity of visual appearance otherwise found in nature. The color scheme led to the large housing development being referred to as the "Paint Pot" or "Parrot Estate", which also expresses a reference to a supposedly natural environment.

Along with the planning of new green infrastructures, novel imaginaries of urban nature were developed in the interwar period. New images of nature were the result of a cultural production that manifested itself equally in architecture, literature and theory.[74] Not only the urban park but a whole repertoire of different green spaces was elaborated. In a city like Berlin, centuries-old forest stands combined with green roof terraces and building facades. Sometimes architecture is the stage for new green spaces, sometimes green spaces enclose selective architectural interventions; urban green infrastructures can belong to nature as well as to culture. With his competition entry, Hermann Jansen speaks of the "sea of stones in Greater-Berlin", to which he contrasts "a different system than before":[75] a new green Berlin. According to the architectural historian Katharina Borsi, Jansen understood in a particularly convincing way how to translate the new demands for hygienic living conditions for the working class into a language of architecture and urban planning. "Questions 'external' to the domain of architecture, such as, for example, those of health and hygiene, were transposed onto the surface of the drawing where they were experimented with by questions 'internal' to architecture, such as density, adjacency or proximity. Here the drawings are not understood as mute representations, but as surfaces of engagement between questions 'inside' and 'outside' of architecture and urbanism."[76]

→ fig. 86   Democratizing Urban Nature   85

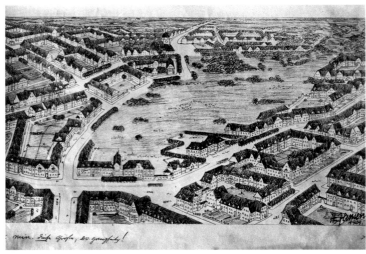

85
Hermann Jansen, Type of a small settlement in Buckow-Rudow (Berlin).

86
View of the Riemeisterstrasse at the forest housing estate Zehlendorf (Onkel Toms Hütte), Berlin.

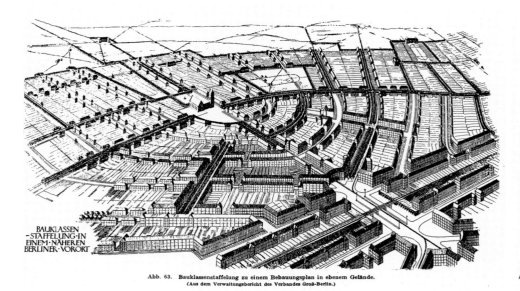

87
Varying building classes for a development plan. Source: Heiligenthal 1921.

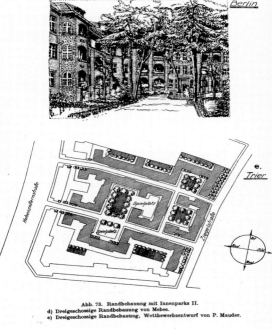

88
Peripheral development with indoor parks. Source: Heiligenthal 1921.

European Developments before 1945

89
Aerial view of the settlement Lindenhof, Berlin-Schöneberg, 1918–1921.

90
Martin Wagner and Leberecht Migge, settlement Lindenhof, Berlin-Schöneberg, 1924.

Democratizing Urban Nature

3.2 Access to the Sun: Metropolitan Architecture

Alongside Berlin, other European metropolises were home to a new microclimatic experimentalism. The tension between urbanization and de-urbanization, between affirmatively promoted urbanization and the overcoming of the city as a way of life, shaped not only the architecture of the time but also thinking about the climate in cities. As early as 1903, Georg Simmel's *Die Grossstädte und das Geistesleben* ("Big Cities and Intellectual Life") referred to many of the challenging aspects of urbanized societies. Especially in the Expressionist architecture of Bruno Taut and Hans Poelzig, but also in the organic architecture of Hugo Häring and Hans Scharoun, the ambivalence of these highly condensed living conditions described by Simmel was reflected not only in the celebration of the big city but also in anti-urban reflexes and imaginations; the first Expressionist housing estates in Berlin, such as Taut's Hufeisensiedlung, were built from the mid-1920s onwards. The discourse on the man-made climate is a mirror of this ambivalence, which points to superordinate social conflicts and debates.

The metaphor of the "big city" shaped the cultural and social imagination of the man-made climate (and more broadly the architectural production) of those years. With his essays and book publications on *Groszstadtarchitektur* ("Metropolitan Architecture"), Ludwig Hilberseimer provided a blueprint for interpreting man-made microclimates as components of new typologies, programs and technologies of the 20th century.[77] "Metropolitan Architecture" in Hilberseimer's work means not only residential buildings but also high-rise buildings; buildings for commerce, traffic, and industry; as well as hall and theater buildings. What to Karl Friedrich Schinkel had still appeared ambivalently as a dystopian chamber of wonders—the new industrial landscapes—was with Ludwig Hilberseimer elevated to the new emphatic guiding framework of the "New Building"—and its microclimates. The modern way of designing, building and living was accompanied by new microclimatic conditions and the need to experience and design them consciously.

3.2.1 Solar Geometries: Graphical Methods of Heliomorphic Design

Solar geometries—or, as they came to be called a few decades later, heliomorphic design (see chapter 6)—radically increased the importance of the sun for the generation of architectural form in urban contexts. Under solar auspices, designing in groups of buildings generates a differentiated interplay of reciprocal shading and illumination. In this context, the inseparability of "illumination and heat retention"[78] must be emphasized.[79] While the empirical penetration of the big city was developed by meteorologists, architects and urban planners tried to make the influence of the sun and the mutual shading of buildings fruitful for the design of buildings. In contrast to the graphical character of heliomorphism, urban climatology is first and foremost an empirical science. However, heliomorphic design has to be conceived in the context of other emerging methodologies of applied urban climatology.[80]

Urban heliomorphic design comprises two fundamental aspects: 1. it scrutinizes the sun (and other environmental factors) as *form-defining forces*. And 2. it investigates the *thermal interplay* of groups of buildings (based on local insolation) focusing on mutual shadowing. This anticipated and promoted an architectural understanding of the urban climate as generated by the coproduction of building groups. While traditional solar-building research has not been concerned with the neighboring built environment, in the case of cities the built context requires a profound analysis.

The Semiotics of Shadows

In the first half of the 20th century, a number of German-speaking and American architects elaborated heliomorphic models showing the way to "heliomorphic form generation" in the city.[81] A science of urban shadows promoted an understanding of how the sun can be utilized for the heating of apartments. The sun was the starting point of architects' investigations and the main reference for new "solar geometries".[82] Heliomorphic design was an early model for later investigations on passive climate control in cities; it was an important model for investigations on how to make environmental factors fruitful.

Whereas in Europe the discussions were dominated by the rampant tenement districts of the big cities, in the United States it was debates about the incipient high-rise construction that stimulated new heliomorphic approaches. Remarkably, the methodology and

→ figs. 65, 103, 112, 116, 117   European Developments before 1945   **88**

the historiography of heliomorphic design went hand in hand. Authors such as Ludwig Hilberseimer and Henry Nicolls Wright were also instrumental in the historiography of "urban solar access".[83] New processes of scrutinizing and learning were launched, with the Bauhaus as an exemplary institution of this development. Since the beginning of 20th century, there had been research into controlling solar heat gains in cities by considering the impact of neighboring buildings. The shadow always appears here in a semiotic sense as an index of the microclimatic conditions within a group of buildings.

Typologically, solar geometries of the interwar period were characterized by four themes, which can be explained by the background of hygienic concerns: apartments and hospitals as well as courtyards and street canyons became exemplary and "especially interesting point[s] of intersection between climatology and design".[84] Courtyards and hospitals appear as symbols of the modern debate on hygiene, and the streets as symbols of the evolving modern cityscapes at the turn of the century. In contrast to American debates on street canyons of the new *high-rises*, heliomorphic design in Europe was characterized by the *block* with its courtyards. Hospitals, with their nested building tracts and scattered pavilions, in turn became actual microclimatic islands in the city. The form of the volumes; the width-to-height ratio, "the surface materials oriented at a particular angle"; "the sky view whose geometry determines solar access by day and loss of energy by night"; or, in short, the basic principles of passive climate control—heat gains, storage and insulation—are some of the factors that were assessed.[85]

Honeycombs (Alexander Klein)

Due to the upheavals that he experienced as a result of his Jewish background, the German–Russian architect Alexander Klein was forced to live in three fundamentally different climates (Petersburg, Berlin and Haifa), which probably contributed to his sensitivity to changing climatic demands on the architecture of the city. Between 1921 and 1933, Klein lived in Berlin, where he worked intensively on the optimization of apartments ("Studies for Small Apartments"). In the process, he investigated "the utilization of the sun as a source of light and heat".[86] His systematic approach to the subject of the micro-apartment or the apartment for the minimum existence promoted a quasi-scientific way of designing, which was based on an in-depth analysis of apartment floor plans. Klein, like other protagonists of heliomorphic design, argued that the design of a building did not primarily follow artistic methods but resulted from given conditions, including climate.

To this end, he developed a "graphical method" to evaluate floor-plan formation and spatial design;[87] which must be considered as being a significant contribution to a theory of heliomorphic design. Klein, with his sublime drawings of the future housing of the European working class, was also an eminent visualizer of potential microclimates in the medium of line drawing. He strove to "illustrate the invisible qualities" of a design; in his drawings, floor plans are analyzed under the aspects of circulation, areas of movement, visual relationships, spatial proportions, and the effects of light and shadow. Later on, further criteria, by means of which microclimatic qualities are assessed, were added. Based on the construction of cast shadows, Klein developed "a graphical method of investigation with which he could determine which areas of the room were illuminated by the sun at a certain time of day and year".[88] In this way, he follows on from the methodologies developed by William Atkinson. The thermal regulation of the apartments goes hand in hand with a question of the most desirable orientation of the rooms; the conduction of heat through the external wall is still understood as a significant thermal contribution. "It is desirable, if possible, to have the bedroom group facing east and the living room group facing west, so as to have morning sun in the bedrooms and afternoon sun in the living rooms, so that the occupants are able to take most advantage of the sunshine. Also, locating the bedrooms facing west is unfavorable because of the late sunset in summer, as the heated exterior walls gradually emit their heat to the indoor air, with the temperature of the latter rising at night when bedtime arrives."[89]

In particular, Klein demanded a "reduction in the depth of the building" from the prevailing 12–13 meters to only 8–10 meters in order to make the lighting and ventilation of the apartments possible.[90] In doing so, he experimented with new typologies: "The [...] variety of new development forms and building types arose from the simultaneous consideration of building climatic and economic requirements." Klein proposed "folded row developments of two-span houses", a "sawtooth-shaped row development with four-span houses", a "honeycomb-shaped development based on the addition of a Y-shaped house type" (1927) and "designs for arcade houses", some of which were realized in the large housing estate of Bad Dürrenberg (near Leipzig, with about 1,000 housing units) in 1929.[91] In the case of the honeycomb development, courtyards of a width of 43.3 m and 1,300 m² in area were provided, with a "terrain ratio per dwelling" of

→ figs. 91, 93, 94   Democratizing Urban Nature

117 m². "Due to the sloping position, most of the dwellings also had southern exposure, while the others are either east-west or purely south-facing."[92] Three apartments were accessed by a common stair landing, and one apartment offset by one half floor. Despite intensive efforts to save space, lateral "access corridors" were introduced into the floor plans to provide direct sunlight and heat to the stairwell and the hallways of the other two apartments. This shows how heliomorphic analysis leads to new typologies and floor plans. The search for new heliomorphic large-scale typologies was accompanied by a systematic optimization of the interior spaces. Klein devoted himself intensively to the reduction of interior circulation areas, which led, among other things, to free-standing H-shaped buildings and to the aforementioned arcade houses. The modernist row is subjected to numerous transformations as part of Klein's heliomorphic analyses: "The problem of optimal building geometry is also crucial for adapting a building to distinctive climatic conditions."[93]

In addition, studies on the mutual shading of single-family houses began in 1934, shifting the focus to the thermal design of small groups of houses. In the context of this study of the optimal single-family neighborhood, "shade constructs"[94] function as indices for the desirable arrangement and materialization of the buildings. On the basis of different times of day and year, Klein made an in-depth analysis of the changing shadows cast by three buildings (in Potsdam). The investigation showed that even in the case of "the worst positioned" house "on the shortest day of the year, the living spaces of the house were—though briefly—lit by direct sunlight".[95] Such considerations were combined with possible innovations in the area of access to the plots and their architectural materialization. The analysis of the "streetscape" shows a play between the "opposing elements" of "stone" and "planting." While on the "shady side" "light walls" are shown, on the "sunny side the same wall [...] is heavily overgrown."[96] Here, the heliomorphic analysis leads to specific forms of materialization in the field of tension between architecture and landscape architecture. Horizontal and vertical surfaces are formulated in a *vegetal-organic* or *mineral* manner; "smooth, vertical wall surfaces [are combined] with horizontal lawn, and on the other side vertical vegetation wall with the horizontal stone surface of the sidewalk".[97]

In a 1942 study, "The Influence of Climate on the Organic Design of Floor Plan and Elevation", a comparison of the urban architecture of Haifa, Tel Aviv, Berlin and Oslo is made on the basis of generalized

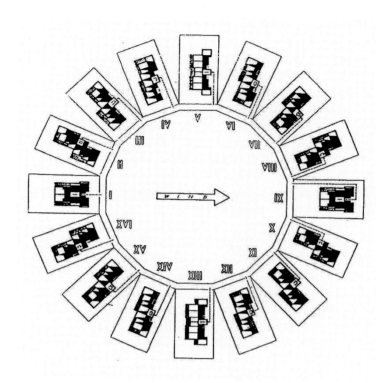

91
Alexander Klein, 1942 study on how a building floor plan must be varied depending on its position relative to the winds in order to ensure optimal ventilation. Ventilated areas are shown in white, non-ventilated areas in black..

climatic factors of influence related to the floor plan and facade of a residential building.[98] The object of study is a "three-story, two-apartment plan with six two-bedroom apartments for a north-south facing row development". Klein adapts this type "to the different climatic conditions" of the four cities mentioned. The study "examines the effect of climate on building form, room height, window placement and size, arrangement of uses, formation of balcony parapets and window details".[99] Although Klein's approach is in line with the thinking of bioclimatism, his investigation of explicitly urban architecture is remarkable and speaks to Klein's comprehensive awareness of urban problems.

Pyramids (Peter Behrens)

A patent submitted June 25, 1930 by Peter Behrens deserves special attention. In his patent *Baublock bestehend aus mehr- oder vielgeschossigen Einzelhäusern* ("building block consisting of multi-story individual houses"), the scales of the building and of the city are brought together in a meaningful manner.[100] Behrens, who had been professor at the Academy of Fine Arts in Vienna since 1922, experimented

→ fig. 95   European Developments before 1945   90

92
Alexander Klein, 1942 comparative study on the cities of Haifa, Tel Aviv, Berlin and Oslo (from top to bottom): in order to ensure a comfortable indoor climate (by passive means), modifications of the same building type (under different climatic conditions) are necessary.

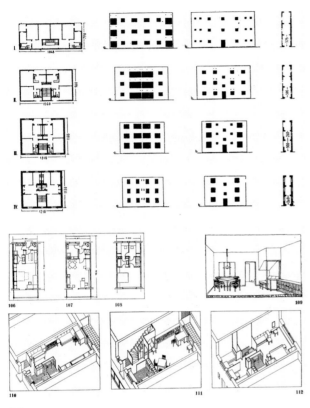

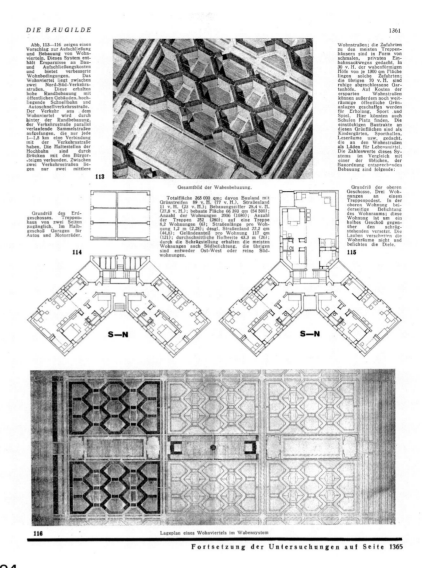

93
Alexander Klein, views of the living space, studies according to the program of the Reichsforschungsgesellschaft, Berlin 1927.

94
Alexander Klein, honeycomb development, studies according to the program of the Reichsforschungsgesellschaft, Berlin 1927.

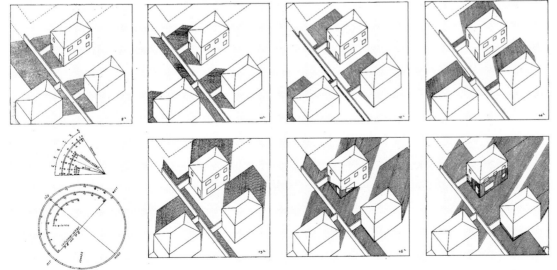

95
Study on insolation: spring and autumn. Source Klein 1934.

Democratizing Urban Nature

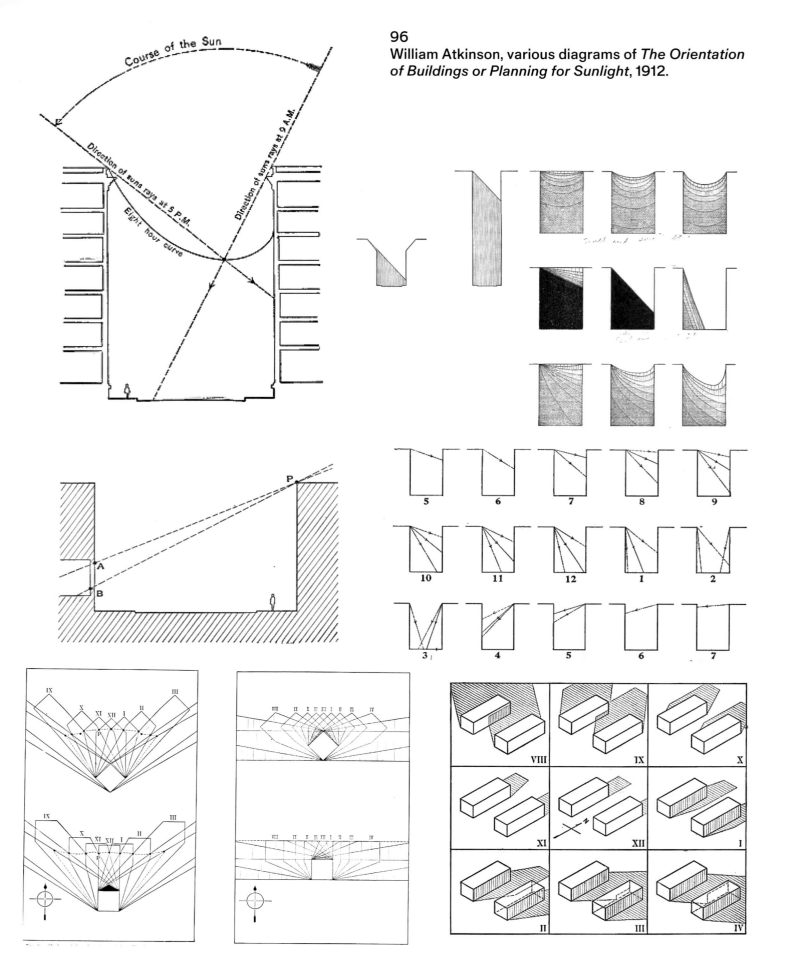

96
William Atkinson, various diagrams of *The Orientation of Buildings or Planning for Sunlight*, 1912.

European Developments before 1945

97
Otto Wagner (layout and church) and Carlo von Boog (pavilions), psychiatry "Am Steinhof" (Niederösterreichische Landes-, Heil- und Pflegeanstalt für Nerven- und Geisteskranke), Vienna 1907.

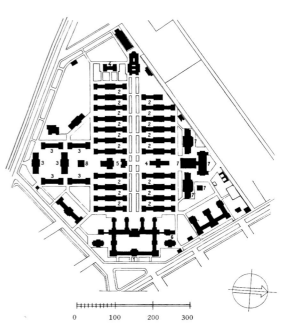

98
Ludwig Hofmann, site plan Rudolf Virchow Hospital, Berlin 1906.

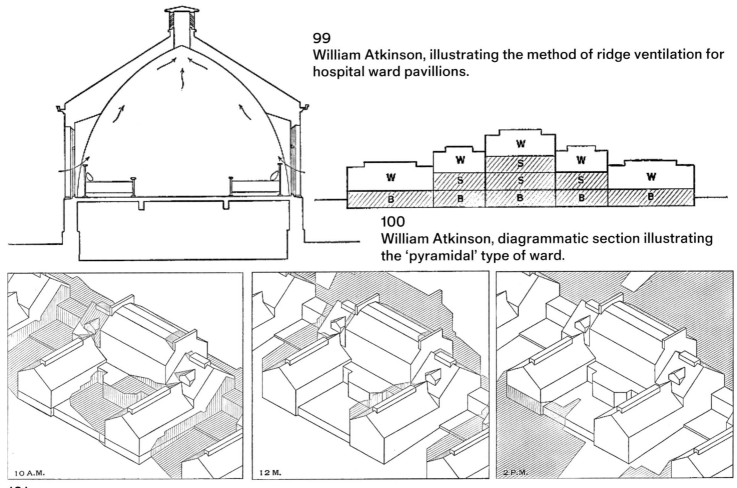

99
William Atkinson, illustrating the method of ridge ventilation for hospital ward pavillions.

100
William Atkinson, diagrammatic section illustrating the 'pyramidal' type of ward.

101
William Atkinson, pyramidal type of ward unit.

Democratizing Urban Nature

102
Hans Poelzig, House of Friendship, project for Constantinople (Istanbul, Turkey), 1917.

103
Hans Poelzig, competition for a convention center in Hamburg, 1925.

European Developments before 1945

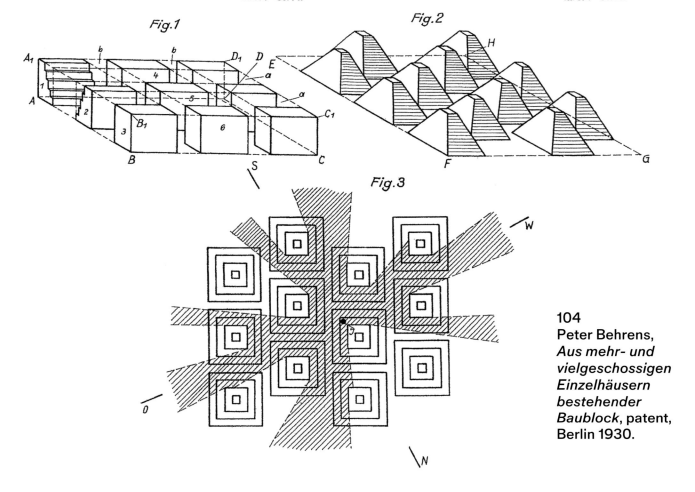

104
Peter Behrens, *Aus mehr- und vielgeschossigen Einzelhäusern bestehender Baublock*, patent, Berlin 1930.

105
The principle of the terrace house.

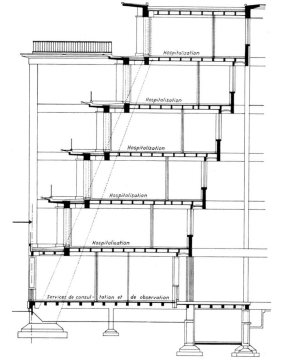

106
Adolf Loos, rendering of an urban terrace project, combining two-family houses.

Democratizing Urban Nature

extensively with the terracing of buildings and also tried to introduce terracing into mass housing construction. In addition to the *Terrassensiedlung* of 1915 and the *Terrassenhaus für Arbeiterwohnungen* of 1923 (both for Berlin) and the *Terrassenhaus* realized in 1927 in the Weissenhofsiedlung near Stuttgart, the *Terrassenhochhausstudie* developed in 1929 (together with Alexander Popp) is also noteworthy as a design exploration of this typology. The patent was the culmination of these investigations over the previous 15 years.[101]

The patent describes a city made of pyramid-shaped structures, promoting the idea of the terraced house from a heliomorphic point of view.[102] The idea of the urban terrace house was extended to the scale of a city, with shifted pyramidal structures considering the mutual influences of the buildings on insolation and shading. Due to heliomorphic analysis on an urban scale, the rectangular arrangement of the terrace houses was abandoned in favor of a setting of displaced pyramids. In light of the rapidly growing cities of Berlin and Vienna, the patent was aimed at new urban quarters, which were intended to form a hygienic alternative to the high-density tenements. Behrens was intensively involved in Berlin and Vienna in overcoming the tenement as a building type.

According to Behrens, the "modern metropolitan housing estate" was engendered by three triggering factors: the "needs of the residents", the "existing building regulations" and the "economic efficiency of the development" to be achieved. The proposal's real objective was to combine a conceivable maximum "of illuminated area" of individual dwellings with a minimum of shadowing "both on the footprint and on the other houses".[103] In this way, Behrens anticipated considerations that would be continued some 40 years later by Ralph Knowles in California (see chapter 6). Following this preliminary consideration, "the arrangement of the individual structures must be done in such a way that one building casts less shadow on the other, or that each building itself casts as little shadow as possible".[104] This solution lies in the "design" of the houses and "their position in relation to each other". Each house is "terraced on all sides" (resulting in "truncated pyramids"), which enables "each dwelling [to have] an outdoor, air-circulated living space that is immediately accessible from the dwelling". The all-round stepping of the individual buildings is combined with offsetting "their axes against each other", so that a maximum field of view is opened for the residents and thus also the desired sunlight.[105]

In addition to the almost prototypical heliomorphic exercise, Behrens also pointed out the thermal implications of his urban-planning approach. It is "desirable to create surfaces from which solar heat is reflected back onto the building walls, so that even with low solar irradiation, especially in transitional periods and also in winter, a certain heating of the building by the solar heat takes place, at least during the midday hours".[106] Thermal added value–the heating of the cold, at that time largely unheated, apartments–is time-specific for the winter and the transitional periods, while the foreseeable overheating in the summer is concealed or conceived as heliotherapeutic gains. The aim of his patent was the greatest amount of incident sunlight, even though the scientific analyses of hygienists at the time warned against the overheating of apartments.[107] However, in contrast to the widespread primacy of favoring certain directions for the orientation of buildings, Behrens' proposal was an urban design approach of staggered axes that would lead to new urban orders. He viewed this kind of urban terraced house as the epitome of a modern way of building and living, and proposed that it should shape the way cities were built in Europe.

Pavilions (an Excursus on William Atkinson)

The planning of hospital complexes offered the opportunity for heliomorphic investigations that were precursors to the consideration of complex projects in urban planning. This was exemplified by the systematic considerations of the American architect William Atkinson. In Atkinson's 1912 study *The Orientation of Buildings*, hospitals, together with street canyons, are prominently highlighted. Sunlight and shade, as they appear on and in hospitals (and in streets), are the main concern of his heliomorphic studies. Atkinson, a Boston-based architect, was able to combine his rigorous geometric analyses with architectural history and material research. His stringently structured book has four chapters: The first two provide the geometrical foundations of heliomorphism–"The Astronomical Data" and "Shadow Diagrams[108]–while chapters 3 and 4 address "Hospitals" and "Streets". The central treatment of the geometric formation of hospitals refers to public-health objectives while the last chapter deals with the mutual solar influence of buildings of a city.

The two main concerns of the publication were the reformation of hospital construction, as it was shaped in the late 19th century, and the insolation of city buildings. In large-scale hospital buildings of the 19th century, the microclimatic problems of the big city–overheating, cooling, mold, etc.–had made

→ figs. 96, 104   European Developments before 1945

themselves felt in a particularly detrimental way. In addition, new "requirements of modern medical treatment" led to a new program of hospital design. For example, outdoor spaces in the form of balconies, loggias, terraces and roof terraces (for sunbathing) were now part of the basic provision of a hospital facility. "The program has changed and a new type of hospital construction must be devised to meet it."[109] The decentralization of hospital buildings to accommodate patients was a common approach at the turn of the century, leading to large-scale hospital facilities in Berlin and Vienna, for example; Atkinson makes special mention of the "Rudolf Virchow Hospital" in Berlin, which was equipped with a central avenue and built in pavilion style by Ludwig Hofmann in 1899–1906. At the same time, "the Lower Austrian Provincial, Therapeutic and Nursing Hospital for the Nervous and Mentally Ill, Am Steinhof" was built in Vienna, which was planned by Carlo von Boog and revised by Otto Wagner. The hospital was opened on 8 October 1907 after three years of construction.

With these architectural developments in Europe in mind, Atkinson develops his recommendations for the reform of hospital construction around 1900, including the Newport Hospital in Newport (near Boston, USA), designed by Atkinson himself in 1910, which is also explained in his publication (p. 95). In addition to questions about the sensible orientation of the building wings (wards), the building distances and building heights, he devotes himself to the arrangement of the windows and the solar protection ("The ward windows should always be provided with blinds, shutters, or shades.")[110] If the building spacing between one- or two-story buildings is 1½ times each, the two-story solution is the more economical; nevertheless, Atkinson advocates single-story solutions that allow vertical venting ("ridge or monitor ventilation") because of the more efficient air conditioning of the building wings. "In all respects, the one-story pavilion is the best for the patient, but especially because it is best adapted to ventilation by natural means. [...] Artificial or forced ventilation, which in our climate is a necessary evil during a large part of the year, may be successfully adapted to a building of superposed stories, as well as to a subway or a mine."[111]

Atkinson examines the access and insolation of individual pavilions and of entire groups of pavilions. The balance is between hygiene and efficiency. Shading from service rooms and neighboring buildings must be avoided as much as possible. Symmetrically arranged groups of buildings, which have very different sunlight conditions, are rejected. "Pavilions may be grouped in various ways. [...] There seems to be no

good reason why the pavilions of a general hospital should be placed any farther apart than is necessary to secure adequate sunlight and a free passage of air between them. Compactness saves steps and helps toward economy of administration."[112] To connect the individual pavilions, Atkinson proposes "covered corridors".[113] The in-depth interest in the geometric control of shading is coupled with the thermal optimization of the buildings (via heat gains). Heat gain must be controlled by means of adequate ventilation, as a cross-section to his hospital design also shows.[114] As early as 1894, he was confronted with heliomorphic issues when planning a hospital building. In search of rational criteria for the orientation and related natural air conditioning of the buildings, Atkinson studied the course of the sun; window arrangements on the east and west sides proved particularly disadvantageous, "since they received lots of sun in summer and little direct solar heat in winter".[115]

In order to do justice to his proposals, Atkinson farsightedly remarked, a new conception of architectural representation was also necessary. Atkinson criticized the prevailing conventions of plan representation, which prevented a representation of the heliomorphic aspects of the site. "A careless habit is engendered of regarding the architectural drawing as an end in itself, while actual conditions of site, surrounding and exposure are lost sight of."[116] In particular, he said, shading, which had been limited in its representation to elevation drawings, needed to be extended to drawings in plan. "The study of shades and shadows [...] is seldom applied to the rendering of the plan, although it is here that its greatest usefulness is found, especially in the study of groups of buildings, or for the representation of landscape work."[117] Heliomorphic analysis uses the casting of shadows to account for the lighting and thermal control of interior spaces. It is not surprising, then, that Atkinson, among others, chooses (the then uncommon) axonometric representation to make clear the mutual shading of wards. Heliomorphic design has always depended on specific and new methods of representation, which allow a thermal "study of groups of buildings". In this sense, Atkinson is an important thinker of passive climate control in urban contexts; the different scales of his heliomorphic analysis (involving the city) are noteworthy.

Negotiating Research and Design

Hospital construction proved to be particularly sensitive and permeable to urban climatic considerations. In the first half of the 20th century, the hospital underwent a strong, also structural, reformation in which

→ figs. 97, 98, 99, 100, 101   Democratizing Urban Nature

the climate–the sun and the winds–became a decisive influencing force. In the dispositive of modern "metropolitan architecture", as theoretically conceived by Ludwig Hilberseimer, the modern hospital appears as an exemplary typology, anticipating the architectural impulses of urban climatology. The imperative of maximum hygiene favored a comprehensive study of solar exposure. This concerned not only the orientation and architectural formulation of the individual buildings; the arrangement of buildings in larger groups also necessitated a reflection on reciprocal shading. The spatial complexity that already characterized hospital buildings in large cities at the end of the 19th century explains the heliomorphic examination of this building type.[118]

Nikolaus Pevsner emphasized the importance of "terrace building" especially for Germany and Switzerland, which he acknowledged as a regional development that influenced hospital building in particular.[119] In 1926–29, Richard Döcker built a structure that served as a model for modern hospital construction. In Waiblingen, in the outskirts of Stuttgart, the new approaches to the terraced house came into play. From 1903 onwards, the physician David Sarason had anticipated with his "open-air house" central considerations for modern terraced hospital construction with his patented "Terrace System". Referring to Sarason, Atkinson, like Peter Behrens later, emphasizes the advantages of terracing hospital buildings. Atkinson speaks of "the pyramidal type of ward construction", which allows natural air conditioning of the rooms.[120] "The successive reduction in area of each story, makes it possible to place the buildings at the minimum distance apart."[121] Parallel to the emergence of terraced sanatorium and hospital building, experiments with terraced residential structures were undertaken. In particular the city of Vienna, one of the points of origin of modern urban climatology, must be emphasized. These impulses were taken up by Adolf Loos and his pupil Rudolf Schindler, among others. Loos, who had designed in 1912 the terraced Scheu House, developed his interest further on the basis of multi-family structures; his residential building for the municipality of Vienna from 1923 should be mentioned. Also noteworthy is Oskar Strnad, who also excelled in 1923 with the design of a large-scale curved terrace housing development. The classic row or perimeter block form was abandoned here in favor of a freer orientation. In 1928, Rudolf Schindler, also a pupil of Loos, designed the Wolfe House in Avalon (CA, USA). The house is located on a neglected, very steep slope. In this context, one can also think of the so-called *Wohnberg* "Residential Mountain" envisaged

by Walter Gropius in 1928 (and officially exhibited in *Visionary Architecture* at MoMA New York 1960).

The geometric examination of the interplay between solar radiation and urban building constellations created a field of activity that dealt with the core competencies of the architect and thereby opened up a methodically conscious approach to the urban climate. The rigorous exercises that were applied in this process–one thinks of Alexander Klein, Peter Behrens or William Atkinson–were characterized by their high degree of abstraction. It is geometry itself that formed the central agency here, although the investigation referred to the shading and thermal effects of neighboring buildings. In this respect, heliomorphic design could be described as the actual driver of a renegotiation of the "relationship between research and design". In the context of Ludwig Hilberseimer's heliomorphic research, Ute Poerschke speaks of "evidence-based" and "data-driven design". Climatological knowledge combined with hygienic insights to form a novel basis for design. "Their designs resulted from collected and analyzed data and expert input rather than historical reference or creative talent."[122] However, there was a tension between the abstraction of solar geometries and the empirical nature of urban climatology, weighing the microclimatic conditions of the city. The architects' solar geometries remained committed to architectural autonomy rather than their scientific legitimacy.

3.2.2 Indoor Nature: Climatic Scenery in Large Halls

In his 1931 book *Hallenbauten* ("Hall Buildings"), Ludwig Hilberseimer astutely pointed out that the ever-increasing proliferation of large-scale halls was a recent phenomenon, related to the industrialization of construction and the modern metropolis.[123] Driven by the new structural possibilities that emerged from reinforced concrete and steel construction, numerous uses–not only production but also trade and leisure–were accommodated in large volumes. Hilberseimer unfolds a whole typology of large halls as part of the then-modern big-city architecture: "commercial buildings" such as department stores; offices and business premises; hall and theater buildings for cinemas, exhibitions and trade fairs; and "transport buildings" such as railway stations, large garages and airports. In the introduction to his book, we read, "In the past, hall buildings were relatively rare and therefore extremely significant. It was through them that the constructional-technical as well as the architectural-artistic problems developed. On the other hand, no closer attention

→ figs. 102, 103, 105, 106   European Developments before 1945   **98**

was paid to special needs such as good viewing or listening, as is required by the various purposes of the halls today."[124] Hilberseimer thus understood large halls in the modern era not only as a problem of construction but also of perception. In the case of hall buildings, the architectural construction itself cannot be separated from the perception that takes place within it. Ensuring the desired *perception* is a cognitive act of *construction* in itself, to be achieved both by the buildings architect and the users.

Mass and Air

While the consumption of (horizontal) surface area in large halls is due to functional reasons, the multisensory requirements that integrate hygienic, climatic and perceptional aspects of architecture involve a vertical dimension (and, thus, volume). Large halls, as they evolved in European cities from the beginning of the 20th century, set up new sites of publicity for, and representations of, urban societies—recognizing themselves in the idea of the "mass" (as, for instance, described by Siegfried Kracauer or Elias Canetti).[125] Through their principal gesture of enclosing air, large halls became part of the symbolism of the crowd and the large volumes of enclosed air contributed to the emerging—proto-fascist—"heroic style" in European architecture.[126]

With their wide-span supporting structures of iron and reinforced concrete, the market halls built by Richard Plüddemann and Heinrich Küster between 1906 and 1908 in Wrocław (Poland), by Tony Garnier between 1909 and 1914 in Lyon (France) or by Martin Elsaesser between 1926 and 1928 in Frankfurt am Main (Germany) are more reminiscent of cathedrals or palaces than conventional market halls. In a market hall, the enormous enclosed volume overhead forms an architecture that transfigures a vast volume of air into a kind of mass ornament. It is no coincidence that during the interwar period in Europe the term 'palace' was used for the (self-)aggrandizement of new halls dedicated to the consumption and leisure of workers and employees; there was talk of a "Sports Palace" or a "Spa Palace". With Siegfried Kracauer, the large air space of the new halls can be understood as contributing to an urban proxemics that evokes a feeling of hygiene above the heads and between the bodies of the individuals. The revues, such as those performed in the Haus Vaterland (Fatherland House) or the Grosse Schauspielhaus (Great Theatre, in 1947 renamed Friedrichstadt Palace) in Berlin, epitomized the new *Körperkultur* ("physical culture")[127] for the masses, which also coined the proxemics of everyday urban life and in which, despite the high density in the

cities, a regulated separation of people was maintained. In densely "packed stadiums", Kracauer writes, "performances of the ever-same geometric precision take place".[128]

Architects such as Hans Poelzig, Mies van der Rohe, Lilly Reich and Leo Nachtlicht as well as landscape architects such as Georg Piower and Herta Hammerbacher explored, in the course of their design work in Berlin—and at the same time as Kracauer—the everyday rituals of modern life. Exhibitions, theatres, night clubs, restaurants and public baths, located in large halls, opened up spaces for new visions of design during the 1920s and 1930s. As I will now show, not only the structure and building services but also the vertical surfaces and cladding—whether painted, woven or vegetal—played a critical role for the *sensory imaginary* in large halls. Vertical surfaces served as amplifiers of perception—and thus as promoters of a new "good viewing"[129]—in that a mass of air, for instance, either sets textiles in motion or is set in motion by them. Relying on the agency of surfaces, such a symbolism of large air volumes contributed to a *climatic imaginary*, which converged in the notion of man-made climate.

Installation of Freely Hanging Fabrics: The Velvet and Silk Café

The topological interlocking of interior and exterior spaces was at the center of urban climatological thinking in the 1920s and 1930s. Accordingly, climatic *interactions* and thermal *transitions* played a critical role in the climatologists' investigations and the design appropriations made by architects. A subtle reflection of this is the Café Samt und Seide ("Velvet and Silk Café"), developed by Ludwig Mies van der Rohe and Lilly Reich[130] in 1927 for *Die Mode der Dame* ("Ladies' Fashion") fair on behalf of the German silk industry.[131] In the project, textiles were used to create a filigree space of self-representation in which all the heaviness of architecture seemed to be suspended. Freely hanging textiles formed highly sensitive surfaces that made the invisible air space of the large halls recognizable. The air volume of the hall was turned into a sensitive space of resonance of the man-made climate. Over an area of 300 m², silk and velvet fabrics hanging at different heights formed various textile wall segments. As a spatial installation, they anticipated the sculptural minimalism of the American artist Richard Serra. In addition to two semicircular walls, the café was structured by textiles arranged transversely and longitudinally to create a number of different zones. The design potential of free-hanging fabrics was evident in the shapes of the

→ figs. 112, 113, 116   Democratizing Urban Nature   **99** walls and the selected colors and textures of the fabrics, as well as in their optical overlapping and their movements in the room.

Depending on the line of sight and the fabrics, different overlaps and alternating shiny and matt color spaces were evoked, thus creating a dynamic play with color, materials and textures of silk and velvet fabrics. According to Philip Johnson, the only person who was later able to recall the colors of the fabrics used, the velvet fabrics were black, orange and red, and the silk fabrics gold, silver and yellow, while the floor was finished in white linoleum.[132] The color palette had a strong contrast as much as it evoked 'national appeal'–a gesture to acknowledge his clients' desires that was not untypical for Mies.[133] Although the lighting in the café area softened the contrast between the different colors, it further enhanced the powerful simplicity of the spatial installation as a whole. Frames were installed along the long sides of the café where the light sources were installed, distributing the light diffusely.[134] The fact that the light sources were hidden, again served to emphasize the staged character of the café.

Temporary installations, such as the Velvet and Silk Café, created spatial subdivisions of huge interiors that are reminiscent of the dazzling scenery and staffages of a theater stage. With the sociologist Erwin Goffman, one can speak of "front" and "back" stages, which characterize this kind of temporary architecture in exhibition halls. The Café in Berlin can be conceived as a blueprint for the open floor plan and a feel for materials used later in the main living room of the Villa Tugendhat in Brno, which Mies was designing at the same time as the Café and completed three years later, in 1930. The textiles used as room dividers; the nickel-plated steel and brass tubes of the suspensions; and the chrome-plated surfaces of the cantilever chairs, stools and glass tables were part of a new modern vocabulary. In the villa, the principle of "scenery-like walls"[135] was applied in a permanent version with onyx marble, an exemplary application of Gottfried Semper's principle of "material change" (*Stoffwechsel*) in the transition from "textile" (*Gewand*) to "wall" (*Wand*).[136]

A Courtyard with Images of Natural Vegetation: Gourmenia House

Urban microclimates–from smoky coffee houses to stuffy workplaces, from unheated apartments to overheated subway stations–made the modern metropolis such as Berlin a focal point of new climatic experiences in which air pollution, heat and cold went hand in hand (see chapter 1). The ambivalence of the man-made climate of the city became subject of the design appropriations made by architects.

One remarkable project related to this ambivalence of urban climates is the Haus Gourmenia,[137] a building (erected in 1928–29 at the Hardenbergstrasse in Berlin-Charlottenburg) with a "pioneering interior design",[138] which housed several restaurants. In the case of the so-called "Traube wine restaurant", a spiral staircase connected the garden with the galleries of a three-story hall. The idea of creating a "glass-roofed inner courtyard" with a "natural garden" was the brainchild of the building's architect, Leo Nachtlicht, and landscape architect Georg Pniower.[139] As Pniower remarked, this courtyard gave the interior a "subtropical character". In addition to plants, including palm trees and cacti, the garden of the Traube restaurant featured water basins and fountains as well as a complex topography that included bridges and changes in materials reminiscent of traditional Chinese gardens (see chapter 5). The artificial subtropical climate was created in the Haus Gourmenia by means of mechanical air conditioning and artificial light for the plants. It was Pniower's main concern to "tightly combine the horticultural with the architectural in order to create natural images of vegetation" within the restaurant.[140] Such an amalgam of nature and culture formed a central design and planning idea of the interwar period in Berlin. The intensive greening of the big city, as had been demanded by architects and planners, not only included streets, squares and facades but also interiors. In this sense, the restaurant comprised vertical images of vegetation.

The Haus Gourmenia has since been regarded as one of the "outstanding projects of Pniower in the Weimar period".[141] The garden developed here formed the first culmination of Pniower's experiments with conservatories and flower windows, which he designed for private clients in Berlin during the Weimar Republic.[142] Johannes Reinhold emphasized the visionary significance of this project, in that it showed, among other things, Pniower's "scientific way of working". "For the first time, work was carried out here with an air-conditioning system and the plantations were maintained under artificial light. The experience gained there […] was a valuable contribution to cultivation techniques in horticulture. With this work, Pniower also provided many creative suggestions, e.g. regarding the inclusion of animals."[143] At the Traube wine restaurant, not only plants but also parrots enriched the room acoustically and visually.

→ figs. 114, 115   European Developments before 1945   100

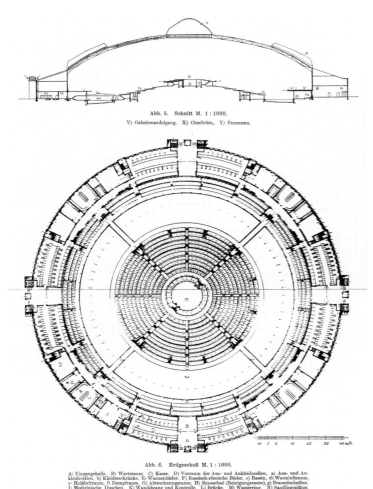

108
Spa Palace Berlin, view of the 12-m-high mural.
Source: Goldmerstein and Stodieck 1928.

107
Spa Palace Berlin, section and ground floor.
Source: Goldmerstein and Stodieck 1928.

109
Spa Palace Berlin. Source: Goldmerstein and Stodieck 1928.

Democratizing Urban Nature

110, 111
Hans Poelzig, Spa Palace Berlin, interior and exterior view, 1928.

112
Martin Elsaesser, Wholesale Market Hall, Frankfurt am Main, 1928.

European Developments before 1945

113
Hans Poelzig and Marlene Moeschke-Poelzig, The Great Theater Berlin (Grosses Schauspielhaus), designed in 1919 for theatre director Max Reinhardt. The building was originally a market hall designed by Friedrich Hitzig.

114
Lilly Reich and Ludwig Mies van der Rohe, Café Samt und Seide, Berlin, 1927.

115
Leo Nachtlicht and Georg Pniower, wine bar "Traube" in the House Gourmenia, Budapester Strasse, Berlin, 1930.

116
Tony Garnier, slaughterhouse, Lyon (France), 1906–32.

Democratizing Urban Nature

Panopticon of the New Physical Culture: Spa Palace

The democratization of society striven for during the short era of the Weimar Republic was also reflected in the desire to build climatic infrastructures for broad sections of the population; climate control was still seen as part of public urban infrastructure rather than a private matter. With the *Thermenpalast* ("Spa Palace")–an (unbuilt) project for an extraordinary public bath in Berlin–the empirically experienced climate of Berlin was countered by a deliberately created man-made climate. Based on the new technical developments in air conditioning, the design developed in 1927–28 was an exemplary blueprint for an artificial climate, worked out in every detail. Goldmerstein and Stodieck,[144] whose brainchild the Spa Palace was, entrusted the architect Hans Poelzig with the architectural design, and, with the assistance of numerous experts from the "building and bathing industry", a precise "evaluation of the feasibility, cost and profitability"[145] of the project was elaborated. The design, with its scenery-like interior, is reminiscent of the Berlin Grosses Schauspielhaus designed by Poelzig in 1918–19—actually a conversion of a former market hall—and the sheer scale of the dome is comparable to the revolutionary architecture of Etienne-Louis Boullée or Claude-Nicolas Ledoux. The aim was to present a generic construction project "that [could] be executed in all cultivated countries and under all climatic conditions".[146]

The interplay of a 150 m dome, a huge ring-shaped water basin and artificial topography reflected the desire to create an environment beyond the conventional scale of a building, in which the dichotomy of inside and outside was suspended.[147] "In the large, airy and light-filled room, the visitor has the sensation of being outside in the open air."[148] In contrast to Buckminster Fuller's famous *Dome over Manhattan*, the Spa Palace had an explicitly socio-political dimension, since a democratic concern was the driving force behind the project. The Spa Palace was intended to create a climatic counter-world to the unhygienic atmosphere of the living and working environments of the general population. The project initiators spoke of "about four hours [...] of relaxation after the day's tiring work". Between 20,000 and 30,000 daily visitors were expected.

With its multimedia approach, which combined technical and artistic aspects, the Spa Palace would have represented a total work of art, in which a mural was intended to provide a counterpoint to the interior-exterior dichotomy. Comparable to natural–historical dioramas, the outer infrastructural ring of the bathing facility (with its serving rooms) was to merge seamlessly into the landscape of an Arcadian island world.[149] The project description explains this in detail: "The periphery of the hall will be artistically equipped with all the aids of modern stage technology, both sculpturally and visually, in such a way that the visitors think they are in an unlimited space with a wide panoramic view. The 12-m-high mural, which closes the hall like a panorama, represents friendly, sunny landscapes. Cold, rain and all unpleasant natural phenomena will not bother our guests."[150] Huge areas of skylight of 8,000 m² (40 percent of the hall floor area) was to provide a sun-flooded atmosphere, which was further enhanced by an "artificial sun".[151] The heating systems also conveyed novel microclimatic sensations on the ground. "All berths are heated. The bathers should definitely feel as if they were on a beautiful, warm summer day on a natural, sun-drenched beach."[152] Everything was designed to simulate or even augment nature with modern artistic and technical means. The technical means should be entirely at the service of a new type of man-made climate, comparable only with experiences "in the mountains": "The stay in the hall is to be arranged to such a recovery, as one finds it in the extent almost only in the mountains in free, light high-altitude air under the rays of the mountain sun. The benefit of the mountain sun shall be transmitted to the visitors of the hall by artificial high-altitude sun rays. [...] The apparatuses are quite new, they do not only emit ultraviolet rays as the apparatuses often used for healing purposes, but all the rays that the sun sends down to the earth."[153]

The new "physical culture" propagated by the initiators–"the planned palace should not only serve water sports, but any physical culture at all"[154]–was thus overlaid by a panoptic experience that combined the "needs [...] of good viewing" (Ludwig Hilberseimer) with thermal sensations of various kinds. "Everyone should be involved in personal hygiene."[155] In addition to sports activities, the focus was on improving the health of the sick. Accordingly, "hydrotherapy" had a special position. For this purpose, different "thermo-atmospherically defined places" were to be created, such as "bathing cures, drinking cures, therapeutic gymnastics, inhalation, electric baths" as well as "Russian-Roman baths [...] steam baths, warm and hot air treatment [...] and air baths". For the electric baths, "sunlamps, X-ray irradiation, electric sweat baths and the like" were to be used.[156]

With the Spa Palace, an exemplary project of man-made climate was elaborated on an urban scale and in every detail. Like hardly any other project, it stands

→ figs. 107, 108, 109, 110, 111   European Developments before 1945

104

for the architectural, technical, but also political and sociological prerequisites of the discourse on man-made climate. With its interplay of a huge dome and an artificial topography inside, the Spa Palace anticipated the bathing landscapes and huge shopping malls that emerged in the United States from the 1960s onwards. In addition, it also anticipated ideas of the closed biospheres (see chapter 5), promoted by ecologists and scientists since the 1970s. The Spa Palace relied on the idea of being somewhere else and of bringing the climate of somewhere else to the users' own place of living. In this sense it must be seen in line with Walter Benjamin's reading of the new mass media such as the cinema, providing space of distraction from people's own miserable living conditions.

Amplifiers of Perception

German Expressionist architecture represents an important echo chamber of the man-made climate, manifested in a mimetic empathy with the thermal geography of the big city. Bruno Taut's and Hans Poelzig's work of those times was strongly motivated by the ambivalent urban conditions of cosmopolitan cities such as Berlin or Vienna. The sanitary and microclimatic inadequacy of the housing sector of the working class led to extensive debates on the future character of cities. Taut explored a new "Alpine Architecture" (1919) and the "Dissolution of Cities" (1920) during the First World War, launching proto-ecological arguments about the "world" as an expanded form of architecture. Both publications were part of a larger intellectual project to reconsider the city in the light of nature (and vice versa).[157] The mountains appear not only as architecture but as a model of another kind of topographically conceived architecture. The architecture itself is converted into another type of mountain, representing a new fusion of nature and city, topography and buildings, figure and ground. Architecture is now literally territorial architecture, while the mountains themselves become *world architecture*. This topographical reading of architecture, which had matured during the war, was exemplified in Hans Poelzig's 1917 design for a "House of Friendship", a large-scale building complex for Constantinople (Istanbul). The building includes terraces, furnished with gardens and green spaces. In a remarkable way, this "artificial mountain"[158] is an early attempt to fuse the three main dimensions of microclimatology–buildings, greenery and topography–anticipating the idea of the "megaform" (see chapter 4).

As I have aimed to show, climatic imaginaries of the interwar metropolis cannot be reduced to the intellectual framework that emerged in the second half of the century; the broad understanding of urban climate was not congruent with thermal and energy issues. Instead, design approaches to urban microclimates were explored as trade-offs between eccentric *sensations* and biological *needs*. Gandy speaks of a "series of modernist architectural interventions of the time that explored the possibility of creating a new synthesis between nature and culture".[159] As Gandy emphasizes, interwar Berlin was "the capital of European modernism"[160] and an "experimental city",[161] and as such had a major impact on how architects and planners of the time approached the man-made climate. The omnipresent nature of Berlin opened up a "space of urban imagination" that shaped architectural thinking about the man-made climate. The "connection between the city and nature [appears] in the existence of natural spaces within Berlin, such as the Tiergarten or the Landwehrkanal", which inscribed themselves into the design thinking of the "modern and future-oriented city".[162]

Three forms of man-made climate appear in the designs of the interwar period: the (uncontrolled) air volumes in large halls that are made visible, as transitional climates between inside and outside, and as simulated outdoor spaces indoors. Referring to climatic scenery in large halls, three design approaches to man-made climate were presented: freely suspended fabrics forming an installation, vertical vegetation as part of a subtropical courtyard and a landscape mural contributing to the visual regime of a panoptic public bath. These designs stand for the staged character of climate in big cities, in which visual impressions are synaesthetically combined with thermal experiences. The "phenomenon of cladding"[163] played an important role in conveying the microclimatic diversity of the city, hitherto underestimated in the historiography. With large halls in big cities, a modern symbolism of air volumes evolved that connected the indoor with the outdoor climate and static structures with movable surfaces. Climatic scenery was used to amplify the experience of climate in large halls—in the process revealing a broader spectrum of how climate was conceived and represented, meandering between (thermal) control and (visual) simulation. Mechanical air conditioning was combined with pictorial representations of a man-made climate. The notion of man-made climate comprised the full ambivalence of the industrialized metropolis, which was also conceived as a thermal ambivalence.

→ fig. 102   Democratizing Urban Nature

3.3 Access to the Winds: Zoning

Modern urban-planning law begins with the deliberate climatic separation between industrial and residential areas. Factories were evaluated according to their emissions and their distances to residential areas[164] In Vienna (September 1909) "the determining factor in setting up the zoning districts was the prevailing winds. The object was to take the smoke away from the city instead of toward it."[165] According to Ludwig Hilberseimer, "buildings laws and zoning laws" had not succeeded in getting a grip on the hygienic deficits of the industrial city: "They have been concerned chiefly with individual lots instead of large integrated urban areas."[166] "Instead of improving housing conditions, zoning laws have at times served to legalize the exorbitant exploitation of the land."[167]

While the sun unfolds its form-defining influence primarily at micro-scales (solar geometries), the winds inscribe into the structure of the city especially on macro-scales.[168] Designing with the winds is directed against the contaminated conditions of inner cities or aimed at the (passive) thermal regulation of residential areas. The new modern housing estates in Germany as well as the city foundations in the Soviet Union were the test areas where such urban climatic considerations were applied.

3.3.1 Urban Ventilation

In particular, the Bauhaus and the authorities in Stuttgart were pioneering the practical dimensions of climate in urban design. The city of Stuttgart established the tradition of urban climate mapping and the associated regulations in their publication *Städtebauliche Klimafibel* ("Climate Booklet for Urban Development"), which was widely accessible and applied in the second half of the 20th century.[169] However, although Stuttgart was the most prominent case, it was not the only city to promote cooperation between urban planning and meteorology in order to develop new regulations and building codes: already in the 1940s, the German meteorologist F. Linke mentioned that various "towns have town meteorologists in their staff who investigate these things and on whose consultation town planning measures are based".[170]

Dessau

Dessau, which since 1925 had been the new location of the Bauhaus and home of the famous Junkerswerke, was the site of a comprehensive socio-climatological analysis. Initiated by the seminar of Bauhaus teachers Hannes Meyer and Ludwig Hilberseimer, three Bauhaus graduates–Hubert Hoffmann, Wilhelm Hess and Cornelius van der Linden–developed a study on the urbanization of Dessau for CIAM 4. While Meyer promoted a comprehensive urban analysis of Dessau, Hilberseimer aimed, from an urban climatology point of view, in particular at the problematic relationship between areas for living and those for working. "Although Hilberseimer himself did not contribute to the CIAM analysis for Dessau, his theoretical support for it is indisputable."[171]

The 48-page leporello begins with the natural features of Dessau. Chapter 1 examines the "soil conditions" while the three following sheets are dedicated to "climatological" aspects and the "wind diagram." These climatological considerations are resumed in chapter 11 on "Advantages Disadvantages of Location." In the illustration dedicated to Dessau, considerations are given to "living in unfavorable location."[172] The schematic diagram shows the double threat of wind and water to the residential areas of the lower-income strata: the location of the working and residential areas leads to excessive air pollution in the residential neighborhoods and the proximity to the river basin to recurrent flooding. Thus, by simple means, problematic hygienic conditions are addressed that here, already, receive a clear urban climatological reading. "Conceivably detrimental to residential neighborhoods is the location of industry to the west. Almost all residential neighborhoods are bothered by smoke, soot and emissions. In the middle of the industrial line there are chemical factories (refinery, yeast, tar, breweries), whose unpleasant smelling exhaust gases mix together and are spread over the residential quarters by the predominantly north-westerly and southwesterly winds. They cause adverse health effects (nausea, malignant eye inflammation, etc.). The streets located near the industry are often completely covered with soot. Very unfavorable is also the location of the large hospital complex next to the tar and roofing felt fabric and the odor zone of the yeast factory."[173] From the analysis of the disadvantageous arrangement of industrial and residential areas, proposals for new Trabant housing estates were elaborated, which assumes a comprehensive rehousing estate of the population and demolition of the existing tenements and old town areas. The focus was on a new *Trabantenstadt*, a school of thought that had found its way into German modern architecture from Unwin through May. For the workers of the Junkerswerke, an independent housing estate for 10,000 residents was proposed.

→ fig. 122   European Developments before 1945

Although the study was not used for CIAM 4, the analyses found their way into the exhibition *Housing, Working, Traffic, Recreation in the Contemporary City* shown in Amsterdam (in the Stedelijk Museum) in 1935. Furthermore, two schematic drawings by Hilberseimer dating from 1932 were to be used in revised form as representations of the "European Industrial City" in *The New City* of 1944.[174] The work of Ludwig Hilberseimer is exemplary of how central the urban climate should be in the conception of new (modern) cities. He paid the greatest attention not only to the influence of the sun in determining urban densities but also to the influence of the winds in the arrangement of urban districts. In *The New City,* suggestions are made for the establishment of new cities based on wind regimes and the reorganization of existing cities is examined; Hilberseimer also described the "American industrial city."[175]

Stuttgart

With its metropolitan region of 2.6 million inhabitants, Stuttgart was the first city in the world that systematically made urban climatological knowledge fruitful for urban regulation; in this respect, it can be called the first modern climate city. The reasons for the climatological investigations and the associated urban regulations lie in the topographical situation of Stuttgart and "the lack of adequate air exchange which prevailed here. Thermal stress and sensitivity of the effects of heat are common in combination with the mild climate of a winegrowing region. Stuttgart's second handicap in respect of climate and air hygiene is due to its lack of wind, namely the episodic rise in air pollution."[176] Being situated in the steep valley of the River Neckar and subsequently suffering from low wind speeds and poor circulation of air masses, its city government had been concerned with air-quality problems and heat stress since the beginnings of its industrialization.[177] Due to the meso-climatic situation and the manufacturing sector in and around Stuttgart (with Daimler, Porsche, Bosch plants), air pollution was a greater concern.[178] The first document to regulate urban planning with regard to air quality was issued in 1935 in the form of an urban construction byelaw. In 1938, "the City of Stuttgart decided to appoint a meteorological graduate to perform studies of the climate conditions and their relationship to urban development".[179] During the Second World War, experiments were conducted—in conjunction with urban winds—with artificial fog in urban areas in order to reduce the accuracy of the Allied bomber squadrons. The insights gained into the wind conditions in Stuttgart's basin can be regarded as veritable urban climatological experiments under real-life conditions; in the post-war period, these played a decisive role in regulating urban development.[180]

However, already in 1926, the government architect of Stuttgart, Alfred Schmidt (Dr. Ing.), had outlined the path taken by Stuttgart in his essay *Die Bedeutung von Sonne und Wind für den Stadtkörper. Ein Beitrag zu zeitgemässem Städtebau* ("The Importance of Sun and Wind for Urban Figures. A Contribution to Contemporary Town Planning"). In this essay, which to a certain extent builds on case studies of southern Germany and Stuttgart, principles for an urban climatologically informed urban planning are discussed that, in particular, place in the center winds, topography and a strategy for zoning derived from them. His explanations sound like the blueprint of Stuttgart's later program of integrating urban climatology and urban planning. Schmidt saw the creation of hygienic living conditions as a great challenge. "It is necessary to consider the differences between seaside, inland, mountain and valley towns, the characteristic properties of the wind, its changes in direction and speed, the relationship between the shape of the terrain and the air flow, and its influence on streets and squares. The effect of mountain and valley winds must be considered, the value of natural ventilation channels in the form of rivers and canals must be recognized and, above all, the necessity of air renewal in valley towns must be recognized if conditions that are truly conducive to health are to be created. Then we will understand again the rules of the ancient city builders and understand that the wind can have a favorable or harmful effect for dwellings, depending on the direction, strength and connection with dust, precipitation and soot, and depending on whether it brings air from healthy areas or not. The direction, length and width of the roads would have to be determined from these points of view. Then sunny living and working spaces could be created on dust-free and yet well-ventilated streets and yards without any special effort."[181]

Schmidt, anticipating the institutionalization of urban climatology in planning, pointed out the importance of climatic factors in urban development. The control and regulation of the urban climate appears, with full clarity, as a new aspect of urban planning. Both indoor and outdoor conditions, air pollution and overheating are taken into account. The urban climatic analyses suggest—less than a rigid determination of certain street directions—a new kind of zoning, which thus further differentiates the zoning plan. "One could now, by compiling the street directions advantageous for sunlight and wind, create a scheme for an ideal city plan. But such a scheme would be of little help to urban

→ figs. 117, 118, 119, 120, 121, 123, 124

planning. It seems more important to me to develop the idea in a different direction and to examine the building site, whether it is favorable for the construction of streets advantageously situated to the sun and wind or not, whether it is particularly suitable for closed, open or only for extensive building. This would result in a natural division of the urban area into zones according to the type of building and use (residential, commercial, industrial), as well as open spaces."[182]

3.3.2 Linear Cities

The design of new cities in the USSR oscillated between concepts of urbanization and de-urbanization. Like Mikhail Okhitovich, Nikolay Milyutin's concept of *sotsgorod*[183] helped shape the debate on the "dissolution of cities". Bruno Taut had already worked on the possibility of overcoming the city during the First World War. However, while in his case the mutual interpenetration of the urban structure and the natural territory was the driving idea, in the case of the Soviet town planners it was rather economic considerations that motivated strategies of dissolution.

The dissolution of the city, as it was discussed in the young Soviet Union until the beginning of the 1930s, allow numerous insights into the new modern urban planning principles in combination with the new urban climatological knowledge. The already fully developed aesthetic principles of Constructivism were combined here with social theorizing and a rationality of the winds. Albert Kratzer writes in reference to the cities of Magnitogorsk and Stalingrad, "Here, for the first time, the urban climate as such, together with the wind direction, is elevated to the status of a city-forming element."[184] Kratzer emphasizes the preeminence of the new cities and the inexhaustible space available. Besides the influence of the winds, it is the greening of the new city that gained special importance. Urban climatic approaches are evident in the setting of the new city facilities, their basic mix of built-up and greenery. The consideration of the main wind direction is carried out in an exemplary way within the framework of these new urban formations. "The residential quarters are always against the wind, the industrial quarters only behind the residential quarters. A line drawn perpendicular to the wind direction through the center of the city divides each city into a climatically most favorable windward side and a climatically disadvantaged leeward side. From the windward side, all industrial plants and freight stations should be banned. This proposal has now found a very logical fulfillment in some Russian cities."[185]

The Soviet examples show how the dual aspect of wind in the city—as an environmental and thermal regulator—came into play and predisposed the zoning of the new cities. Numerous documents survived from Ernst May's time as head of an urban-planning group of German architects in the Soviet Union. The central role of the geographic–climatic conditions in the conception of these new cities is obvious in these descriptions. In a newspaper article for the *Frankfurter Zeitung* of January 24, 1931, May describes the processes and the speed with which these German urban planners developed their designs for cities such as Stalingrad. The plan was for a parallel striped arrangement of functions: A: Railroad; B: Area of production (industry and industrial schools); C: Park; D: Highway; E: Residential area; F: River Volga. The new cities have to be built from scratch. After a day of inspection of the site a first draft for the new city is sketched that very evening: "Clearly we see before our mind's eye the city crystallizing with compelling logic out of the geographical conditions, the city as a ribbon on the Volga River. The railroad road will run along the river bank, thus enabling direct transfer from the barge to the railroad along the entire city. A kilometer-wide industrial belt will run along the riverbank. The band of housing estates will extend at the foot of the steep slopes to the second terrain terrace. A strip of green several hundred meters wide separates the industrial and residential areas. Following the ravines drawn by nature, at larger intervals, at the same time separating the different categories of industry, it sends green strips down to the Volga, on the cool banks of which the inhabitants of the city will seek refreshment in the blazing heat of summer."[186]

Magnitogorsk was chosen as the "first model city of socialism".[187] Its planning followed the ideas of the ribbon development.[188] However, this primacy of economics was repeatedly at odds with urban climate considerations. Already "Lenin had formulated the necessity of locating industrial centers in the vicinity of raw material deposits."[189] This gave rise to specific urban-climatic issues, such as avoiding, or at least mitigating, the enormous air pollution through adequate zoning.[190] After a large urban-planning competition, in which questions of the communal house type were in the foreground, the German architect Ernst May and his team were appointed; besides May, Hans Burkhart, Karl Lehmann and Mart Stam also took part.[191] The excursion to Magnitogorsk took place between October 26 and November 8, 1930. The residential quarters of Magnitogorsk were designed as a Garden City, with special attention paid to the "self-sufficiency of the individual neighborhoods".[192] While the first design variant showed

→ fig. 42   European Developments before 1945   **108**

117
Ludwig Hilberseimer, *The New City*, 1944.

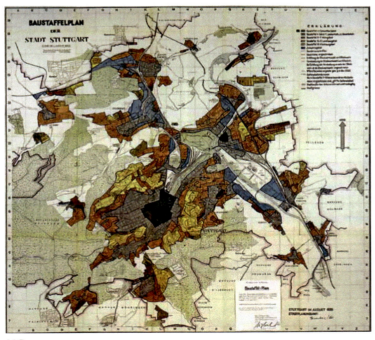

118
Bye-Law Plan of the city of Stuttgart, 1935.

119
Hillside development in Stuttgart.
Source: Döcker 1929.

Democratizing Urban Nature

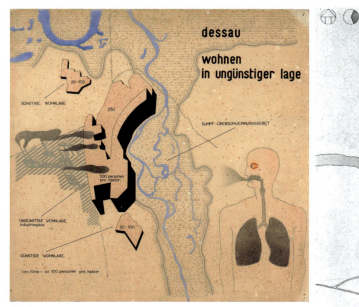

120
Dessau: Living in an unfavorable location.

121
Ludwig Hilberseimer, plan of the city of Dessau (Germany), announced as "European industrial city. Diagram of the present state and condition".

122
Hubert Hoffmann, Wilhelm Hess, Cornelius van der Linden, 48-page leporello on Dessau's climatic, social, and economic conditions, as part of the arbeitsgruppe dessau der int. kongresse f. neues bauen, 11, 1932.

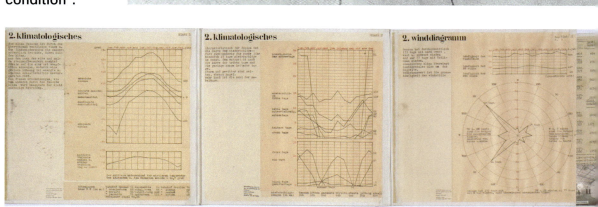

European Developments before 1945

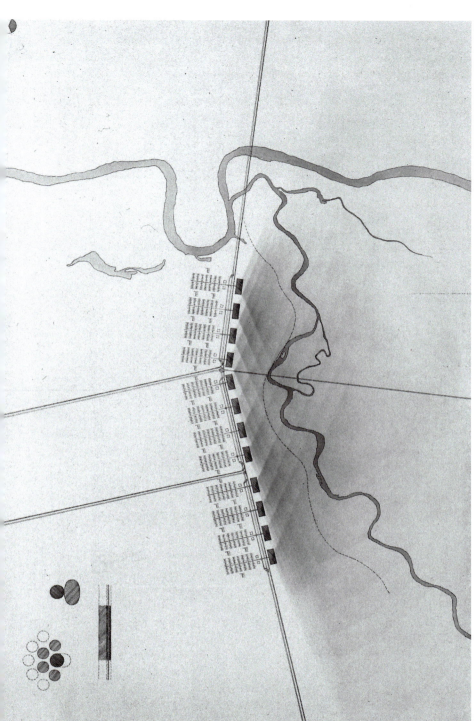

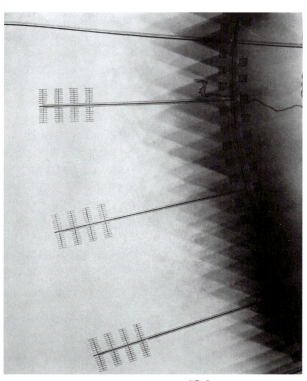

123
Ludwig Hilberseimer, plan of the city of Dessau (Germany), announced as "European industrial city. Diagram of its proposed replanning, 1933".

124
Ludwig Hilberseimer, "Wind conditions necessitate a separation of residential areas from industrial areas." Source: Hilberseimer 1944.

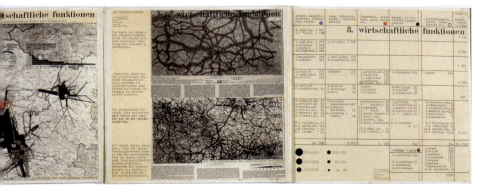

Democratizing Urban Nature

"parallels to the development plan of Ernst May and Walter Schwagenscheidt for the garden city Goldstein in Frankfurt am Main", the second variant, decisively influenced by Mart Stam, was "a rather schematic juxtaposition of rows planned in a north-south orientation".[193] The large distance of 40 m planned between the three-story rows allowed for a strong greening of the residential quarters ("four times the height of the houses"). Each quarter was also to have a "central green space for play and sports activities". In keeping with the new urban-planning paradigm of the "satellite system", the residential complexes were to be "directly connected to the surrounding nature, with interior and exterior spaces merging, thus bringing the functions of living and recreation closer together".[194] A forest was to be planted as a buffer "between the plant and the city", which connected the two areas with each other as well as taking on a "protective function". The background to this novel sensibility was the principle of closely linking production and living. The workplaces had to be connected with the residential areas and new green recreational areas. Urban climate considerations regarding the interaction between the industrial combine and housing played a role: "In addition to favorable light conditions for the residents, this arrangement was also intended to provide a certain degree of protection for the houses in the southern part of the city from the influence of the combine. However, the consistent row development did not correlate with local wind conditions."[195]

European Developments before 1945

Global Adaptations after 1945:
Three Design Metaphors

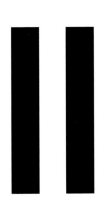

**Thermal Heritage
Microclimatic Islands
Energy-Synergy**

Thought Styles of Man-made Climates

An Introduction

In the 1930s, "man-made climate" was the term for a consciously designed microclimate, inspired by the ambivalent atmospheric character of big cities. The central focus of urban climatological knowledge of the interwar period was the reciprocal interaction between inside and outside, brought together in the collective practice of climate control. After the Second World War, numerous traces of urban climatological thinking could be found in architecture and urban design that continued this approach. The dialectic thinking of outdoor climate and indoor heating and cooling demand, launched as science in the interwar period, remained relevant, especially in the immediate post-war period, due to the insignificant distribution of building services in the housing sector. Martin Wagner opens his 1951 publication *Wirtschaftlicher Städtebau* ("Economic Urban Planning") with a chapter on the relevance of the urban climate for urban planning. Referring to Brezina and Schmidt's 1937 publication, he recalls the continuity between interior and exterior spaces: according to Wagner, the "unintentional influence of the urban climate" is contrasted with "the planned influence, which today is limited only to the very easily influenced climates of the interior, but which may become of the greatest importance for human life and activity in the future. After all, the interior space is only the continuation of the exterior space, and the exterior space only a continuation of the interior space. The feedback effect of both spaces on each other should certainly not be overlooked by those planning these spaces."[1]

Wagner was by no means alone in his interpretation of the urban climate. The arrangement of the urban fabric (which inflence urban microclimates) was still significant for technological thinking on heating and cooling. As the meteorologist Ferdinand Steinhauser pointed out with regards to Vienna, "the wind strength accompanying certain weather phenomena plays a role, for instance, the winds in cold weather with regards to the heating of living spaces".[2] In the middle of the century, geographers such as the Briton Charles Ernest Brooks or the East German Günter Grundke were still equally concerned with both natural and mechanical forms of climate control. The interior was still heavily dependent on external climatic conditions, and the meteorologists who pioneered the scientification of urban climate remained convinced of the practical relevance of their research.[3] The meteorologist Werner Strempler pointed out that "special emphasis should be placed on the interaction of indoor and outdoor climates". In a report written for the City of Berlin on the use of urban climatology in urban planning, Strempler called for "the recapitulation" of all existing "city-owned and third-party" studies, with the goal of "creating a bioclimatology of the city of Berlin".[4]

115

Global Receptions and Transnational Entanglements

The growing internationality of urban climatic research after the Second World War has been extensively and adequately recorded.[5] In Tony Chandler's 1970 bibliography prepared for the World Meteorological Organization,[6] as well as in Bob Frommes' series of publications[7] released from 1978 onwards, one finds bibliographies listing publications from Germany, the USA, Japan and many other Western countries. Socialist countries are also represented, with numerous contributions from the USSR, East Germany (GDR) and Yugoslavia. The paradigm of state control in the socialist countries played into the hands of large-scale planning approaches with a higher-order perspective that transcended the individual building.

As far as the German-speaking world is concerned, one can speak of a certain scientific continuity or continuity of scientific careers. Scientists and architects dealt with urban climate issues in the context of reconstruction and (post-fascist) nation building. In both West Germany and Austria, as well as in the GDR, lines of thought such as those that emerged at various universities before the war were continued. In the GDR, for example, a number of scientists and planners carried on the legacy of the pre-war decades; explicitly, for example, in the case of the geographer Günter Grundke, the meteorologist Wolfgang Böer and the landscape architect Friedrich-Herman Pfützner. Böer published the small paper *Klimaforschung im Dienste des Städtebaus* ("Climate Research in the Service of Urban Planning") in 1954. Here, once again, the whole expectation of a "close collective cooperation of planning architects and meteorologists" is set out.[8] In the chapter "The Application of Climatic Values", for example, an approach is outlined that fits in with the planning needs of the post-war period in Germany. The idea of interdisciplinary collaboration, which had been envisioned for decades, is now the subject of socialist construction in labor collectives. However, while in West Germany, urban climatology from the 1960s became oriented along the lines of the emerging ecological movement and the critique of environmental pollution, the initial efforts of urban climatology in the GDR were led *ad absurdum* by a rampant air-pollution problem resulting from overemphasizing economic development. In considering both parts of Germany, one has to consider what were increasingly strong negative associations of the term "urban climate".

As far as the USA is concerned, specific lines of reception must first be mentioned that were not least the result of contributions made by emigrant European scientists and architects. A few figures in particular were instrumental in conveying research published in German to the Anglo-Saxon architectural and scientific community immediately after the Second World War: the two Hungarian brothers Victor and Aladar Olgyay, the German climatologist Helmut Landsberg and the American architectural publicist Jeffrey Ellis Aronin. *Design with Climate*, published in 1963 by Victor Olgyay, presented a synthesis of thermodynamic insights at different scales of architecture and urban design. The Olgyay brothers were able to comprehensively integrate and process the meteorological and structural–physical state of knowledge of their time. For the concepts underpinning *Design with Climate,* the Olgyay brothers, who evidently could read German, significantly relied on the research of Albert Kratzer, Helmut Landsberg and Rudolf Geiger,[9] thus critically

Global Adaptations after 1945

contributing to the reception of this research in the Anglo-Saxon world and hence integrating passive strategies of climate control in the field of architecture and urban design. The architectural publicist Jeffrey Ellis Aronin must also be emphasized. With his publication *Climate and Architecture* (1953), he provided a careful bibliography that includes numerous references to German-language urban climatology. His descriptions of collaborations between architects and climatologists, for example in the planning of the capital city of Chandigarh (India) and the industrial city of Kitimat (British Columbia), deserve special mention. Helmut Landsberg, a German climatologist who emigrated to the United States, worked on both projects as a key consultant to the New York architectural firm Mayer & Whittlesey.[10]

The reception of European-influenced urban climatology, as it occurred among American landscape architects in the 1960s and 70s, took place in light of the new ecological paradigm of the era. Anne Whiston Spirn proclaimed that "a fresh attitude to the city and the molding of its form is necessary".[11] What was lacking was the translation of the enormous amount of fragmented interdisciplinary studies on urban nature and urban climatology into a coherent design perspective in order to make urban climatology applicable. "I discovered that a wealth of information about urban nature did exist, sequestered in specialized scientific journals, in conference proceedings, and in technical reports. This book arose out of my frustration in failing to find a volume that summarized that knowledge and applied it to urban design."[12] Spirn addressed the central theoretical problem of my publication: the translation of scientific knowledge into a design methodology. As much as this knowledge aims at application, it has not yet been made fruitful for "molding the form of the city—the shape of its buildings and parks, the course of its roads, and the pattern of the whole".[13]

The paramount importance of the concept of "urban heat islands" cannot be justified from the history of architecture and urban planning examined here. Rather, the reduction of urban climate to problems of heat islands has contributed in good part to giving urban climate a marginal importance in the discipline of architecture. The concept of urban heat islands, which has been strongly emphasized since the 1970s, has contributed to the disappearance of the complex relationship between inside and outside that governs the many microclimates in urban environments.

From the Hygiene to the Ecology of the City

The most striking shift in the architectural understanding of the urban climate in the post-war era was a result of far-reaching social changes, which manifested themselves discursively as a gradual transition from *hygiene* to *ecology*. Instead of *the proxemics of hygiene*, *the system of ecology* emerged, oriented towards superordinate wholes. No longer were the distances between the objects of the city at the center of the analysis but the energetic processes in the system as a whole. Cybernetics and, later, political ecology, formed stages of the new systems-based thinking. While the former aimed at control of the city by means of infrastructures, the latter particularly elicited nature and its energies as components of the city. While applied climatology of the 1920s and 30s was still separate from energy issues (focusing on thermal themes), energy conservation gained increasing importance in the second half of the

→ fig. 125   Thought Styles of Man-made Climates   **117** century, in which the urban environment was considered as an "energy landscape" (Michael Hough). The ecological shift within urban climatology led to the understanding of the possibility of saving energy at an urban scale.

An architectural theory of the urban climate must accordingly unfold in two ways. Firstly, as a history of urban microclimates with *hygiene* as the leading interdisciplinary science. And secondly, as a history of energy landscapes, in which *ecology* is the pivotal concept. Whereas German-speaking architects and climatologists of the interwar period had developed a planning toolkit to use climatic factors such as wind and sun for structuring cities, in the second half of the 20th century everyday perspectives gained importance for the design of microclimates. The figure of the city dweller was recognized as being important for an understanding of the man-made climate. The growing inclusion of everyday perspectives in architecture and urban planning also changed the view of urban climatology.

This development was accompanied by a growing relevance of other media in the interaction of urban climate and architecture: methods of measurement, simulation techniques and computerization showed the central position of media for the recording and assessment of the urban climate; however, they were also components of a new "identity politics" of urban societies.[14] Michael Hebbert and Brian Webb summarized this development as follows: "The science appeared to hold great promise for climatically-informed town planning. Data-gathering was enhanced by new techniques such as vehicle-mounted observation points, high density weather stations on public buildings, balloon mounted sensors for monitoring city air-flow, radar, aerial photography and remote sensing of atmospheric conditions. The state of the art shifted from descriptive research to process analysis of energy exchanges and air circulation within the complex three-dimensional geometry of urban landscapes, using physical and then numerical models."[15] The *internationalization* of German-speaking urban climatology was accompanied by an *institutionalization* in the form of new laboratories and methods of data collection. In the following chapters, the new, epistemically relevant interconnections between empirical–ecological studies in the field, simulations in the laboratory and the designs in the offices of landscape architects and architects will accordingly be shown. The collaboration of architects with landscape architects and ecologists sharpened the multi-scalar imagination of those involved. What emerged was an understanding of the multi-scalar contextuality of buildings that clearly extends beyond the immediately visible.

With modernity, the nature of the city may indeed have lain fallow as a resource, as numerous authors have critically noted. However, the process of reintegration, which has accelerated with the emergence of urban climatology, had already begun in the interwar period and found a continuation in numerous variants after the Second World War. Traces of an architecture-related urban climatology can be found in both the West and East, which anticipated the explicitly ecological thinking found from the late 1960s onwards and developed an urban perspective. In this respect, one can speak of a proto-ecology or an ecology *avant la lettre* in the field of architecture and urban design. An integrated view reveals an ecological history of architecture and urban design that does not

Global Adaptations after 1945

follow the narratives of the anti-urban American ecology movement (with its focus on individualism and single buildings).

Thought Styles of Urban Microclimates

To conceive urban climatology detached from *the practice of design* relies on the erroneous assumption that its application is a purely techno-scientific project, which does not always involve the traces of architecture and urban development. Michael Hebbert and Fionn Mackillop characterized "urban climatology applied to urban planning" and architecture as "a postwar knowledge circulation failure".[16] "The application of climatic design principles, […] remained localized and little-known until the rise of concern over global climate change prompted fresh interest in the present century."[17] By subordinating urban climatology to the causality of applied science, this statement relies on one-sided epistemological assumptions: the "application" itself appears as a technological problem rather than one of *design methodology*.

However, urban climate is a classical example of a "wicked problem" as outlined by Horst Rittel and Melvin Webber in 1973.[18] "In dealing with wicked problems, the modes of reasoning used in the argument are much richer than those permissible in the scientific discourse."[19] Wicked problems such as urban climate rely on "a model of planning as an argumentative process in the course of which an image of the problem and of the solution emerges gradually among the participants".[20] Rittel and Webber point to the "plurality of publics" as critical social contexts of planning.[21] The epistemic boundaries of the "problem" are not clearly definable since "a wide array of social problems" and actors are involved.[22] The man-made climate has to be conceived as a "wicked problem", which can't be equated with the "well-tempered environment" as promoted so widely in the second half of the 20th century. The notion of man-made climate conceives microclimates as human artifacts, which in a significant way bring together nature and society. Hence, urban climatology represents a *scientificity* that is steadily haunted by metaphors intertwining research with design.

125
From hygiene to ecology: design approaches to urban climates.

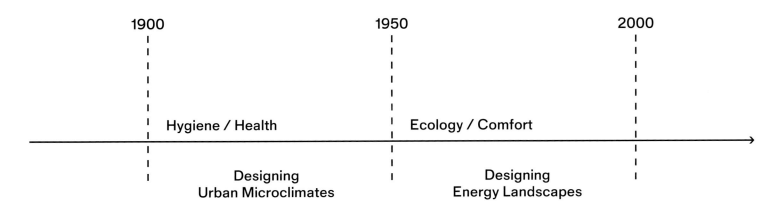

Thought Styles of Man-made Climates

The urban microclimate is hence an entity that is based on findings from various disciplines, but it has always had obvious connections to architecture and landscape architecture. As an exemplary hybrid topic, its appropriation was coined by specific "thought styles". Ludwig Fleck defined the "thought style" of expert communities as a "directed perception with the corresponding mental and factual processing of what is perceived".[23] Thought styles of urban microclimates refer to different forms of appropriation of the microclimate concept as they were practised in the second half of the 20th century. The focus was on the translatability of the scientific object into a design artefact by means of architecture and landscape architecture.

Following Giambattista Vico's "logic of the imagination",[24] philosopher Hans Blumenberg emphasizes the autonomy of what is "self-created" by man: "the world of his images and constructions, his conjectures and projections".[25] With Blumenberg, it must be assumed that metaphors are in many cases the only way to help the recognized truth to break through into reality. With regard to the "transferring speech" of metaphors, the "translatio" has an inescapable role in the formation of the reality.[26] Following such an understanding, "imagination and logos"[27] are inseparable also in the case of the appropriation and implementation of urban climatological insights through architecture and urban design. Establishing links between scientific knowledge and design methodology requires imagination and a sense that responds to hidden clues and indirect signs. Not all of the designs discussed in this publication were created under explicit urban-climatological auspices, although they offer significant points of connection to urban climate. Metaphors play a central role in the agency of urban climatic knowledge in architecture: this begins with the metaphor of "man-made climate" itself and is particularly evident today in the metaphor of the "heat island". Hence, the methodological approach chosen here is based on the uncontrollable semantic richness of the metaphors that can be found in the history of ideas of urban climatology.[28] Concepts such as the "urban landscape" changed the way architects thought not only about the structure of the city but also about its urban microclimates. With the pictorial turn, urban planning itself evolved in the direction of "urban design" and thus increasingly addressed the visual qualities of the city.

Against all odds, the notion of man-made climate continues to be thought about on a global scale in the form of new approaches and themes. Contrary to the assumption of a "postwar knowledge circulation failure", the second part of this book opens up methodological and imaginative attempts that further differentiate the approaches of a collective practice of coping with urban climates in architecture and urban design. Three urban metaphors in particular represent the process of globalization and transformation of German-speaking urban climatology:

↘ thermal heritage

↘ microclimatic islands

↘ energy synergy

Global Adaptations after 1945

4 Thermal Heritage

Collective Memory and the Natural History of Cities

Writing in 1945, shortly before the end of the war, in an article in the Swiss architectural journal *Das Werk*, the Swiss–Austrian architect and urban planner Ernst Egli declared the climatically adapted city to be a genuinely European project, whose universal legitimacy, however, had been cast into doubt by the war and would be put to the test by the pending reconstruction. "Will Europe have the strength, in the midst of its own trial, or perhaps because of it, to think once again for the whole world?" In the face of heavily damaged cities such as Warsaw or Berlin, Egli wondered what principles would be used in future to consider the climate in cities. It was "obvious", he said, "that urban design science is not ready".[1] A few years later, in his book *Climate and Town Districts* (1951), Egli coined the metaphor of "the city as a house" in order to address the architectural program of urban climate and its consideration in architecture: "people will doubtless build their towns as they would build a home."[2] In light of the coming reconstruction, he argued it was no longer individual buildings but entire city districts that needed to be adequately air-conditioned. Accordingly it was the city—and no longer individual buildings—that formed the new thermal interior of society, which implied a fundamental shift in scale in the field of climate control.[3] For this task, nature in the city provided the guidelines.

In the face of the obliteration of entire urban districts during the war, architectural reconstruction was bound to the natural history of these cities. Such a "collective memory"[4] included not only what had been destroyed—the built artifacts—but also the system of natural forces in the space of the city: the quasi-natural constants. These were to be formed into a new type of *infrastructure* possessing both aspects of technical artifacts and technologies of "mnēmosýnē";[5] collective memory played a critical role in the conception of these climate-related infrastructures. The new urban arrangements were derived from the memorial recordings of the natural conditions, representing the thermal heritage of a city and its region.

4.1 Post-war European Districts (Hans Scharoun, Revisited)

Throughout the 1950s, publications appeared that dealt with war-destroyed cities and the necessity of a reconstruction that was sensitive to the urban climate. For this, the 2nd edition of Albert Kratzer's book published in 1956 (originally published in 1937) may be seen as representative. The renewed and expanded edition was influenced by the obliterated European urban districts and the view that urban climate was an important design parameter in the reconstruction process. In the preface to the 2nd edition, Kratzer writes, "The need to rebuild cities that had sunk into rubble provided an opportunity to put urban climatological knowledge to practical use in planning the reconstruction of the city."[6] The main reason he gave for this was the loss of cultural connections that buildings once represented: given the extensive destruction of architectural and infrastructural artifacts, the nature in the city was a critical *surviving context* to which architects and planners could refer. In addition, there were extremely *sparse resources* available, which led to an increased consideration of the natural forces in the city.

The Second World War engendered destruction on an unprecedented scale, and the experience of extinction became part of the urban climate discourse. What was needed were design parameters beyond culture that would foster an altered culture of building. In reconstruction projects, attempts were made to tie in with the nature of the city by paying particular attention to the urban climate and the natural geological conditions that influence it. "Progressive climate monitoring will certainly make it possible to find ways to favorably influence the urban climate. As cities are rebuilt, there will also be ample opportunity to eliminate stone deserts that retain heat and block the winds from entering. Green spaces should take their place, and residential areas, hospitals, and schools should be moved to the outskirts."[7]

4.1.1 Deep Structures: Collective Memory in Urban Environments

According to the philosopher Maurice Halbwachs in his volume on *Collective Memory*, published posthumously in 1950, places and spaces possess, "equivalent to a language", a "meaning": "Even if they do not speak, we nevertheless understand them, since they have a meaning which we playfully decipher."[8] Collective identity and spatial realms, Halbwachs argues,

→ figs. 126, 127, 128, 154

121

126
Scenes of destruction, Berlin, 1945.

127
The destroyed district of Kreuzberg in the area of Alexandrinenstrasse, Berlin, 1945.

Global Adaptations after 1945

128
Stuttgart (Germany), 1945.

129
Hiroshima (Japan), 1945.

130
Scheme for determining the damage groups in the 'Guidelines for Statistics and Presentation of Damage in Destroyed Cities', issued by the German Reconstruction Planning Staff in July 1944.

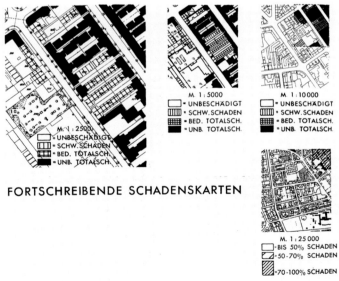

131
Sample for updating damage maps in the 'Guidelines for Statistics and Presentation of Damage in Destroyed Cities', issued by the German Reconstruction Planning Staff in July 1944.

Thermal Heritage

essentially coexist, together constituting the totality of "the ideas, traditions, and remembered histories of particular groups", which he calls "collective memory."[9] He sees such a collective memory as having a life of its own, in the form of its own history and effectiveness.

Halbwachs emphasized the preeminent significance of space as a carrier of memory. The meanings shaped by collective memory fundamentally rely on places and territories, which always include the natural elements of the city. Urban nature represents a resource that is not only to be harvested, refined, formed and packaged but also to be remembered. The central prerequisite for this is a collective memory that is capable of recalling the nature of the city and its historical transformations. The process of remembering makes the shaping of new, climate-related infrastructures possible.

Zones of Permanence and Change

Halbwachs pointed to the temporal implications of space: he argued that space in all its manifestations—within the horizon of processes of rapid modernization—represents the epitome of "permanence" and "continuity". It is these aspects that epistemically bring space close to nature: "The different parts of a city and the buildings within a district have a fixed place and are as strongly anchored in the ground as trees and rocks, as a hill or a plateau. From this it follows that a group of city dwellers does not appear to change as long as the appearance of the streets and buildings remains the same and there are few social formations that are at once more firmly established and more permanent."[10] Space undermines social change by cushioning its impact through "inertia". The (more) persistent space formed by "images" unfolds its effect. As a memory carrier, space represents a premise of human action. The relationship that the inhabitants of a city have with their environment forms its own history and thus its own meaning in a longer perspective. The term *adaptation* is a paraphrase of this very process.

Halbwachs developed his argument along two intersecting axes of thought: both the identity of "groups" and the meaning of "spaces" are generated in the area of tension between *memory and perception* as well as *permanence and change*. Depending on the level of social change, the two poles prove to be indissolubly contradictory: in smaller towns with enduring traditions, for example, the collective memory of this community largely falls into line with what can be perceived on the ground—clearly familiar houses, clearly

familiar people. Likewise, the transformation of this community proves to be attuned to local traditions on the ground. This is quite different in the modern city; here, social change is overshadowed by the patterns of collective memory: "While the group evolves, the external cityscape changes much more slowly. Local habits resist the forces that seek to change them, and this resistance allows us best to perceive the extent to which collective memory relies on spatial images in such groups."[11] Under these auspices, it is impossible to refer exclusively to the (perceptually structured) meaning of a newly created spatiality. According to Halbwachs, perception remains overshadowed by a pre-existing understanding of that place that identifies individuals as members of a particular collective. The collectively produced meaning of a place is an inescapable part of dealing with that very place and leads collectives to try "to regain their former equilibrium in the new circumstances."[12] This is true even where, as in Tokyo (1923) or Hiroshima (1945), an entire city is destroyed.

Kon Wajiro's[13] project of an ethnographic survey of everyday life in Japan ("modernology") represented a new kind of research into the built environment following its total destruction (by an earthquake). The sketches of the post-disaster conditions after the earthquake in Tokyo bear testimony to the collective memory. Hiroshima was destroyed even more instantaneously with the dropping of the atomic bomb on August 6, 1945. As a city, it was arranged densely over a small area of 10 km², with about one million inhabitants. After the bomb was dropped, it was assumed that no trees or other plants would grow for decades. But just the opposite happened; immediately, growth began again.[14] The power of urban nature contributed significantly to the regeneration of the city. The rapid "regrowth of [...] Hiroshima's radioecology"[15] shows how "the making of landscapes" follows "multiple temporalities", as anthropologist Anna Tsing has emphasized.[16] According to architect Hiroshi Sambuichi, this was due to urban nature such as the wind flow as well as the seven rivers flowing through the city.[17] The interaction of winds, water bodies and the sun had for centuries led to a balanced urban climate in Hiroshima.[18] While the water of the rivers comes from 1,300 meters above sea level, in the Seto Inland Sea area, the water of the sea flows deep into the delta of the city at high tide. According to Sambuichi, the way clean water and air circulate in Hiroshima was why the city recovered so quickly.[19]

→ fig. 129   Global Adaptations after 1945   **124**

Urban Environments

Already in the immediate post-war period, there were signs of a new type of city-related environmental research in both Germany and Austria, in which the urban climate was negotiated as part of an urban natural history that made use of novel methodologies. In Vienna, the Institute of Science and Art[20] organized "a series of ecological lectures" entitled "Problems and Findings of Environmental Research", in which "the manifold and complicated relationships between organism and environment" were discussed from interdisciplinary perspectives.[21] Early attempts to understand the city and the environment in their continuities and interactions can be seen in these approaches. The "big city" appeared as a multidisciplinary subject, as shown by the "bioclimatic working group" of the Viennese *Forschungsgemeinschaft für Grossstadtprobleme* ("Research group for big city problems"). Also in Vienna, the journal *Wetter und Leben* ("Weather and Life"), first published in April 1948, was dedicated to the countless interactions between the two. The editors speak of the "weather-based influences on life",[22] which were to be the subject of the new publication.[23] The journal explicitly inherited the *Bioklimatische Beiblätter* ("Bioclimatic Supplements") of the *Meteorologische Zeitschrift*, which had been discontinued in late 1943.[24] The economic tint of *Wetter und Leben* distinguishes it from the climatic and cultural studies of the first half of the century. The journal emphasizes the relevance of meteorological research to civil engineering, construction, urban planning and economics in general. The authors stress the agency of "climate" for very different spheres of life, with construction as the fulcrum.

Klima und Bioklima von Wien ("Climate and Bioclimate of Vienna"), a synthesizing publication of decades of weather measurements in Vienna clearly demonstrated the pioneering character of this meteorological project: a product of the "Central Institute for Meteorology and Geodynamics in Vienna, Hohe Warte", it was published in three volumes between 1955 and 1959. The subtitle directly addressed the intended audience of these publications: "an overview with special attention to the needs of urban planning and construction". "In particular, the climatic data of the Central Institute also provide valuable bases for the purposes of civil engineering, industry and various branches of the economy."[25] The three volumes relied on an unusually long-standing series of measurements made in Vienna, and initiated a paradigm shift in meteorology that had been emerging since the 1940s—away from the statics of unchanging climate data to dynamic approaches, i.e. to the study of climate change within shorter periods of time.

F. Steinhauser, one of the authoritative protagonists of the pre- and post-war periods, emphasized the importance of an exact "insight into the climatic conditions of Vienna", especially as a "basis which can and should be used as quantitative measurements for various kinds of planning and other practical purposes".[26]

Anticipating the "ecological" arguments of the late 1960s, in the third volume natural plant growths are conceived as indicators of urban climatic conditions. For example, "the distribution of bark-inhabiting lichens in Vienna is interpreted as an indicator of the effect of the urban climate". Such lichens are viewed as being particularly sensitive to chemical "air quality", the "microclimatic conditions" such as drought and "reduced dew formation", and to a superposition of both factors.[27] While the actual city center is lichen-free, a certain transitional zone stands out with low lichen content, also in comparison with the distribution zone in the surrounding area. The authors consider the chemical qualities of the air to be of greater relevance than climatic influences; in this respect, the "map-like representation of the distribution can be considered as a supplement to the results of the [...] air pollution" as such.[28] The authors mention the exhaust gases of a power plant, which—despite "uniform climatic conditions and the presence of the same tree species"[29] in the area in question—led to the absence of lichens in the catchment area of the power plant.

In an innovative way, the nature of the city is here understood as an indicator of the quality of life in the city; the lichen growth is made emblematic as an "indicator of the effect of the urban climate." Lichens are particularly sensitive to the specific climatic conditions of a place, indicating where the interplay between air pollution, greenery and housing stock becomes visible. Subtle changes in arrangement and positioning can lead to different growths in the same street. The prevailing wind direction is central to the distribution of lichens. "In the Viennese urban area, trees with rough, cracked bark, especially maple species, elms, lindens, poplars and oaks are preferred. Horse chestnuts and walnut trees are less frequently colonized, conifers almost never."[30] The comparatively small number of lichen species in the city appears to demonstrate the "complicated character of the area requirements for the site".[31] Here, anticipating an argument later made by Anne Whiston Spirn, the city viewed as nature or natural space appears as an interplay of myriad urban ecologies; the microclimates favoring lichens are also desirable microclimates for people. In particular, to address the serious

Thermal Heritage

problem of urban air pollution, relief can be provided through a consideration of the prevailing winds in conjunction with adequate zoning. "For purposes of urban planning and zoning, of all climatological elements, wind is of the greatest importance. The wind transports air pollutants generated in industrial areas and traffic centers, the harmful effects of which on public health depend essentially on the spatial arrangement of individual functional urban areas in relation to climatological conditions. In recent times, increasing attention is being paid to this aspect in the planning and reconstruction of war-damaged cities."[32]

Deep Structures

The natural history of the city as a reference space for collective memory can be further deepened with an ecological conception made by the American landscape architect Anne Whiston Spirn (see chapter 6), building on the work of Ian McHarg, who had posited that there are "two systems within the metropolitan region": "the pattern of natural processes" and "the pattern of urban development".[33] McHarg emphasized the autonomy of the natural processes and their role as starting points in design processes: "Rather than propose a blanket standard of open space, we wish to find discrete aspects of natural processes that carry their own values and prohibitions."[34] Following McHarg, Spirn distinguished "deep structures" (the natural history of a city) from its "surface structures" (the cultural history): "Deep structure expresses the fundamental climatic, geomorphic, and biotic processes in a particular place. Deep structure is the product of these processes operating and interacting across vast scales of time at the scales of large regions and at the microscale."[35] Spirn's central ecological idea is that the natural and cultural history of a city must be in conceptual congruence; otherwise, an unwanted energetic, economic and communicative loss occurs through friction. "Deep structure can be masked, but it cannot be erased. When surface structure obscures or opposes deep structure, it will require additional energy, materials, and information to sustain."[36] In contrast to the surface structure, deep structure cannot be obliterated; however, its neglect leads to compensatory measures, including additional expenditure of energy, resources and data. "In modern cities, where the surface structure of buildings, roads, sewers, and parks seem to obliterate deep structure, the danger is that this more enduring natural environment may be ignored."[37] In Halbwachs and Spirn's considerations, it can be seen that the planning of cities can refer to both surface structures and deep structures while contradictions can never be

completely eliminated: The spatial order of the city remains in tension with the societal order. One can speak of an "inherent logic of spatial configurations" related to collective memory.[38] Spirn argues that urban development, however, must take the inextinguishable significance of the deep structure into account. "Traditions, values, and policies may change, but the deep structure of each city remains an enduring framework within which the human community builds."[39]

The Barbican Estate in London (1956–82) can be viewed under the conceptual premise of collective memory, foregrounding urban nature. The estate was designed in the mid-1950s by the office of Peter Chamberlin, Geoffry Powell and Christoph Bon. The vast area is situated in central London; its site was a heavily bomb-damaged area from the Second World War. While the design process was completed in 1959, the construction lasted throughout the 1960s and 70s. The buildings are arranged to bring sun to enclosed and wind-protected squares. The three-dimensional complexity of this city within the city is remarkable, permeated by numerous green spaces and water basins. The three main towers are placed on the northern perimeter, allowing the sun to heat the lower-situated buildings. A series of ponds and plants help to cool down these zones with evaporating water in the heat of the summer. Generally, these measures help balance the temperature in the estate. Part of the original design proposal was to promote a pedestrian zone in the city, elevated from urban traffic and noisy streets.[40] The living conditions include a walkable zone in which inhabitants and cultural activities are within a reasonable distance from one another.

4.1.2 Megaforms: CIAM and the Territorialization of Architecture

After the Second World War, CIAM representatives developed a growing interest in the uneven density distribution of the territory. The character of the territory became critical for the arrangement of new buildings and urban services. The design of architecture as man-made topography or greened space shaped architectural thinking about the climate-adapted city. The City of Stuttgart issued the very first guidelines for the influence of the urban climate that addressed topography (see chapter 3). Urban-planning strategies with a topographical character were also used by José Luis Sert and Paul Lester Wiener (Town Planning Associates) in their planning activities in Latin America immediately after the Second World War.

→ figs. 136, 137   Global Adaptations after 1945   126

The urban plan they prepared for Chimbote (Peru, 1946) had numerous courtyards sunk into the ground as a response to the extreme aridness and cold found in that region.[41] Here, the topography became an important urban structuring element, which always possessed urban-climatic implications.

The transformation of architecture into a kind of man-made topography (or an artificial nature) became one of the major thematic foci of designing urban climate. Topography came to be understood as architecture of territorial dimensions, and architecture was transformed into a topography—thus drawing more heavily from nature, exemplary in the case of terrace houses. Either infrastructures were conceived as territorialized architectures or architecture was reinterpreted as an infrastructure—with numerous implications for the deliberate influencing of urban climate. Buildings, green space and topography were now in a state of mutual affinity and metamorphosis.

Plants, Topography, and Buildings

During the Second World War, Walter Gropius and Marcel Breuer had already explored how topography and buildings—in interaction with the climate—could be understood as a field of forces, in their project for the so-called *Aluminum City Terrace* in New Kensington (near Pittsburgh, Pennsylvania). The settlement, comprising 250 units of apartments in a hilly location outside of town, was designed in 1942 for American war-workers. The urban arrangement contradicts all the aesthetic and heliomorphic guidelines as developed in pre-war Germany. According to Arnold Whittick, this Aluminum City was "sited irregularly according to the sun, and with complete disregard of any formal or geometric layout. They represent the culmination of twenty years' progress in this matter."[42] The orientation of the buildings followed different fields of forces, as Gropius explained in an interview with Jeffrey Aronin, discussing the project: "Gropius has told to the writer that this project is an illustration of what has to be considered in factors of climate and character of the land. He gives great importance to these factors, but notes that there are many and that they have to be studied carefully. He says that the southern exposure is always superior from the living point of view on account of the steep angle of the sun during the hot time of the year, but, if the buildings have to be put on rolling ground, the orientation of the buildings has to be adjusted, so the southern exposure cannot always be a one hundred per cent selection. Furthermore, he notes, the views from the house and the wind direction of the region in question have to be considered as well; and when all these factors are balanced out, an optimum is reached for that location."[43]

In an article published in 1949 in the *Journal of the Royal Institute of British Architects*, geographer Gordon Manley addressed the still unclear disciplinary location of microclimatology. The original relationship of this field of research to botany (and thus to the world of plants) seemed clear. However, Manley, addressing the architectural community, foregrounded three aspects of a design-influenced microclimatology: in addition to the microclimate of the plant population, it is the microclimate caused by the topography as well as the microclimate of the buildings that architects must pay special attention to. "Quantitative measurement of these local differences leads to the study of microclimatology. Unfortunately the scope of the study is not clearly defined. Botanists with some justice prefer to restrict the term to the discussion of the small but significant differences found within the zone occupied by the plant cover; within a ripening wheat field, for example, there are appreciable differences between the prevailing conditions of temperature and humidity at the level of the ears and beneath. [...] Others have for long used the term with regard to the local differences that we readily perceive to be associated with topography. The differences arising from variations of altitude, aspect and shelter due to relief come first to mind. The proximity of the water bodies, the character of the soil and the extent to which it is drained, and the prevailing vegetation cover introduce notable modifications. Lastly, replacement of the vegetation by buildings gives rise to further modifications. It might be wise to describe all these effects under the term 'local climatology'. But we can also learn much from the more detailed work of the botanist whose technique and results may often be of interest to the architect engaged in the study of small spaces, or the problems of what happens at the surface of his materials."[44] Gordon Manley's triad—vegetation, topography and architecture—formulated a toolbox that would in fact manifest itself twice in the course of the 20th century: first, as the consciously designed topology of microclimates, and second, as the reciprocal metamorphosis of this triad. Fostering the ever-stronger merging of this triad was one of the pivotal points of thinking on urban climate in the second half of the 20th century. An urban theory of climate control draws its scientific vocabulary from all these three fields of inquiry—plants, topography and buildings. They have to be conceived in close relation to climatic parameters such as the sun, the winds or the humidity.

→ figs. 138, 139   Thermal Heritage   **127**

132
Jerzy Soltan et al., Sports complex Warszawianka, Warsaw (Poland), 1954–72.

133
The sports complex Warszawianka as man-made topography.

134
The megaform providing a microclimatic diversity.

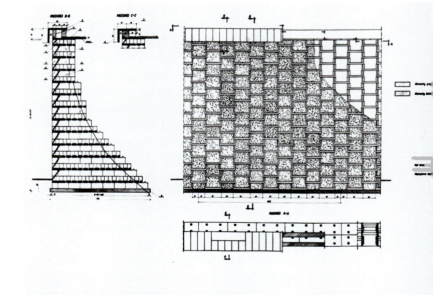

135
Interlocking concrete blocks (designed by Lech Tomaszewski) that could be laid as chains, were applied as stabilizers of the slopes, forming terraces across the complex.

Global Adaptations after 1945

136
José Luis Sert and Paul Lester Wiener (Town Planning Associates), master plan of Medellin (Colombia), 1948.

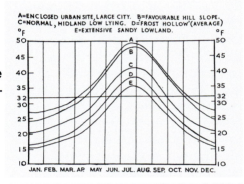

138
Impact of the topographical and urbanistic conditions on the temperature profile. Source: Manley 1949.

137
José Luis Sert and Paul Lester Wiener (Town Planning Associates), view of future Chimbote (Peru), 1946.

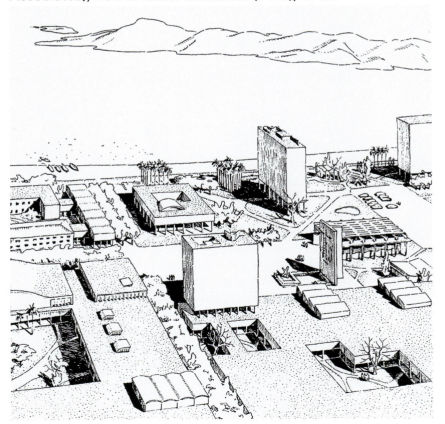

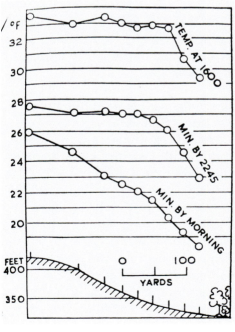

139
Temperature profile over the course of the day. Source: Manley 1949.

Thermal Heritage

The Warszawianka Cooperative Sports Club

The city of Warsaw was largely destroyed during the war. According to the historian Christhardt Henschel, "Warsaw is a metropolis whose urban DNA is indelibly inscribed with the consequences of the German occupation during World War II. At the beginning of the war, the city officially had about 1.3 million inhabitants, including more than 360,000 Jews; in 1946, the official figures were still just under 480,000 and 18,000, respectively. In addition to hundreds of thousands of people, entire neighborhoods disappeared during the occupation. The mountain of rubble left behind by the Germans is estimated at 18-20 million cubic meters."[45] Urban nature such as the hydrological system, the system of winds and some of the green spaces remained therefore as important survivors, which represented outstanding parameters for the reconstruction of the city. The Warszawianka Cooperative Sports Club complex (1954–72) "set into the gentle escarpment of the Vistula river valley" can be considered an exemplary megaform with a microclimatic diversity (in Gordon Manley's sense).[46] Under the harshest post-war conditions and in view of a largely obliterated inner city of Warsaw, the natural forces of the city were in this case activated using the least amount of material possible. The main elements of the sports center included "earthen structures, plants, and multifunctional mini-pavilions."[47]The historical reconstruction of the site unfolds, as the Polish architecture and research collective CENTRALA (Małgorzta Kuciewicz and Simone De Iacobis) emphasized, the natural incentives of this project. CENTRALA speaks of a "multi-element, conceptual sculpture, which is founded on the great memory of the earth and harnesses the forces of nature" in the city of Warsaw.[48]

Due to the very limited resources available after the war, Jerzy Soltan, a member of CIAM (Team X) and coordinating architect of this complex, decided to work with the territory and the natural forces of the city. Soltan led a multidisciplinary team of the "Art and Research Unit" of the Academy of Fine Arts of Warsaw, comprising artists (sculptors) and landscape architects, among others. Rather than proposing buildings in a strict sense, they formed the urban territory. The layout of the sports complex followed the main water and wind corridors of Warsaw, in this sense anticipating Anne Whiston Spirn's idea of the "deep structure" of cities, based on their natural history forming the *longue durée* of urban environments. The water and wind corridors corresponded to the existing green corridors. The subterranean waters of the valley were critical for the microclimatic manage-

ment. "The terraces, embankments and hollows of Warszawianka direct the course of the water that has been eroding the slopes of the Warsaw Escarpment continuously since the end of the last glacial period. On both sides of the body of the stadium, ducts and cascades lead to a body of water which is level with the lower terrace of the Vistula River. Here, its surface is stirred up by the wind into mist. Rather than being negated, the movement of water has become an ephemeral component of the architectural design."[49] By forming the topography consciously, it provided a play with different microclimates. After the war, people had a highly developed sense for microclimatic conditions and the ways in which to deal with changing atmospheres.

The limited resources available led to the absence of reinforced concrete and foundations, reducing concrete consumption "by 40 percent".[50] Instead, interlocking concrete blocks designed by Lech Tomaszewski, which could be laid as chains, were applied as stabilizers of the slopes, forming terraces across the complex. "Unlike deep-foundation walls, which resist the pressure of the earth, the Warszawianka retaining walls move together with its masses, without interrupting the circulation of water or the movement of animals."[51] "As planned by landscape designers Wanda Staniewicz and Alina Scholtz-Richert, mosses and lichens were planted in the earth that was deposited inside the open-ended self-locking modules in such a way that that the roots of the species selected would withhold the slope's pressure and prevent the soil from spilling over."[52]

It is no coincidence that Jerzy Soltan is quoted at the beginning of Kenneth Frampton's famous essay on the "megaform".[53] In contrast to the infrastructural *megastructure* (see chapter 6), as emphasized in particular by the Japanese Metabolists, the *megaform* in Frampton's definition is developed from topographical conditions. Frampton emphasizes the fact "that while the megaform may display certain megastructural characteristics, the large-scale manifestation and expression of the megaform's intrinsic structure is not its primary significance. What is much more pertinent in the case of the megaform is the topographic, horizontal thrust of its overall profile, together with the programmatic, place-creating character of its intrinsic program."[54] The megaform has a natural-historical trait that distinguishes it from the technological definition of megastructure. In this case, architecture does not primarily become *technology* but rather an artificial *territory*.

→ figs. 132, 133, 134, 135   Global Adaptations after 1945

4.1.3 Urban Landscapes: Post-war Berlin as a Model

While in the Berlin of the Weimar Republic the democratization of urban nature had constituted the field of forces that had produced new solutions, it was the war-ruined Berlin, with the beginnings of an economic boom, that guided the search for innovative planning solutions. Post-war Berlin, largely flattened by bombing, was characterized by new intermediate zones (brownfields). As a result of its intensive bombing in 1944 and 1945, the city now had huge open spaces at its disposal, which gave the once highly dense inner city a completely new character and which were to be used for reconstruction from an urban climatic perspective, as Albert Kratzer emphasized in the second edition of his book. "The destruction of the Second World War confronted the administrations of many cities in Germany and Europe with the task of, to a greater or lesser extent, rebuilding them anew. A whole series of publications dealt with this subject and endeavored to take advantage of the hardship of the destruction and to incorporate the resulting rubble fields into generously planned green belt. I can only refer to the list of publications."[55]

Berlin again proved to be an incubator for planning principles that took the fragmentation of the city as a starting point for a new urbanism, weighing the interplay between the urban fabric and the renaturalized areas of the city. During his time as Berlin's Councilor for Building (*Stadtbaurat*) in 1926–1933, Martin Wagner had been unable to implement a proposed system of new green spaces for Berlin's densest quarters in the inner city; finance was limited by the constraints of civil law. It was not until the bombing raids in the war that large swathes were torn through the residential districts and new green areas subsequently became imaginable. On May 23, 1945, Hans Scharoun took over the newly created "Department of Construction and Housing" in the Magistrate's Office of the City of Berlin.[56] Together with a small team[57] of colleagues selected by him, he drew up the so-called "Collective Plan" for the reconstruction of the city.[58] Among other things, the plan was "based on the damage mapping of 1945."[59] It showed the areas that had been preserved (cross-hatching), partially destroyed (diagonal hatching) and predominantly destroyed (no hatching).[60] In January 1946 at the request of the Red Cross, Scharoun estimated the extent of the destruction as follows—about a third of the city was completely damaged. Charlottenburg and Friedrichshain were the most heavily damaged districts of the city. "Of the 224,917 buildings that stood in Berlin in 1939, 30,000 are totally destroyed,

47,000 are moderately to severely damaged. In terms of dwellings, this means: of 1,562,000 dwellings before the war, 266,000 are completely destroyed, 410,000 moderately to severely damaged."[61] These numbers outline the massive task of reconstruction but also the potential it offered for introducing new parameters for urban planning.

1946: The Exhibition *Berlin Plans. First Report*

Against the background of extensive destruction of the urban structure and in declared dissociation from Albert Speer's urban planning proposals, the natural and topographical conditions of Berlin became the central "point of reference" for the new "Collective Plan", which was referred to as "urban landscape". The result was a combination of city and countryside, which was characterized by new kinds of in-between zones. The topos of the "lack of history" was combined with a reading of the "topography" as a guiding instrument for urban development.[62] The new city was laid out between the hills along the Spree. With "urban landscape", a scintillating concept was introduced that was to some extent intended to replace the old notion of "garden city" or at least extend it into the hinterland. Immediately after the end of the war, the landscape architect Herta Hammerbacher, in consultation with Hans Scharoun, conducted a series of lectures on "Open Space Policy, Urban Landscape and Regional Planning" at the Institute for Urban Planning at the Technical University of Berlin, in the context of which the concept of the "urban landscape" was decisively introduced.[63] The "Berlin urban landscape"[64] became a new major planning paradigm of the second half of the 20th century, also affecting design approaches to urban climate.

In the summer of 1946, as part of the exhibition *Berlin plant. Erster Bericht* ("Berlin Plans. First Report"), the first main outcomes of the Collective Plan were presented to the public in the White Hall of the Palace of Berlin. The radical proposal, written entirely in the spirit of "moral and political renewal", left only a few "museum souvenirs in the center", denying the remaining urban structures any "commemorative value".[65] In its reference to natural-historical and quasi-natural conditions, Berlin was conceived as a renewed city foundation at the moral null point of 1945. Scharoun argued that since the war had produced a "mechanical loosening" of the city, "the possibility of creating an urban landscape" was to be used and "a new living order" created from nature and buildings, from low and high, narrow and wide.[66] Scharoun and his team referred to the category of the

→ figs. 130, 131, 140, 145   Thermal Heritage

131

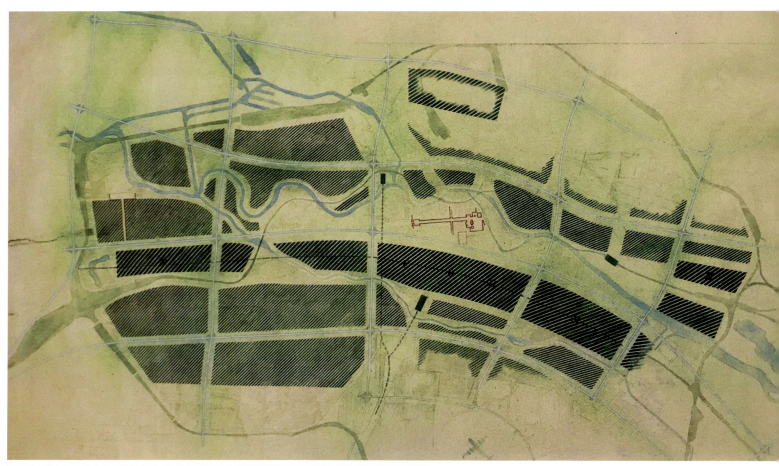

140, 141
Collective plan prepared by the planning collective, 1945/6. Original scale 1:10,000, 95 × 195 cm. The "green mist" that covered the entire central plan was meant to be a clear indicator of the complete "interpenetration" of built and green structure.

The transportation concept of the "Collective Plan" largely ignored the existing radial structure of Berlin; instead, a grid-like scheme was proposed, based on the course of the *Urstromtal*.

142
The architects referred to the so called *Urstromtal* ("Valley of the Urstrom"), which was to form the natural-historical basis of a new "ribbon city".

Global Adaptations after 1945   132

143
Principles of the envisaged new city using the example of Charlottenburg. The model was shown in the exhibition *Berlin plant. Erster Bericht*, Berlin 1946.

144
Planning collective (Selman Selmanagic), redering of the reconstruction of Berlin, 1946.

145
View from the entrance of the exhibition *Berlin plant. Erster Bericht*, Berlin 1946.

Thermal Heritage 133

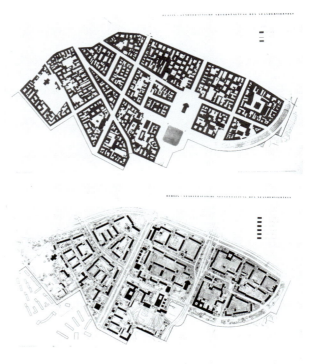

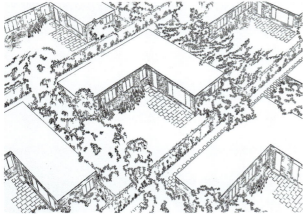

148, 149
Layout of *Neanderviertel*, Berlin, 1939.

Hermann Henselmann and Werner Dutschke, *Heinrich Heine Quarter* (former *Neanderviertel*), Berlin, 1957.

146, 147
Hans Scharoun, the principle of the *Wohnzelle* ("residential cell").

Ludwig Hilberseimer, the principle of the *Wohnzelle* ("residential cell").

150
Aerial view of the *Heinrich Heine Quarter*.

151
Alvaro Siza,
Block 121 of IBA
1984/7 in Berlin.

152
Model of Block 121 of IBA 1984/7 in Berlin.

153
Green courtyards in Kreuzberg, Berlin. Source: Fisch, Maass and Rating 1984.

Thermal Heritage

Urstromtal ("Valley of the Urstrom"), which was to form the natural-historical basis of a new "ribbon city".[67] The central plan suggested an east–west expansion of the new city, with Spandau and Köpenick two of a few specific locations. Notably, the "fine green mist" that covered the entire central plan was meant to be a clear indicator of the complete "interpenetration" of built and green structure, striving for an "adaptation to natural conditions".[68] While the faint hatching represented the "housing", the darker band, located more centrally, included the different "functions of work", such as administration (V), electro-industry (E), printing (D), garment manufacturing (K), mechanical engineering (M) and trade (H). Finally, the buildings and zones that were to be considered as "cultural heritage" in the reconstruction of the city were marked in red; the Charlottenburg Palace in the west and Unter den Linden in the center. The clear separation of work and living in the form of the ribbon city indicated a return to the existing planning concepts of the pre-war period, which also always had urban climatological implications. Kratzer referred to Magnitogorsk and Stalingrad as exemplary ribbon cities (1937), and Hilberseimer to Milyutin's concept of the ribbon city (1944) (see chapter 3).

The transportation concept developed in the Collective Plan largely ignored the existing radial structure of Berlin; instead, a grid-like scheme was proposed, based on the course of the *Urstromtal*. Highways were planned in a grid pattern as the main access routes for the new urban landscape, replacing the radial urban concept of the pre-war period. As Jörn Düwel noted, the trained mathematician and architect Peter Friedrich must be considered as being central to this approach.[69] Immediately after the war, Scharoun had at his disposal one of the "few surviving copies" of the book of urban-planning theory, *Die gegliederte und aufgelockerte Stadt* ("The Structured and Dispersed City"), which had already been carried out during the war as a response to the bombing of German cities on behalf of Albert Speer by the Deutsche Akademie für Städtebau, Reichs-, und Landesplanung (German Academy for Urban, Imperial, and Regional Planning) and printed in January 1945.[70] Geist and Küvers describe the Collective Plan as "a first concrete application of this theory using Berlin as an example".[71]

The discussion launched in the interwar period on the relationship between density and urban microclimates continued in the discourse between "building forms and open spaces". Even in the planning collective's designs for a new cinema complex, there was an interpenetration of the landscape and complex. Of particular importance, however, was the so-called *Wohnzelle* ("residential cell") and the associated mixed-use design, which Ludwig Hilberseimer developed during his time as a Bauhaus lecturer and published prominently in *The New City*. Hilberseimer propagated different types of dwelling—row houses, semi-detached houses, detached houses, apartment buildings—which were also adopted by Scharoun and his team. "We can have both the one-family house with its gardens and the apartment house with a free view over these gardens. This new type of settlement is a form of housing which meets the actual needs of man. It gives him complete freedom to choose the kind of dwelling he prefers."[72] In his book, Hilberseimer advocated L-shaped single-family neighborhoods that combined quality of life and high densities with sufficient privacy. And he made proposals for a "mixed type of settlement", such as four-story row buildings (with "40 people to an acre"). The relatively large spacing between the rows, allowed for the insertion of single-story buildings in addition to intensive planting.[73] This combination Hilberseimer considered to be the settlement approach of his time. The suburbanization envisioned was later developed into an overall approach to regional planning that was to lead to the de facto dissolution of the urban–rural dichotomy that also appears in Scharaoun's notion of urban landscape.

1962: Row Development versus Block Development

The areas of the *Luisenstadt* (in today's *Kreuzberg*) and the *Neanderviertel* heavily destroyed by Allied bombing, became the testing grounds for a new form of urban development—both in the West and in the East—that considered urban climate requirements. In the doctoral dissertation of the East German landscape architect Friedrich-Herman Pfützner, urban climatic improvements in the cities of the Soviet occupation zone under reconstruction are praised as socialist achievements.[74] With direct reference to the earlier work by Martin Wagner, Pfützner examines the effects of urban green spaces on heat build-up in cities. Extensive greening, as already envisaged by Hermann Jansen in his groundbreaking competition entry of 1909 (see chapter 3), was assumed to be the new norm of East German urban design some 50 years later. The real goal of such urban design was to achieve "a city climate approximating the open countryside",[75] an idea which was widespread in those times.[76]

→ figs. 42, 141, 142, 143, 144, 146, 147

Berlin's former "Neander District" (*Neanderviertel*), renamed *Heinrich-Heine-Viertel*—located between Köpenicker Strasse, Heinrich-Heine-Strasse and the sector border—serves as an example how socialist design principles met the urban climatic requirements. Pfützner compares the district's state in 1939 with the plans of the newly rebuilt quarter in the eastern part of the city. The Entwurfsbüro für Hochbau I ("Design Office for Building I"), with its lead architects Hermann Henselmann and Werner Dutschke, had developed the plans for the urban redevelopment of the *Neanderviertel* as part of the first series of three-, four- and five-story models of development of the GDR. "If one looks at a typical area of old buildings, it becomes clear how right Brezina and Schmidt were in their observation that the climate of such an area is similar to that of a stone desert. The depiction of the same urban area, rebuilt according to the principles of socialist urban planning, makes it obvious that here the comparison with the climate of a stone desert can no longer be maintained. The larger horizontal section results in a lower thermal load. After all, the development, the position of the buildings, the traffic areas as well as the design of the open spaces cause, among other things, a change in the radiation conditions, the wind speed and the air temperature in comparison to the open countryside. Thus, in such an area, one can speak of a climate approximating that of the open country."[77] The cooling of warmed cities is postulated as a planning objective here, and thus a principle that had also gained validity in the West with the concept of the urban landscape. However, Pfützner emphasizes the fact that the mere reduction of the density of buildings is not a sufficient guarantee of "optimal thermal conditions" in urban development.[78] It is necessary to draw the right conclusions with regard to the "building placement" in the "new cities and urban areas in summer"[79] as well as the placement of the existing buildings. These contribute to shading, but also lead to reduced ventilation. "The more the climatic situation of a place tends to produce clustered warm days—for example, cities in basin locations—the more every opportunity for ventilation must also be used."[80] The solar exposure of dwellings and the ventilation of outdoor spaces are in conflict and accordingly require periodic rebalancing.

Row development versus block development was a widely discussed, ideologically charged dichotomy—not least in the post-war socialist societies—which also had strong urban climatic implications. To consciously design microclimates appeared also as an urban-planning and socio-ideological debate about the correct typological character of the city. Climate and ideology were in a reciprocal relationship,

reinforcing each other. "On summer days, row development is clearly superior to block development", Pfützner states emphatically.[81] He presents examples from Berlin, Rostock (Hoyerswerda) and Eisenhüttenstadt—the authoritative socialist housing projects of his time—and examines the projected tree populations in the new developments. According to Pfützner, climatic considerations had been incorporated far too little into the new planning of the cities; it is worth noting, for example, the criticism of the GDR's main representative avenue, Karl-Marx-Allee, as well as the adjacent Alexanderstrasse in East Berlin. The primacy of the architectural would have prevented the necessary thermal design measures by the planning of adequate green. "To what extent might the architect's fear of obstructing the view of the facades of the buildings have been the driving force here?"[82] In the case of the new socialist model city of Hoyerswerda (Housing Complexes II and III), the adequate design of the areas of open space is generally emphasized. The planting of shrubs along the housing blocks and the basement stories is emphasized, which led to an increase in the amount of cooling. On the other hand, additional hedges and bushes in the case of a block development in Eisenhüttenstadt are negatively evaluated, as they led to an increased reduction of wind speed, while additional plantings along the facades are evaluated as positive. And in the case of a "sand playground" in the small town of Schwedt, a decreased wind speed and an increase in radiation was determined, the closer playing children are to the ground. Rather than additional hedges, shading would be the adequate response.

The shading of open spaces and buildings and their cooling through adequate wind speeds formed a permanent dilemma of planning, which had to accordingly be weighed up. This necessary balancing was taken into account by the parameter of the *Abkühlungsgrösse* ("cooling factor"). With this factor, a quantifiable measure for planning–architectural evaluation was created. In many cases, greening reduced wind speeds, leading to a "lower cooling effect".[83] The aim is to achieve a balance between the different parameters, a constant balancing of the interdependencies and the affected uses. In terms of landscape architecture, "smaller fluctuations" and "higher minimum values" of the cooling effect should be striven for.[84] In West Germany, Erwin Neumann provided similar insights: "In order to maintain the effect of moderate winds for air renewal, this must be taken into account in the development [of the city]. Green areas enclosing the city center should be kept low so that the city air can flow away over them. Urban reconstruction provides an important tool in

→ figs. 148, 149, 150   Thermal Heritage

this regard. By prohibiting construction in the most appropriate places, its allows the wind to re-enter as a renewer of the air."[85]

The "ideal" of "grouping the buildings into housing clusters" is postulated as a dialectical overcoming the contrasting typologies of urban development. "In its effects on the human heat balance, this building arrangement occupies an intermediate position between block and row development. The openings between the blocks of houses still permit sufficient ventilation, so that no such unfavorable effects as with the block development are to be expected, and on the other hand a reduction in wind speed can be expected even in the case of strong winds due to the separation of the buildings alone."[86] In the new ideal of "grouping the buildings into housing clusters", not only can one recognize the socialist debate on the organization of neighborhoods as "microdistricts", which had been launched in the West by the idea of the "neighborhood unit";[87] moreover, an approach had been established to the deliberate microclimatic conceptualization of urban neighborhoods. The thermal conception had shifted from the scale of the building to that of the neighborhood (see chapter 2).

1984/87:
The International Building Exhibition (IBA)

The Collective Plan, as elaborated by Scharoun and his team, contrasted sharply with Max Taut's proposal for the reconstruction of Berlin, which was oriented to the city's existing (although invisible) and largely undamaged infrastructures.[88] Instead of a complete reconstruction of Berlin, Max Taut proposed an urban development following existing infrastructures such a street grids and sewage-water systems. As theorized by Anne Whiston Spirn, "deep structures" and "surface structures" indeed form a field of tension; however, the infrastructures themselves could become a deep structure, as Taut's proposal indicates.[89] While Scharoun foregrounded a natural-historical rereading of the urban landscape, Max Taut emphasized the primacy of the surviving infrastructures, which became the new main guideline for urban development after the war.

While the 1957 *Hauptstadt Berlin Wettbewerb* ("Capital City Berlin Competition")[90] consummated the international planning debates on the notion of urban landscape, the *International Building Exhibition* (IBA) of 1984/87 extended the approach by Max Taut, prioritizing the existing built structures of the city and thus historical-renewal planning strategies. The contributions by Hans Scharoun (2nd prize) and Alison

and Peter Smithson (3rd prize) in the frame of the 1957 competition, revisited and continued themes of the pre-war Modern Movement, as they emerged in particular in Berlin as the "capital of European modernism".[91] The questions concerning urban nature in architecture and planning were once again taken up by Scharoun under the auspices of the urban landscape. In his contribution, he combined the contemporary requirements of the modern city with the greatest amount of green spaces. Scharoun had envisaged a 2-km-long "hill of buildings" that would have extended from Leipziger Platz to *Luisenstadt*.[92] Aspects of urban nature were deliberately placed within the conceptual framework of the superordinate collective. Green spaces experienced a further revaluation as infrastructures that enabled the integration of the disparate parts of the city. In the Smithsons' entry, the separation of traffic and pedestrians on different levels, which in turn were connected by escalators and ramps, came to the fore.

The *Internationale Bauausstellung* (International Building Exhibition, IBA) Berlin of 1984/87 focused on the ecological urban renewal in the West of the city (*Kreuzberg*), not far away from the heavily bombed area of the *Neaderviertel*.[93] According to Margrit Kennedy, the "reconstruction phase" destroyed more buildings in Berlin than had been destroyed in the entire war. Now, "cautiously should be saved what still can be saved".[94] The recognition of urban nature was a decisive framework for promoting this protecting endeavor, excavating the natural-historical traces of the city. In the urban planning and architectural debates on the IBA, one sees again an effort to incorporate urban climatological knowledge into the new conceptions of the settlements in Kreuzberg, based on the current state of knowledge relying on measurements. The contribution of Joachim Schmalz[95] (on "urban climate") as well as Rose Fisch, Inge Maass and Katrin Rating[96] (on "green courtyards") were exemplary in this respect. The greening of roofs, courtyards and facades were central themes of historical and ecological urban renewal.

Joachim Schmalz was particularly interested in making urban climatological knowledge (of the early 1980s) fruitful for the urban renewal of West Berlin. The "climate-altering effects"[97] of specific building components and the "improvement of the microclimate"[98] appeared as central design themes, in which air pollution and thermal comfort stood in a field of tension that was often difficult to resolve. The courtyards of Berlin's perimeter-block developments were examined from a microclimatic perspective, and climate and energy were understood as interdependent

Global Adaptations after 1945

138

categories (see chapter 6). The design of individual buildings and urban districts were still an interrelated field of consideration: "Radiant heating of the south-facing interior courtyard facades, which also makes sense from an energy standpoint, also leads to desirable small-scale circulation."[99] The interior and exterior of the courtyards were understood as thermally interdependent and were shown to be of greatest relevance for planning. Special consideration was given to "low exchange weather conditions" without sufficient wind speeds. In this context, the city was "dependent on thermally induced air movements", which could only be brought about by design, i.e. by architectural and planning means. "Larger contiguous green spaces, in conjunction with courtyards and landscaped air movement paths, provide the necessary distribution of favorable microclimatic conditions over an entire contiguous planning area."[100] It was not individual buildings but rather overarching blocks and neighborhood structures that were examined here with regard to the creation of desirable microclimates. Alvaro Siza created one of the exemplary projects of the IBA at the intersection of Schlesische Strasse and Falckensteinstrasse.[101] Block 121, fragmented since the war, was closed with a curved corner building; however, a "narrow strip was left free" at Jakob-Kaiser-Strasse, allowing access to the courtyard area and ventilation from the nearby Spree River.[102]

Divergent interests of the different user groups of the courtyards as well as of the plant biotopes and animal habitats required a careful investigation of the existing potentials for improvements. This is because the existing yards of Kreuzberg were characterized in many cases by poor ventilation and heavy shading. Especially in the eastern zones, there was a lack of comprehensive greening. "To improve the urban climatic and air-hygienic situation", an "increase in vegetation mass, especially also on large roof surfaces and on south-exposed walls" was proposed, combined with an "increase in infiltration and evaporation areas". In addition, an "improvement of the vegetation structure" was proposed, especially avoiding "grove- and forest-like structures in narrow courtyards", as well as an "improvement of the sunlight conditions" through a "partial gutting".[103] From an architectural point of view the "sealed, asphaltic courtyard surfaces" were criticized. "It is only through finely structured materials and textures that livability is created. The more private the use, the softer the courtyard surfaces."[104] In contrast to the immediate post-war discussions, the focus of the IBA 1984 was on existing buildings. Particularly noteworthy were the plant-biological inventories, which were paired with those on the material aesthetics and architecture of the

courtyards. The strengthening of the existing biological, social and structural diversity was clearly at the center of the urban-climatic considerations of the 1980s.

4.2 Future Asian Cities (Ernst Egli, Revisited)

In the middle of century, the notion of the climate-adapted city was a widespread idea. The combination of empirical assessment and reasoned deduction informed the extrapolative elaboration of the climate-adapted city. In those years, climate-deterministic figures of thought were still self-evidently connected with empirical measurements and physical principles, for which the publications of Victor Olgyay are exemplary. In *Design with Climate* (1963), for example, there is talk of a "hot-humid zone housing layout"[105] or of a "community layout in New York-New Jersey area with winter sheltering effect and summer breeze penetration".[106] Until the mid-1960s, this methodological hybridity shaped architecture's approach to climatologically informed urban design.[107] While the effects of the materiality and arrangement of buildings for the urban climate was scientifically known, there was a lack of design methodologies that translated climatological insights into architecture and urban design.

In addition to empirical approaches relying on measurements, *deductions from global climate zones* and the *decoding of thermal allegories* played a central role for translating urban climates into design proposals. The latter two aspects in particular led to the inclusion of architecture and urban-design history, which made the empiricist inquiry too limited. The future city could only be adequately explored through such a multi-perspective approach. In his 1946 lecture series *Theorie des Städtebaues* ("Theory of Urbanism"), Ernst Egli said, "The great […] but also sad […] time that we live [in] has, with a brutality peculiar to history, put cities in heaps, cities without number, cities that shelter millions of people, cities that had been the pride or, nevertheless, the constant worry of their inhabitants, cities that […] plunged uncounted people with them into destruction: This period has at the same time confronted us with the task of building new cities in place of the old ones."[108]

→ figs. 151, 152, 153, 154, 155   Thermal Heritage

4.2.1 Deductions: Anticipating the Subarctic Town (Russia)

While new empirical environmental research in the post-war period was particularly indebted to the great task of European reconstruction, more globally oriented architectural publications aimed at the theoretical interconnection of urbanization and climate—and thus remained committed to climate-deterministic thinking as it had developed since antiquity. In 1958, Gordon Manley spoke of *The Revival of Climatic Determinism*.[109] Such thought strove for the modernization of the city without, at the same time, falling prey to the ubiquity of generic global solutions. This ambivalence is expressed particularly in the works of the architect and urban planner Ernst Egli and the climatologist Charles Ernest Brooks, who both published their books on urbanization and climate in 1951.

The arrangement and materiality of buildings, which had once depended on climatic conditions, were increasingly being superimposed or even superseded by the new mechanical technologies of indoor climate control. In the middle of the century, climatic considerations were already clearly under the auspices of the globally ascendant air conditioning, which led to increasingly generic solutions. Hence, Egli asked himself whether "the basic town structures of the different climatic zones [were] unchangeable" or could they "be changed by technical means of our day", and "in the end be replaced by a standard town type"?[110] What Egli was criticizing was a type of modernization that invariably leads to generic solutions: "Modern technical means of air-conditioning and of insulation would make it possible to build cool and agreeable houses of the European or American type even in the glaring desert."[111]

Climate Zones

In *Climate and Town Districts*, Egli not only provides climate-deterministic reflections[112] on the relationship between climate and planetary urbanization but also models for future climate-related cities for different climate zones of the world. Precisely because he dares to make, as he says, *strong simplifications*, such models became conceivable. An attempt to develop architectural principles for the climate-adapted city from climatic conditions becomes apparent. While Jeffrey Aronin, for example, placed his architectural and urban-planning cases under basic climatic factors of influence—sun, temperature, wind, precipitation and "other climatic factors"—Egli derived the future patterns of urbanization from climatic zones. In contrast to the new empirical environmental research and the widespread view that assumed the global triumph of mechanically air-conditioned individual buildings, Ernst Egli examined the possible "deductions"[113] to be drawn from the climate zones for the reconstruction and design of future cities worldwide. Further architectural developments have to be oriented towards the environments of these zones.

Deductionism, as manifested by Egli with his references to numerous historical examples, makes possible an imaginatively rich theoretical approach to urbanization processes, which lies in sharp contrast to the empiricism of novel urban climatology. Abstract modelling and reasoning allowed urbanization to be examined from a supposedly logical point of view (committed to the architect's expertise) by conceiving of technology, geophysics and architecture together in relation to one another. What resulted were a series of developments that are in contrast to the empirical data and present themselves accordingly as rather utopian and future-related. According to Egli, climate-zone maps alone do not represent reality sufficiently; (empirical) measurements of local conditions are also necessary. However, it is precisely the division into climate zones that opens up a comparative approach to different solutions that point beyond individual location. In *Climate and Town Districts*, Egli uses the comparative rationale of the Modern Movement in his own way to study cities cross-culturally as climate-dependent human organizations in order to develop appropriate solutions for the future. The comparative approach seems to be the real asset of the deductive (climate deterministic) method. "It is always the actual conditions that are important in any determined spot of the earth, and these conditions can be measured, controlled and exemplified. Maps alone are never sufficient to express reality, all the more so if they are maps giving a large and general view. Nevertheless, we must learn how to classify the towns of our earth according to their climatic zones. We must forget the conception of town we were born into, and which we sometimes regard as the conception valid for every town, and teach ourselves to imagine in a living and realistic way the town structures typical for other climates."[114]

What is remarkable about Egli's argument is that he insists on the European legacy of climate determinism and, equally, calls for it to be overcome. Or, to put it another way, climate determinism leads as much to the cementing of cultural stereotypes as it does to the creation of unprecedented future cities. Alongside the Second World War, the end of European colonialism appears as an important context for Egli's reflections.

→ figs. 159, 160, 161, 162   Global Adaptations after 1945

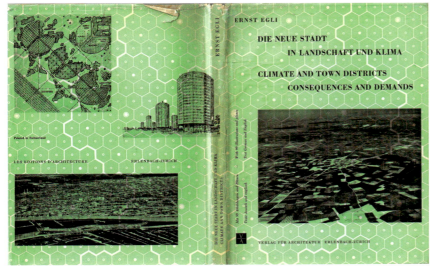

154
Ernst Egli, cover of the book *Climate and Town Districts*, 1951.

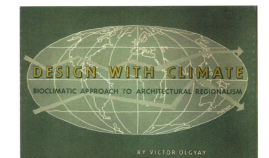

155
Victor Olgyay, cover of the book *Design with Climate*, 1963.

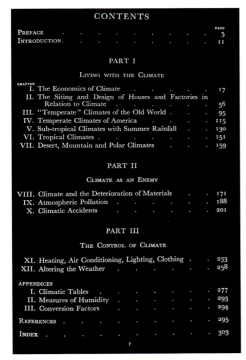

156
Charles Ernest Brooks, contents of the book *Climate in Everyday Life*, 1951.

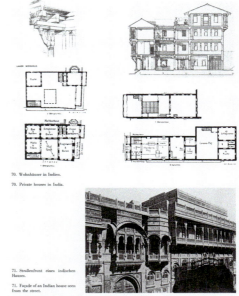

157
Urban cultures of passive climate control, examples from India.
Source: Egli 1951.

158
Gandria (Ticino, Switzerland) Source: Egli 1951.

Thermal Heritage

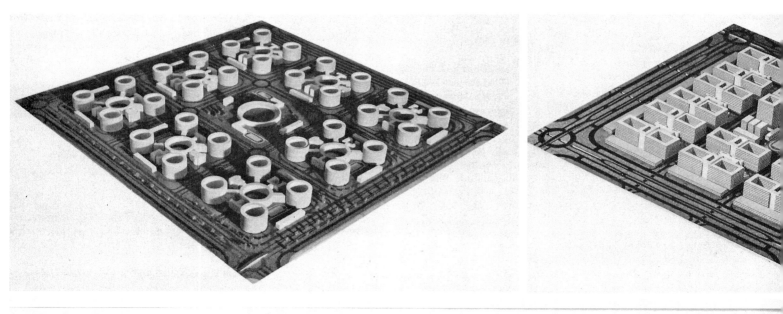

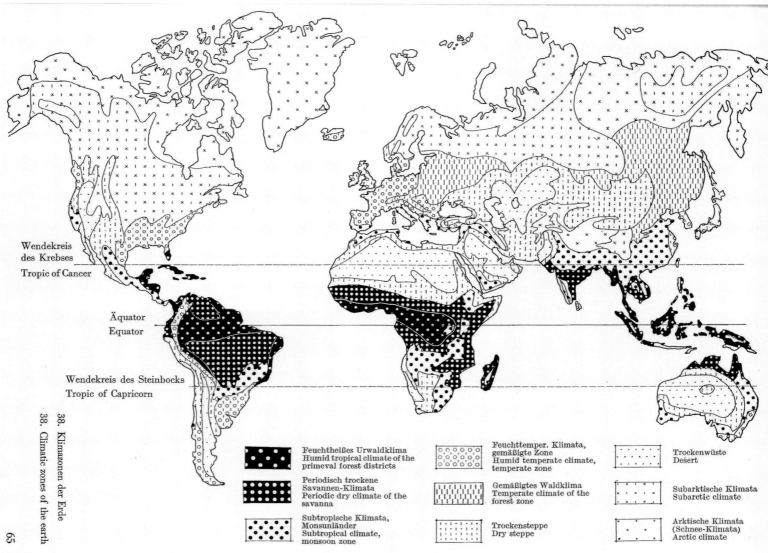

159
Are "the basic town structures of the different climatic zones […] unchangeable" or can they "be changed by technical means of our day", and "in the end be replaced by a standard town type"?
Source: Egli 1951.

Global Adaptations after 1945   142

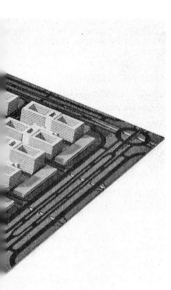

160
"Deductions" of different climate zones: A town model for the subarctic zone (left) and for the dry zone. Source: Egli 1951.

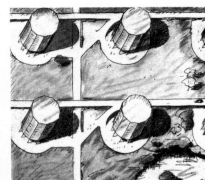

161
Comparative overview of cities in different climatic zones and topographical locations. Source: Egli 1951.

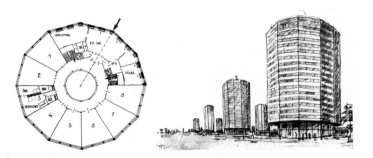

162
Comparative overview of different climate zones and related climate adaptation measures. Source: Egli 1951.

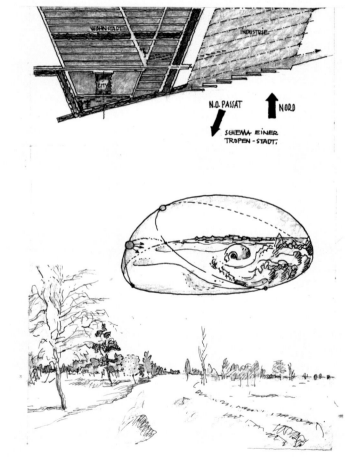

163–164
Ernst Egli, future houses and urban layout of the subarctic zone, taking the thermal heritage of vernacular architecture into account.

Thermal Heritage

"New cities of European spirit [will] no longer dominate the rest of the world as models of urban development."[115] "The time seems ripe for the idea of the European city itself to be transformed. It is dawning in European urbanism."[116] "Europe's old cities are dying."[117] According to Egli, part of the European heritage of "thinking for the whole world"[118] is understanding cities in different parts of the world—and not just those in Europe—in the context of historical knowledge. Egli was emphasizing the role of culture and tradition as the basis of future climate-responsive cities. For him, climatic determinism and historical knowledge were two sides of the same coin.

Everyday Life

In the second half of the 20th century, man-made climate became an everyday problem of allocation by different technical infrastructures. It was this development that inspired Charles Ernest Brooks to entitle his 1951 book *Climate in Everyday Life*, which from today's perspective combines in a completely unthinkable way climatic zones with domestic technology, outdoor with indoor conditions, deductive with empirical reasoning. Brooks's publication integrates the perspective of subdividing the globe into climate zones with that of comfort research of building services. The structure of his book is quite telling. It consists of three parts: Part 1, "Living with the Climate", describes the economic and cultural dimensions of the climate and of microclimates; Part 2, "Climate as an Enemy", focuses on the destruction of materials and air pollution; Part 3, "The Control of Climate", shows structural–physical, architectural and building-installation-related strategies of influencing the climate inside buildings. With the titles of the three parts of his book, Brooks neatly summarized the modern challenges posed by urban microclimates for architecture and urban development.

Human "efficiency" under different climatic conditions is the crucial concern of the book.[119] Different climate zones related equally to the dangers of climatic threats (e.g. rainfall, drought) and the requirement for climate control inside factories, where the top priority was to increase or maintain productivity by means of adequate environmental conditions (see chapter 1). Brooks refers to Ellsworth Huntington, to highlight the contrasting importance of European and Asian areas for the world economy.[120] He relies on differential perceptions of warmth between genders, (ethnic) groups and individuals in different work situations, comparing the "comfort levels" of "Europeans" and "Asians".[121] The colonial gaze is evident: "Deterioration climates. These are regions of high humidity combined with steady high temperatures which, however, are not extreme. Metals corrode rapidly and leather goods, clothes, paper, etc., soon go mouldy. The native populations lack energy and initiative, and white immigrants cannot maintain their efficiency for many years without occasional recourse to a cooler, more stimulating climate."[122]

What is striking however, is the integration of various scales, proving a planetary perspective on climate. Brooks brings a decidedly economic view of climate into play shortly after the war, which is remarkable in its integrative character. The incipient global economy of the post-war period appears as a driving agent to mediate between the different scales of climate. An economic perspective examines all scales, thus following the logic of global industrial capitalism. With the historian of science Deborah Coen, one can speak of an art of "scaling", a methodology that connects the world with the city and the home.[123] The scales of climatic determinism comprise towns, districts, typologies, building, rooms and even furniture; they thus integrate outside and inside perspectives. The inside is still connected with the outside, thus promoting an inter-scalarity in climatic thinking. The way climates are made inside buildings appears still in a conceptual connection with the climatic zones of the world.

Unprecedented Cities in the Subarctic Zone

As a well-travelled architect and town planner, Ernst Egli was not only familiar with pre-industrial cities, especially those of Turkey and the Middle East; he was also aware of their increasing modernization. In this context, he particularly emphasized the unprecedented character of future cities in the different climatic zones: "The basic structure of the towns in the different climatic zones" has yet to be found;[124] for example in the subtropical zone, where "new landscapes are coming to life, and in them, new peoples and new towns".[125] It is precisely this lack of imaginative models that cannot be overcome with the empirical methods of the meteorologists.

Central to Egli's theoretical argument is the so-called "subarctic zone". Winter cities were important epistemological testing grounds of the 1950s. The absence of urban models in the case of arctic and subarctic climates explains the great boom of proposals that emerged in modern architecture between the 1930s and the 1970s. From Ernst May to Ernst Egli, and Ralph Erskine to Frei Otto and Kenzo Tange, the subarctic city provided the subject for creative work on the means and methods of urban design itself, sharpened to climatic challenges. "In the arctic zone, the

→ fig. 156   Global Adaptations after 1945   **144**

closed-in, warm housing type could be developed to a town structure where all the single elements close in around a warm centre sheltered against the wind. For the necessity to keep the warmth and to find shelter against the wind would have to influence the great form of the town just as it influences now the single living unit."[126]

Egli was particularly convinced of the novel character of Soviet urban planning in Siberia. After the tsars had tried to import mainly European urban typologies to the region, the time had now come to develop genuine urban formations that were based on its prevailing climatic conditions. Like Albert Kratzer, Egli also emphasized the unprecedented character of these new cities. "No new town structure, conceived and developed from their immediate surroundings, has as yet been found at all in the artic, or better in the sub-arctic zone, let us say Siberia. Here, the Soviets are offered a unique possibility of methodological research and of daring realizations, or, if necessary, of large-scale experimentation with few settlements forms. For before, during the reign of the Tsars, the only way of procedure had been to copy imported plans for the unpopular Siberian colonies, and, in the best case, to let a shadow of European splendor shine in them. But once roused, this great and powerful country Siberia will develop its own town plans for the future, not only in a logical, but in a creative way. Let us say this in advance against the possible objection that here, in any case, is a great climatic zone that has hardly a sign of authentic creativeness to show, at least in town-planning."[127] Up until that point, according to Egli, cities had been built in the wrong way; it was prevailing tradition that had to be overcome in building more appropriately. "Neither the hut of the Laplander nor the hut of the Samojede will be used as the elements of a future Siberian or Alaskan town, but new creations, developed from the fundamental attitude of the inhabitants of those regions towards their surroundings. [...] Thus, basic structures are not absolute, but can be changed and developed."[128] This anti-anthropological, yet historically informed, approach to environmental transformation opens up possibilities for thinking about climate-adaptive urbanism.

Egli proposed round, high-rise apartment buildings with a warm core area, whose circular form would capture the most solar radiation. Protection from winds and absolute minimization of direct contact with the outdoors are further requirements of the future winter city. The schematic city layout follows the pre-war paradigm of adequate separation of residential and industrial areas, taking into account the pre-

vailing winds (see chapter 3). These are additionally blocked by "areas of planted trees."[129] There remained, however, a need for comprehensive work in translating these ideas, as Egli's schematic town models show. The town model of the flat country of the arctic and of the flat country of the dry zone prove Egli's intention to develop new design methodologies. "Modern technical evolution makes it possible to begin town-building colonization in these zones. In what form these town settlements will ultimately be realized cannot yet be told."[130] With his "Sub-Artic Habitat" (developed from 1958 onwards), the Sweden-based British architect Ralph Erskine was literally making real the ideas of Ernst Egli. As a talented architect, Erskine was able to lead the general climatic concepts towards site-specific architectural and urban projects. One could speak of an elaboration of an architectural grammar for polar latitudes. His project integrated the urban scale and the building scale, taking into account orientation, form, structure, material, along with the elements and details of the buildings.

With their references to different traditions of building, Egli's climate-deterministic urban models proved to be more open-minded towards questions of design than empirical urban climatology. In Egli's view, a climate-deterministic interpretation of architecture would open up a reservoir of (sophisticated) traditional architectural solutions, such as those found in Kashan (Isfahan, Iran) or Gandria (Ticino, Switzerland), which the future city should draw from. The interest in non-European forms of urbanization therefore also formed the prerequisite for the new European city of the future.

4.2.2 Empirical Knowledge: Interdisciplinary Collaborations in Chandigarh (India)

Urban climate-related design should, as the landscape architect Friedrich-Herman Pfützner wrote, "be characterized by an impeccable method".[131] Such an application of scientific knowledge involves notions of *causality*, which often lead to an opposition (to be reconciled) with artistic and creative approaches in architecture. Jeffrey Aronin, who drew both on urban climatology and the urban-planning projects of CIAM members, spoke in 1953 of "a city plan laid out on scientific sunlight principles".[132] Such a plan strove to transfer scientific insights into architectural design. In his 1953 publication *Climate & Architecture*, Aronin attempted to bring together the empirical understanding of urban climatology, as it had emerged in the interwar period (see chapter 2), with the new

→ figs. 158, 163, 164   Thermal Heritage

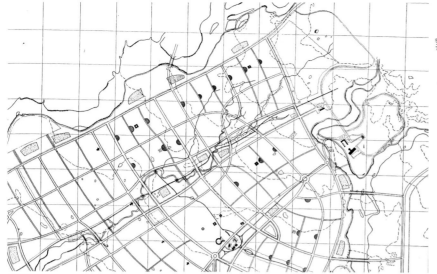

165
Layout of Chandigarh, Punjab (India).

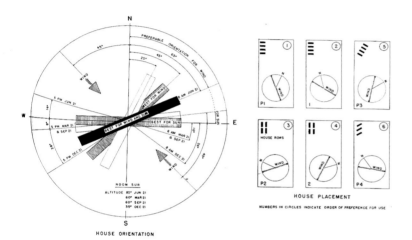

166
House orientation in relation to sun and prevailing winds, Chandigarh (India).

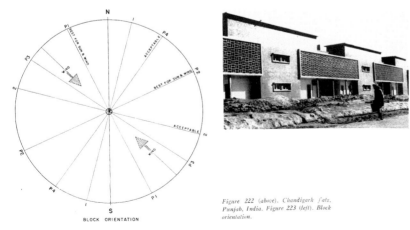

168
Block orientation in relation to sun and prevailing winds, Chandigarh (India).

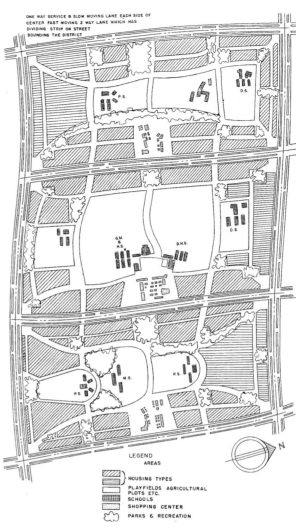

167
Three neighborhood super blocks, Chandigarh (India).

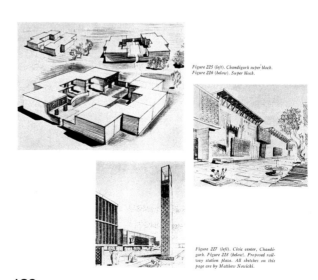

169
Matthew Nowicki, different building types of Chandigarh (India).

Global Adaptations after 1945   146

urbanistic principles of CIAM. He programmatically highlighted the dialectic between the urban climate as a form-giver *and* a product of architecture. In this context, the difficulty of translating high-level scientific insights into high-level architecture and urban planning appears. The solutions were to be found in the field of tension between methodological rigorousness and architectural autonomy.

Interdisciplinary Collaborations

Aronin addressed the necessity of collaboration between climatologists and architects. For example, he highlighted the collaboration between German climatologist Helmut Landsberg, who emigrated to the United States (in 1934), and the New York architectural firm Mayer and Whittlesey. Landsberg worked prominently as a consultant during the elaboration of the master plan for Chandigarh (India) by Mayer and Whittlesey. Climate considerations played a critical role in the framework of the master plan. This collaboration is less known than Le Corbusier's collaboration with French climatologist André Missenard, despite the fact that the initial master plan formed the conceptual basis that was later continued by Le Corbusier.[133]

The project began following the decision that Chandigarh should become the new capital of the state of Punjab (India). The master plan was to be drawn up for 500,000 people, "built from 'scratch'".[134] The site, consisting of approximately 70 square kilometers, comprised "a gentle sloping plain"[135] and a "prominent elevation in the north-east quarter as well as a river near the north and east boundaries".[136] The project relied on a multi-scalar approach, combing the territory with the other scales, such as the buildings and the streets. The planning concept linked the exterior with the interior: "It has been stressed that the line of direction to follow in site planning is first to consider the sun, then to consider the units of a building with respect to the sun, thirdly to consider the massing of the house with respect to the sun, other buildings, and open spaces, and finally to plan the streets around these structures."[137] In preparation for his book *Climate & Architecture*, Aronin was able to gather information on the collaboration between meteorologist Landsberg and the firm of Mayer and Whittlesey. One sentence begins with: "Landsberg and the architects came to the conclusion that",[138] indicating the relevance of this relationship. Mayer and Whittlesey considered comprehensively "the nature of the terrain and the accompanying macro- and micro-climate. It is notable that Mayer and Whittlesey recognized the tremendous importance of designing

Chandigarh from the climatic point-of-view, because in the State of Punjab, as in many regions of India, excessive temperatures and humidities make life rather uncomfortable [...]. One of the first things this firm did was to consult Dr. Helmut Landsberg, noted American climatologist, and to arrange with him to make a climatic and microclimatic survey of the proposed site. The entire program was influenced by this intelligence, with the result that the master plan [...] was extremely well developed."[139]

Winds and Windcatchers

Both the "placement" and the "layout" of the buildings were based on the information provided by the meteorological consultants. There was a prioritization of the environmental factors, leading to a grid of streets following the prevailing winds. "The architects plotted on a chart various orientations concerning prevailing macroclimatic winds and the most favorable sun conditions. [...] It was determined that the optimum orientation for streets would be greatly influenced by the winds, not the sun. [...] From this data, studies were made concerning the preferred placement for each block and the layout of the housing units inside them."[140] There was a planning strategy with a lot of vegetation "to prevent the building up of solar heat".[141] The "industrial area" was selected in the "west and southwest" considering the prevailing winds. The main goal was "to avoid smoke and odors being carried over the city. Smoke from the railroad and yards on the east side was prevented from drifting over the Capitol in summer by placing a park strip 600 feet [180 meters] wide, with close planting of tall trees as a screen on the windward side of the yard."[142]

According to Aronin, the "relatively small effect" of the "daytime macroclimatic wind" can be compensated for by an appropriate architectural design of the buildings, applying some basic climatological insights. "In Punjab, the cooling to be had from the daytime macroclimatic wind is of a relatively small effect, since the average velocities are only a few miles an hour in the hottest months of the year and seldom greater than 15 m.p.h [24 km.p.h]. However, nighttime cooling slope breezes, micro-climatically induced are important. They are capable of ventilating well any building that are widely spaced; that is, to an extent where there is at least twice their depth between ends and twice their height between fronts and rears facing the wind."[143] At a micro-scale, cross-ventilation played a crucial role, and was achieved partly through windcatchers in the middle of the houses. The superposition of macro- and

→ figs. 165, 166, 167, 168, 169   Thermal Heritage

147

microclimatic perspectives through urban architecture addressed a central insight of urban climatology. "Buildings, especially when grouped together in small areas, will greatly stop the flow of air. This is not always good: a moving breeze is an excellent way to prevent the accumulation of foul odors and unhygienic areas. The wind over a city will be one or a combination of the following: the prevailing macroclimatic wind, the microclimatic winds induced by topography, etc., or the wind set up by the presence of the city itself."[144]

4.2.3 Allegories: Reflecting the Windcatchers of Hyderabad (Pakistan)

The global circulation of architectural concepts and images and the new type of knowledge that is acquired in the process have become the focus of research interest only recently. Such a contemporary "geography of ideas", which picks up where Aby Warburg left off, follows the modern "migratory routes" of motifs and images.[145] Much more so than during Warburg's lifetime, globalized distribution channels and globally operating stock-photo agencies generate and leave traces that themselves produce a new "space of knowledge"[146] and a cross-cultural "collective memory"[147] of architecture. This global circulation of images requires a new "way of looking at things in terms of the history of images",[148] which regards *historicity* as an immanent part of the (circulating) motifs and images.

The roofscape of the city of Hyderabad in Pakistan (province of Sindh), is subject to such an encoding in an exemplary way. Since it was discussed in Bernard Rudofsky's influential 1964 book *Architecture without Architects*, the roofscape with its scores of windcatchers has repeatedly appeared in publications on architecture and the ecology movement.[149] Rudofsky's exhibition catalogue shows three views of the windcatchers on a double-page spread. The recurring reference to the *roofscape shown by Rudofsky* has imbued these images with a twofold referential character, which follows the interpretation of Hyderabad that was created by the New York MoMA exhibition. Rudofsky's exhibition and accompanying publication have constituted the framework for an interpretation of the circulating images of Hyderabad's roofscape ever since.[150] We might ask, with Roland Barthes, what constituted "the power of attraction" that this "specific photo"[151]—the motif of Hyderabad's roofscape—exerted on Bernard Rudofsky and the many architects that later made

reference to him. What was the "fascination,"[152] even the promise, that this motif exuded? The iconic character of this city view can best be compared with that of the Uthman Katkhuda Palace (c. AD 1350) in Cairo, which Hassan Fathy introduced into the research literature. This palace, also with its striking windcatcher, was part of a region in which Islamic civilization extended from the Maghreb countries to South Asia. And it, too, was used as a code for a different kind of energy supply in the discourse that communicated modern architecture to a wider public.

Stock-photo Agencies as Mediators of the World

The photographs that have generally been used in the global distribution of images of the roofscape of Hyderabad since Rudofsky are those made by the Zurich-based photographer, author and publisher Martin Hürlimann (1897–1984), who took them in 1927 during the final stage of his two-year trip around the world. Two pictures in particular appear again and again in architectural publications.[153] One shows the roofscape from a slightly elevated position of an old fort; the other one exposes a building with nine windcatchers from a point of view in the street. Hürlimann brought these two photos into circulation just one year after his return to Europe in his book *India*.[154] With this illustrated travel book, he launched a publishing career that led to the foundation of the journal *Atlantis* in 1929, and one year later to the foundation of the eponymous publishing house—a publishing house that increasingly also took on the role of an early modern stock-photo agency.[155]

With his threefold *Atlantis* project (journal, publishing house, stock-photo agency), Hürlimann was not the only broker of global and exotic stock photography of his day. The first stock-photo agencies, in today's sense of the word, had come into being simultaneously in Europe and the US in the 1920s. In this connection, we may mention, e.g. the "phototheque" of the Berlin press photographer Willy Römer or the agency of the New York journalist and photo editor Ewing Galloway.[156] Driven by a new phase in the "structural transformation of the public sphere"[157] and the new paradigm of "technical reproducibility",[158] combined with a popular curiosity about the world, photography attained a novel force of impact during the years between the wars. New forms of publication and presentation took on the role of agents that conveyed the colonial non-European realm of experience to a European and American audience.[159] Accordingly, Hürlimann regarded his *Atlantis* project as an attempt to bring the world—which, as a result of

→ fig. 170   Global Adaptations after 1945

colonialism and economic interrelations, had become increasingly globalized—closer to a European (German-speaking) audience after the First World War. The journal, which existed until 1960, "gave the Germans and Swiss a whiff of the world", as Hürlimann remarked in hindsight. *Atlantis* was one of the first European consumer magazines that was, explicitly in a non-political way, dedicated to the "world", to "countries, peoples, and travelling".[160] It always had to satisfy the expectations of the European readership as well, even though—as the following quotation from Hürlimann's India book from 1928 suggests—the reality observed did not measure up to expectations. The world presented was one based on the expectations the readers were presumed to have, which, despite Hürlimann's actual objective, gave his book an exoticist and orientalist bias: "Rajputana, the area which best matches our concept of medieval India, is colourful and full of proud characters, and still filled with the brilliance of ancient princely courts. On Mount Abu and further west, on the Kathiawar peninsula, are some holy sites of the Jainas. Ahmedabad was once the seat of powerful Islamic rulers, like Hyderabad, the former capital of Sindh, which today is far surpassed by the powerfully emerging trading place of Karachi."[161]

The Town as a House

The photographic representation of Hyderabad that was now beginning to spread was accompanied by the growing presence of Europeans in the city. In the 19th century, Hyderabad constituted the economic and cultural center of Sindh province; lying on the Indus River, the city possessed a privileged trading route to China.[162] Since 1843, it had been subject to British colonial rule—not long after it had been founded as the fort of a regional prince in 1768. The majority of surviving photographs of Hyderabad were taken from this very fort. The gate to the city, which was located outside the fort, also appears in some of Hürlimann's shots. Remarkably, his "Indian" albums contain documentations of urban conditions only from the area that today is Pakistani territory. The central motifs of the three documented cities—Lahore, Peshawar and Hyderabad—are their roofscapes.[163] Yet while in the case of Lahore and Peshawar the image of the city was enriched with portraits of inhabitants, with everyday street scenes and architectural details, the photos of Hyderabad remained strangely stylized, even anemic. Hürlimann pursued that "cultivated ennui"[164] that dominated the European city portraits of the 19th and early 20th centuries; the main photographic motifs were "the fabric of entire cities".[165] Accordingly, there tends to be few

people in his photographs of Hyderabad; the city appears to be nearly desolate.

This puts all the more emphasis on the windcatchers and the roofscape as the true motif of the photographs. The striking quality comes from the overall view of the roofs as thermal regulators of the city. Not just privileged buildings but the majority of buildings—from palace to the lowliest cottage—were equipped with windcatchers. One of Hürlimann's pictures shows a building constructed by the British colonial administration whose roof also had a wind scoop. While windcatchers were used not only in Hyderabad but in the entire Sindh province, Hyderabad was widely known as *manghan jo shaharu*, "the city of the windcatchers."[166]

As a general rule, each room had one or two windcatchers. These were triangular flues that funneled the air into the stories below. Due to the thermodynamic principle of the "stack effect", the winds (in the flue) were accelerated, which generated additional cooling effects. All windcatchers faced the same direction: they were oriented towards the incident summer monsoon winds. In the summer months, the monsoon weather brought these winds from the Arabian Sea in the southwest. This made staying inside the buildings more bearable when outside temperatures reached up to 48°C (118°F). In addition to the high temperatures in the summer, Hyderabad typically has extremely high humidity.[167] The flues also included a system of flaps that made it possible to direct the incident winds. When temperatures rose around midday, the flaps were closed for a few hours. Conversely, during the winter months the windcatchers, all of which were aligned in one direction, were protected against cold winds from the opposite direction. In winter, outside temperatures are as low as 0°C (32°F). As the outdoor air is warmed up on the protected side of the windcatchers, it is funneled into the interior rooms, further modifying their temperature profile.

The further spreading of the motif of the roofscape, which was robbed of its historical context, resulted in object and representation becoming indistinguishable; there was no longer a clear separation between *medium* and *reference* (Roland Barthes), since no particular knowledge on the historical context of the windcatchers was included in the distributed photograph. The windcatchers became part of a modern "iconography of the wind", making visible what is in fact an invisible phenomenon.[168] Walter Benjamin saw the allegorical connection between "meaning and sign" particularly clearly in the "antinomies of the

→ fig. 171   Thermal Heritage   **149**

170
The roofscape of Hyderabad (today's Pakistan), by Swiss photographer Martin Hürlimann, 1927.

171
Vertical and cross section of a windcatcher. Source: Rudofsky 1977.

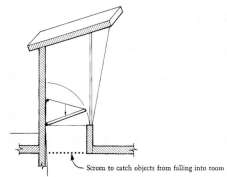
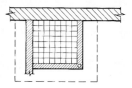

allegorical": "Any person, any object, any relationship can mean absolutely anything else."[169] In the allegory, "all of the things which are used to signify derive, from the very fact of their pointing to something else, a power which makes them appear no longer commensurable with profane things, which raises them onto a higher plane".[170] According to Benjamin's comprehension, the allegory is a "mode of expression" and not just a "conventional system of signs" the way symbols are.[171] Benjamin put photography in general close to allegorical representation.

Modern Knowledge of the Winds

It was architects of Central European extraction who introduced the Pakistani windcatchers into the architectural discourse of the 20th century. As early as 1951, Ernst Egli published a photograph of this city in *Climate and Town Districts*. This was followed in 1963 by Victor Olgyay, who in his study *Design with Climate* raised the previous research on passive climate control to a new level. The publications by Egli and Olgyay, in other words, had determined the direction of the interpretation of windcatchers that Rudofsky started in his *Architecture without Architects*. The Austro-Hungarian upbringing of these three architects (Ernst Egli, born in Vienna in 1893; Viktor Olgyay, born in Budapest in 1910; Bernhard Rudofsky, born in Zauchtel, today's Czech Republic, in 1905)[172] may have contributed to their perceptiveness to culturally connoted forms of knowledge[173] and "the problem of cultural exchange".[174] In any event, it was not until the 1970s (1975 and 1977) that the Englishman Philip Steadman (*Energy, Environment, and Building*) and the American Richard Stein (*Architecture and Energy*) placed the roofscape explicitly into the context of energy, ecology and environmental protection. At that time, the way the roofscape was to be interpreted had long since been established; from then on, it was the modern discourse on energy that determined the interpretative framework for the reception of the roofscape. Examining the different approaches of interpretation of Egli, Olgyay and Rudofsky before the general direction of the interpretation was set is therefore quite revealing. *Climate and Town Districts, Design with Climate* and *Architecture without Architects* represent three alignments in the reception of this modern allegory of the sustainable city. They represent modern forms of knowledge regarding the passive use of the winds in architecture and urban design.

1951: Interdependencies of Culture and Climate in Ernst Egli

In his *Climate and Town Districts*, Ernst Egli pursued a quasi-empirical and cultural approach in order to form links between climatic conditions on the one hand and urban form on the other. Examining the impact of the climate on global urbanization, he applied an environment-related rationality that has concrete historical and cultural consequences. The central architectural "motifs" of the "town of India", for instance, are "garden, oriel, and balcony, airy halls with waterworks, cool and shadow".[175] The "courtyard house" is identified as the most important architectural type of Indian cities, which, according to Egli, repeats on a small scale the spatial arrangement of the cities. It is a characteristic of the courtyard house that "there is a definite turn towards the street, expressed by richly decorated oriels, and the house is built so as to ensure continuous airing of the rooms: ventilators on the roofs, ventilating channels down the house, communicating cool underground living-rooms. The structure of the main story is repeated on the upper floors, but diminishing in size, so that the house is a terraced building, with roof gardens and galleries on the different levels."[176] In Egli's line of argument, Hyderabad is an exemplary Indian town. In the case of Hyderabad, he speaks of an "urban body" that is ventilated inside via airlocks: "View of the town of Haidarabad: many air-channels renew the air in the inner town."[177] In accordance with his book's program, Egli focuses on the city as a whole in his description of the windcatchers.

1963: The Winds' Potential for Expression in Victor Olgyay

Design with Climate combined the scientific description of passive climate control and the objective of a design method. Olgyay's achievement consisted in generating a "universally applicable method for architectural climate control".[178] He possessed the ability to systematically take the interdisciplinary research literature on the link between architecture and climate into account and to synthesize it to meet the requirements of architectural design. Hyderabad's roofscape appears in a chapter on "wind effects and air flow patterns". The optimum shape and alignment of buildings is achieved when solar irradiance and wind effects are considered simultaneously; the weight of each of the two respective parameters must be assessed in relation to the other. "In order to evaluate the specific effects of wind on human comfort conditions, both the annual and monthly variations of prevalence, the velocity, and the temperature of the winds must be

→ fig. 157   Global Adaptations after 1945

analyzed by direction."[179] Olgyay's focus is on the form-generating potential of the climate for architecture. In this respect, *Design with Climate* continues a discursive tradition that examines the tectonic qualities of external and internal forces of architecture. The windcatchers are outstanding examples of this; and modern architecture has yet to reach the level of their expressive power. The winds determine the appearance of this traditional architecture: "The town of Hyderabad is strangely silhouetted with air-shafts and wind catchers standing erect on the roofs."[180]

1964: Other Forms of Building and Lifestyles in Bernard Rudofsky

Rudofsky emphasizes the parallels between primitive and modern man with subtle irony and a rhetoric that relies on dichotomies. In line with his didactic ambition, he stresses the ingeniousness and diversity of vernacular solutions. He recognizes that the central principles and achievements of modern architecture have already been anticipated in vernacular architecture. Rudofsky also sees the natural ventilation of buildings as just another form of mechanized air conditioning. The caption for the three photographs[181] of Hyderabad in his book reads, "The air-conditioners of Hyderabad Sindh."[182] Rudofsky presents the vernacular ventilation system as a technical accomplishment in the modern sense, against which he juxtaposes the "mania for mechanical comfort".[183] The concept of ventilation Rudofsky proposed, as a technical mechanism and contraption, must be seen in the context of the triumph of air-conditioning in the USA. "We learn that many audacious 'primitive' solutions anticipate our cumbersome technology; that many a feature invented in recent years is old hat in vernacular architecture—prefabrication, standardization of building components, flexible and movable structures, and, more especially, floor-heating, air-conditioning, light control, even elevators."[184] The vernacular ingeniousness that Rudofsky suggested possesses a certain timeless character. The contradictory statement that the origins of the windcatchers of Hyderabad are unknown while claiming that they have been used for 500 years reveals his intention to place the windcatchers outside of real time and real space: "Although the origin of this contraption is unknown, it has been in use for at least five hundred years."[185]

Two hundred years after the city's foundation, the windcatchers of Hyderabad had disappeared again. With the construction of an electric power station by the British during the Second World War, they literally lost their significance. "This tradition started to wane with the advent of electricity [sic] during World War II when the British authorities built a powerhouse at Tando Agha. Most of the new houses and buildings have switched to other methods of room cooling, e.g. electric fans, room coolers and air conditioners and, therefore, the windcatchers are no longer numerous over the cityscape. Only few buildings have continued to follow this tradition of having Manghu on their rooftops. These include the newly constructed Civic Centre and the Aga Khan Maternity and Child Care Centre."[186] In other words, right as Rudofsky was postulating the quasi-timeless presence of these windcatchers, their *real distribution* in Hyderabad was already largely a thing of the past. Their *medial distribution* however, which started at that time, took place by resorting to images from the late 19th and early 20th centuries.

An allegorical interpretation of the windcatchers may help to counter the equation of object and representation, which has widely proliferated in architectural discourse, and to recognize vernacular forms of knowledge as mediated and as circulating on a global scale—including thought patterns from different regions and eras. By tracing the photographic tradition of Hyderabad's windcatchers, the historical character of this motif is revealed and thus an allegorical concept of the relationship between architecture and environment, which, given today's focus on sustainability, is becoming increasingly important. "Nature", says Walter Benjamin, "has always been allegorical",[187] thus anticipating an insight of contemporary science and technology studies. The allegorical interpretation does not exclude the technical conception of the windcatchers as a *ventilation system* but even credibly incorporates it.

The Thermal Heritage of the Future City

The spread of the windcatchers in the age of colonialism must have been as much linked to the European sense of comfort as to the architectural traditions of the native population. Their sprouting throughout the city gives the windcatchers an aspect of urban development that Ernst Egli comprehended better than others. Contrary to Olgyay and Rudofsky, Egli was more prone to apply standards of urban design than of architecture—a fact that reflects his experiences with urban planning in Turkey.[188] Even though both Olgyay and Rudofsky had a Central European background, their line of argument was directed towards the technological and cultural environment of the United States of the early 1960s. Both championed an object-oriented and technological interpretation of tradition. This was meant to upgrade the vernacular but in fact resulted in reducing it; grounded in

Thermal Heritage

American individualism, this kind of interpretation ignored the overall urban context by focusing on individual buildings. Beyond the sculptural (Olgyay) and the technological (Rudofsky) interpretation of the windcatchers, the motif directs our eye to a general solution for the entire city, which was gaining importance in the reflections on urban forms of energy supply.[189]

A rereading of the interpretative approaches to the windcatchers reveals the questionable character of the one-sided transfer of modern energetic and technological concepts onto architectural forms that originated in a different cultural and historical environment. Illustrations of windcatchers in modern architectural publications tell us as much about the energy societies in Europe and the USA as about vernacular thermal forms of knowledge in Pakistan. In the case of the windcatchers, both the use and the interpretation of the winds are subject to historical change. Only in its capacity to point to the energy and comfort problems of Western societies did the photography rid itself of its own profanity; the image represented a more sustainable modern architecture.

With its windcatchers, which became widespread in the 19th century, it has left us with the image of a contemporary form of climate control on an urban scale whose incisiveness was only achieved in Buckminster Fuller's *Dome over Manhattan* (see chapter 5). A few years before Rudofsky's exhibition, in 1960, with his photo montage, Buckminster Fuller had created an icon of ideas on urban energy supply. With its ideal form, the dome served as an allegory of New York's citizens forming a community gathered under an order of energy supply (see chapter 5). In the context of the *Dome over Manhattan*, the windcatchers of Hyderabad take on new meaning. The roofscape implies the uniform cooling system of an entire city and appears to be a precursor of a thermal regime that does not leave the energy supply up to the individual homeowners but regulates it city-wide. The task of cooling is transferred from the building scale to the urban scale and from high to everyday culture. The likewise decentralized and city-wide approach of the windcatchers is its paradigmatic model.

→ fig. 191   Global Adaptations after 1945   154

5 Microclimatic Islands

Thermal Spectacles and Closed-system Imaginaries

Although environmental factors such as the sun or green spaces were always critical starting points of architects' investigations on microclimates, another point of reference became increasingly relevant in the design of urban microclimates: the bodies of city dwellers and their capacity to move through the city and to experience selected microclimates within it. Even more than being geometric and spatial entities, microclimates relate first and foremost to thermal perception. Although, among experts, "infrastructure" was considered of paramount importance for the provision of (man-made) climate and energy, in the second half of the 20th century this view was increasingly in tension with user perspectives: the thermal experiences of the inhabitants. Such individual thermal sensations and associated (often cultural) thermal practices formed an important corrective to the non-existent, or at least deficient, thermal infrastructures found all over the world.

From a design perspective there are thus two fundamental starting points to cope with the climate of a city: *top-down* and large-scale, by looking at the environmental factors; and *bottom-up* and micro-scale, by referring to the sensorial experiences and the thermal practices of the inhabitants of cities.[1] The "experiential value" of urban situations, as it was promoted by artists' movements such as Fluxus and Situationism (SI) in the 1950s and 60s, formed the pivot of a new "artificial environment" in architecture and urban design.[2] In this line of thought, the design of microclimates is in the service of the public "spectacle" and the rhetorical force of images is an important concomitant of the man-made urban climate as a new type of spectacle. New media played a growing role, since climate was not only created via built structures and building services but also simulated via new technologies and imaginaries, providing synaesthetic experiences.

5.1 The Archipelago City (Open Systems)

Whereas interwar studies on the impact of climate on the urban fabric were dominated by a negative horizon—avoidance of toxic aerosols in the residential areas, avoidance of overheating in the apartments—a postmodern discourse on "islands" expanded the relational approach of the urban winds. The combating of negative climatic impacts (by considering existing wind regimes) was abandoned in favor of a diversity-focused approach to microclimatic conditions in the city. The recognition of the complex socio-cultural character of "a thermal place" allowed a broader understanding of urban climate phenomena, leaving the purely negative connotation of the so called "heat island" behind.

The notion of the microclimatic island has been echoed more broadly by the "metaphor" of the "urban island", promoting a synaesthetic and multi-sensory understanding of the urban environment.[3] Representative of the diversity of microclimates was the idea of "thermal delight" as elaborated by architect Lisa Heschong or the "climatological seeing" as pioneered by climatologist Karl Knoch. From a climatic point of view, urban islands imply site-specific thermal experiences that must be approached with adequate design methodologies. Thermal differences between these islands are experienced through movement and recorded via mapping and fieldwork. "Climatic change", as the late Martin Wagner had anticipated, appeared as the methodological pivot for the thermal design of the built environment. "It seems as if people never wish to stop fully indulging in the comfort of a climate and stubbornly resisting its inclemencies. But we too easily and too often overlook the fact that it is precisely the change of climate which is of the utmost importance in a world of actions and reactions, and which enables us to live a pulsating life of the greatest potency. [...] The fact that climate wants to be felt, seen and heard in contrasts may still be very 'uninteresting' to our climatologists today, but it must become an axiom for the creative urban planner, who must give this idea a life-forming shape."[4]

5.1.1 Urban Gardens: Rethinking the Picturesque

In her 1979 essay *Thermal Delight in Architecture*, the American architect Lisa Heschong coined the term "thermal delight" for the multi-sensorial and synesthetic experience provided by "thermal places" such as gardens. Emphasizing the exemplary significance of gardens for the notion of thermal delight, the

155

publication's cover is decorated with a gardener's hat. Heschong highlights for instance "the Islamic garden" as "perhaps the richest example of a thermal place with a profound role in its culture".[5] "Islamic gardens [...] offer delights for each sense".[6] Exemplary in this respect are the Moorish gardens, where, as in the case of the Alhambra in Granada or the Real Alcázar in Seville (Spain),[7] the conscious combination of architecture, vegetation, water and topographical measures (such as subsidence) led to a surprising abundance and quality of microclimatic experiences.[8] With the notion of thermal delight a new understanding of thermal sensation was pioneered, transcending the horizon of "comfort". Thermal delight is always linked to multiple forms of experience provided by the body. Overall, this notion introduced "the garden as paradigm" of (climate-related) urban design, highlighting microclimatic islands as critical subjects.[9]

Beauty Without Order

Lisa Heschong's postmodern notion of thermal delight accentuated an idea emerging in enlightened 18th-century Europe, when gardens (and, with them, microclimates) were described as promoters of pleasant sensations. From a historical perspective, it was gardens that pioneered the aesthetics of microclimatology. It is indeed the garden that makes a supreme aesthetic principle of the *transience* that results from the daily and seasonal fluctuations of microclimates. As a result, it is the visual techniques, constantly seeking the ephemeral, which dominate; the blossoming and fading garden is a symbol for this. In the garden and *through* the garden, a man-made *concentration of the climate* is created. At a time when microclimates were not yet measured scientifically, they were already described as part of new emotional landscapes. "The excitement of pleasant sensations will thus be the general purpose of garden art", writes philosopher Christian Cay Lorenz Hirschfeld in his *Theorie der Gartenkunst* ("Theory of Garden Art"), published in five volumes between 1779 and 1785.[10] In this romantic line of thought, the garden is associated with numerous sensations transcending notions purely of beauty. Gardens also provide a *pleasant shiver*, giving the notion of thermal delight an ambivalent note. Thermal delight denotes a comparative mode of perception that builds on the ambiguity of changing microclimatic conditions. Drawing its pleasure from the volatility and difference of thermal conditions makes thermal delight a mode of perception that includes the unpleasant; it may be conceived as "an enjoyment" based on a "mixture of pleasure and pain".[11]

The ambiguity of the thermal sensations provided by gardens has been illustrated by the garden historian Osvald Sirén in his description of the "Chinese treatise of gardening," *Yüan Yeh,* published in the 17th century. Referring to the New Summer Palace near Beijing, the pavilions of the palace's garden appear "intended for the 'retention [enjoyment] of the spring,' and others offering protection against the summer heat".[12] According to *Yüan Yeh*, "here is created an atmosphere that evokes deep feelings".[13] Gardens provide these "deep feelings" of "enjoyment". In them, a space of experience is made that can be described as an emotional geography. The multi-sensory qualities of gardens were made visible in Chinese miniatures with their simultaneous display of vegetal and architectural diversity in the medium of painting. "A successful garden is one where the scenery changes at each step, providing new viewpoints created by intentional framing processes. The winding route of a path helps to guide the walker's gaze and to frame their view. The notion of dissimulation and discovery is essential in the creation of a garden. A Chinese garden only reveals itself as the walker advances, through a subtle operation on space and time."[14] It is not Euclidean geometry but rather topology that forms its epistemic model; noticeably in gardens, it is not primarily the proportion of discrete elements that is all important but rather the gradual and uninterrupted transition from one zone into another.

Gardens are transcultural spaces par excellence. There are, for instance, close conceptual ties between Islamic and Italian gardens of the Renaissance or Chinese and English gardens and landscape parks of the 18th century. The central idea of English landscape parks, "beauty without order", was widely conceived "as a Chinese idea, actually realized in Chinese gardens".[15] Gardens have always been places of reception and recognition of *the Other* and thus of cultural adaptation.[16] Travelers, not least European ones, have taken special consideration of ideal places in foreign countries.[17] The garden—and later the city park—is a symbol of an ideal, even paradisiacal place.[18] In this respect, the garden is indeed, as Michel Foucault pointed out, a "heterotopia"[19] at a small scale, forming a perfect thermal place despite all adverse surrounding climatic conditions. The garden forms a special zone that demarcates itself from the surrounding environment. One (transcultural) "aspect that is common to all gardens is the demarcation from the environment. Through the wall, the hedge, the fence or the moat, the garden becomes a garden."[20]

Global Adaptations after 1945

China's urbanization was indeed characterized by the walls of gardens and courtyard houses, as Osvald Sirén pointed out. "Walls, walls and yet again walls, form the framework of every Chinese city. They surround it, they divide it into lots and compounds, they mark more than any other structures the basic features of the Chinese communities."[21] Traditional Chinese gardens had certain recurring basic elements: in addition to the walls surrounding them, there are various forms of water such as ponds, canals or small rivers; artificial and natural rocks that further structure the garden and give it an artificial topography; and vegetation that depends on the seasons. In this respect, the Chinese garden is a simulacrum of the natural landscape. The architectural historian Andrew Boyd described the vegetal interior of a traditional courtyard house in Beijing (China), providing a sense of the atmosphere in a pre-industrial urban garden: "The outer court [...] was paved with stone slabs. There was a small pool with lotus growing in it near the center of it. A screen covered with morning glory stood in front of the service rooms which backed on to the street and faced north. A 'date' tree (*Zyzyphus vulgaris*) and a crab apple grew in this courtyard, and many flowers in pots were set out around the edge. This was not a mere service courtyard. There were guest rooms and some family rooms in the side buildings. [...] This inner court, encircled by a verandah, was also stone paved. There was an 'artificial mountain' or strangely shaped stone in one corner, and two raised beds of shrub-peonies faced one another, one on each side. The wooden columns of the building and the verandah were all painted red, and the beams and other decorated woodwork other bright colors. [...] In summer the table was set in the open air in the inner courtyard, or, if it rained, on the verandah."[22]

In a cross-cultural perspective, gardens are characterized by two types of thermal environments: on the one hand, green open spaces, structured by plants and filigree types of architecture (pavilions), and on the other hand courtyard houses with interior gardens. In the first case, the architecture is part of the garden; in the second case, the garden is part of the house. The courtyard house cultivates the garden inside the building while the green open space, on the other hand, places the architecture in a floating relationship to the surrounding garden.[23] In both cases, the architectural boundary between inside and outside is dissolved or at least extenuated; the inside unfolds into the outside or the outside is taken into the inside of the house. This constellation of the consciously initiated crossing of borders between inside and outside areas forms the starting point of every architectural theory of microclimates, with gardens

serving as important anthropological references. Gardens are places where horticulture and house overlap. Andrew Boyd provided insights into the heating system in Beijing with great clarity. Due to the courtyard-house structure with several buildings, inside and outside were connected in terms of use. The cold air temperature is mitigated by the incoming winter sun. Boyd speaks of an actual reversal of the modern practice of surviving winter with light clothing and extensively heated rooms; he points to the "greater adaptability" of a system of thermoregulation that relied primarily on warm clothing. His description, published in 1962, shows the relativity of comfort. "The whole window wall of a room on the courtyard side was composed of a panel of windows and doors. Windows were of thick translucent paper; in spring they were rolled up and the rooms opened to the outside air. Wide eaves protected them from rain and from the midday summer sun. [...] The paper windows had a certain amount of thermal resistance but tended to let in blasts of wind. External shutters could also be fitted. The general method of heating was always the portable charcoal brazier, usually in the form of a bowl, prepared by servants outside and brought into rooms in a glowing condition. [...] An important part of the 'heating system' however was undoubtedly clothes. Reversing the modern practice of light clothes and full central heating, the Chinese in winter wore fur-lined or quilted gowns and thick felt-soled shoes in a slightly warmed house, which had the advantage of greater adaptability to going constantly in and out of doors."[24]

In times when microclimates were not yet measured by scientific means, they were described as part of poetry and painting (highlighting the "enjoyment" of gardens) or in the treatises of gardeners and botanists such as *Yüan Yeh* or in such works as the *Theory of Garden Art*. Within these writings on gardens, thermal experiences appear as part of synesthetic sensations. In the sense of Roland Barthes' *Fragments of a Lover's Discourse* (1977),[25] the historical accounts of gardens thus provide a cross-cultural vocabulary for the microclimatic description of urban environments today. Cultural historians and historians of gardens in particular are important mediators of the long history of microclimates. The appropriation of the Chinese garden by Western scholars, particularly in the decades after the Second World War, was conducted in an exemplary manner by the art historians Eleanor von Erdberg (1936) and Osvald Sirén (1949), the historians of science Arthur Lovejoy (1948) and Joseph Needham (1956), as well as the architects Andrew Boyd (1962) and Werner Blaser (1974). The Chinese garden played a central role for

Microclimatic Islands

157

the creation of a man-made nature in Europe, highlighted in the "taste and artistic practice" of the Picturesque (1680–1830).[26] The understanding of microclimates in general has been shaped by the perception of gardens and green spaces, since they integrate both the transient and the periodic dimensions of microclimates. Although microclimatology as a field of scientific knowledge emerged only at the beginning of the 20th century, in parallel with the genesis of the modern European city (see chapter 2), the experience and language of microclimatology rely on the long cross-cultural history of gardens.

Both in the long cultural history of *thermal delight* and the short scientific history of *urban climatology*, gardens and green spaces (in cities) played a critical role. In an anthropological perspective, the garden is the place where sensations of thermal delight are cultivated while in 20th-century urban climatology, green areas received a role mainly as mitigating structures. However, green spaces represent a medium that is able to synthesize the aesthetic dimensions of thermal experience and to make them accessible. A cultural history of microclimates has therefore to be anchored in the context of the thermal experience of gardens; historically, the appropriation of nature took place first and foremost in the form of "gardens, sacred landscapes, and nature symbolism", as Clarence Glacken has emphasized.[27] The experience of thermal differences undermines the insensitivity materialized in the "well-tempered environment" provided by a 20th-century notion of comfort.

Spa Microclimatology
(The Climatological Gaze)

In the context of post-war reconstruction and the construction of new cities, the siting of hospitals led to a comprehensive reflection on the advantage of location in green environments. Historically, "the move of hospitals out of cities to more rural sites was a universal tendency", as Nikolaus Pevsner emphasized in *A History of Building Types*.[28] The construction of new hospitals relied in particular on urban climatic considerations. After the Second World War, hospitals and convalescent homes had to be created in large numbers for the injured and disabled. In *Das Krankenhaus in der Stadtplanung* ("The Hospital in Urban Planning") from 1951, Gustav Hassenpflug derived his recommendations from contemporary urban climatology. Like Werner Hebebrand, Hassenpflug was an architect who specialized in hospital construction in the Soviet Union of the early 1930s.[29] Although the location of hospitals "in the immediate vicinity of residences and workplaces" seemed self-evident, urban

climatic considerations spoke against such a siting. Due to the "reconstruction" of "partially destroyed cities", "dust clouds from the rubble and construction sites [...] will have a highly unfavorable impact on the urban climate".[30] The urban climate of the inner cities was still seen as being so detrimental to health that the hospital of the modern post-war society should ideally be placed in the countryside. "If one considers [...] the siting of a hospital in an urban area with regards to the poor climatic conditions of the urban core, that is, taking into account the poor microclimate of industrial and residential areas, then the location of hospitals outside the actual area of the city is more correct, on the outskirts where the climatic conditions are better."[31] In this respect, the post-war debate on hospitals was exemplary in considering how urban climates could affect the design process at different scales, from XL to S, from territorial planning to the detailing of construction. This meant that "the urban climate can be favorably influenced by skillful planning according to health considerations", as Hassenpflug emphasized.[32] The hospital must be adequately positioned in relation to the city and construction must be geared towards microclimatic conditions.

Increasingly refined methods of studying microclimates were used in the search for ideal sites, for hospitals, spas and other uses. In this context, the vineyard forms a kind of model for the site-specific quality of a microclimate and its measurement; topography and the vegetation itself play a central role for the thermal dynamics of microclimates in the vineyard. In the sense of the notion of *thermal delight*, the thermal assessment provides a sense of the microclimatic diversity that can also be incorporated into the evaluation of urban sites. The climatologist Karl Knoch emphasized the need for integrating quantitative and qualitative, visual and thermal, aesthetic and scientific dimensions of microclimates—a task for which he formed the notion of the "climatological gaze". Knoch referred to the tension between a subjective survey and the measurement of objective data; and he proposed a quick analytical method[33] suitable for the microclimates of vineyards and other locations with pronounced thermal requirements: "The experienced terrain climatologist can assess, for a given purpose, favor or unfavor of a site by sheer sight. He will separate the windy sites from the sheltered ones, recognize the possibility of cold air accumulation in the lower areas (frost holes), the places of preferential fog formation, the duration of insolation, and other aspects. Even if this simple procedure, called 'climatological gaze,' cannot be expressed in climatological numerical values, it is sufficient in special cases to

→ fig. 173 Global Adaptations after 1945 158

separate out the very unfavorable locations, and this is already worth a lot. If, for example, it is a question of creating new spa facilities, such a terrain survey can save one from later disappointments."[34]

Knoch developed his microclimatic arguments in the context of climatic therapies.[35] He provided evidence on how local microclimates could influence the siting and planning of spas; he speaks of "dealing with the climatic peculiarities of the spa".[36] In contrast to already existing old "village cores", "the new spa developments now shooting out of the ground everywhere [...] should go to the unproblematic and above all to the favorable locations proven by the climatic survey and avoid the unfavorable ones".[37] Based on microclimatic assessment, one could go about "the selection of the sunbathing areas, the gymnasiums, the building sites for new health resorts".[38] The evaluation of spas, health resorts and seaside resorts requires a specific "climatological gaze", that should in particular be practiced by architects. The siting and orientation of a building has fundamental effects on its heating and cooling requirements. "The proposed building climate survey is intended to create, so to speak, an inventory of local climatic facts of both favorable and unfavorable nature. Only when such an overview of the whole resort is available, is it convenient to make comparative considerations and to find the most appropriate solution for a given project. There is no doubt that local climatic influences are of decisive importance for spa planning of any kind. [...] Differences in local climates can often be of the same order of magnitude, occasionally even of greater magnitude than the differences of the large-scale climates."[39]

Berlin, the Archipelago City

In the second half of the 20th century, the "heat island" famously became the determinative concept of the discourse of urban climatology. From a design perspective, however, the debate on urban heat islands cannot be separated from the simultaneous discussions on "the city within the city" and the "urban archipelago". The idea of a city structured by fragmented "islands" was, in particular, developed by the architects Oswald Mathias Ungers and Léon Krier. In outlining the concept of "the city within the city" (Krier 1976) and "the archipelago city" (Ungers 1977), these architects provided important keywords for an urban–rural integration, which also included a conscious design of urban microclimates. The fragmentation of the urban landscape as planning principle echoed and intensified the evidence of both the "urban heat island" and the "urban archipelago". It underlined the

idea of a deliberate design of isolated urban microclimates.

In particular, post-war Berlin represented a critical urban territory that allowed the testing of the new concept. One of the topoi of the "Berlin discourse" is to assume that Berlin is a city characterized by islands; evidence of Berlin's "oases" can be traced from the end of the 18th century.[40] Important precursor impulses to the idea of the archipelago city can be also found in the post-war notion of *Stadtlandschaft* ("urban landscape"), promoting ideas for a strongly greened and loosened up Berlin with interspersed city districts (see chapter 4). The event-like character of "the island" and its experience are thus at the center of an urban approach that focuses on difference (rather than identity). What is at stake is the possibility to offer "room for the accident".[41] This is dependent on a specific conception of the city, in which metaphors mediate between the past and the present: "the landscape garden as an urban metaphor" thus emphasizes "the transformation of the existing site in the mind of the designer to form a new entity while retaining its inherent characteristics. Particular views (prospects) were designed as specific 'set pieces', but as intensifications of the existing rather than 'renewal'."[42]

For Oswald Mathias Ungers, the archipelago city represented an entire "urban-spatial planning concept."[43] In the form of a new methodology, Martin Wagner's notion of "sanitary green" is connected with the idea of the green urban landscape of islands—integrating, as Ungers put it, "the urban block", "the urban villa" and "the urban garden".[44] The garden was conceived as an "urban texture" that, conversely, could also be understood as a garden. Ungers, Hans Kollhof and Arthur Ovaska referred to gardens such as those of the monastery of St. Gallen, of Stowe Gardens designed by Lancelot Brown and of Glienicke Palace on the outskirts of Berlin. They derived the idea of Berlin as a green archipelago not least from Schinkel's design of the Havel landscape between Berlin and Potsdam—Berlin as "a gigantic enlargement of Schinkel's Glienicke Palace" (Rem Koolhaas).[45] Ovaska speaks of "this dialectic of reason and sensuality, of past and present, and the multiple, pluralistic view allowing the various symbols to stand for themselves, produced a dialogue between peace of mind and mental stimulation, an active contrast which we might expect in today's cities".[46] In particular the palace complex of Glienicke elaborated by Schinkel and Lénné, with its repeated references to the English landscape garden, integrated in an exemplary manner the ideas of the urban garden and the

→ fig. 172   Microclimatic Islands   **159**

172
Oswald Mathias Ungers and Peter Riemann, proposal for urban islands in West Berlin, 1977.

173
Microclimatic map for planning purposes (indicating sun exposure in kcal/cm²/year), Bad Kissingen (Germany). Source: Knoch 1961.

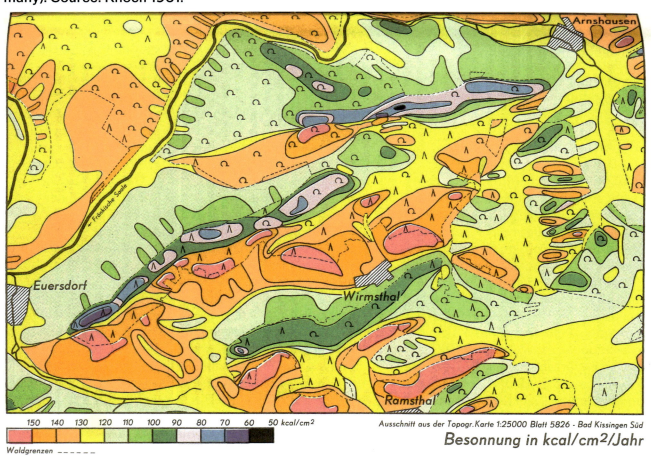

174
Luigi Caccia Dominioni, house at the Piazza Carbonari, Milan (Italy), 1960–61.

175
Mario Asnago and Claudio Vender, house at the Via Andrea Verga, Milan (Italy), 1961–65.

176
Mario Asnago and Claudio Vender, house at the Viale Tunisia, Milan (Italy), 1935–36.

Microclimatic Islands

urban villa. The palace and its associated pavilions and greenhouses demonstrated the island-like character of the urban landscape.[47]

5.1.2 Urban Villas: Revisiting Milan

In the context of the Cornell Summer Academy of 1977 in Berlin,[48] the urban villa became a multi-scalar subject of investigation and a vehicle for addressing different aspects of the urban fabric (blocks, gardens, infrastructures).[49] What emerged historically in cities such Barcelona, Paris and Milan, on the basis of specific urban regulations, appears in Ungers' design approach through a systematic examination of the development of forms and volumes.[50] Recesses, notches, openings and subdivisions were systematically investigated by Ungers in their potential for architectural expression. However, Ungers extended his interest in the urban and architectural form to ecological solutions and prefabrication. With an interdisciplinary team, he investigated "solar typologies"[51] that continued the formal motifs present in the "Roosevelt Island Housing Competition." By proposing a downsized and simplified Manhattan for the group of buildings, Ungers explored issues of typological difference and repetition, which would also play an important role in his ecological approach to the city. There was at the same time a reflection of the thermal requirements of the contemporary city and a play with the man-made character of urban microclimates. Ungers' "solar typologies" for a milking plant in Landstuhl (Germany) combined a "stone house", a "glass house" and a "green house".[52] The urban villa takes the urban garden as a starting point for its formal diversity, which also includes numerous microclimatic situations.

With their highly differentiated facades, attics and courtyards, the buildings of the Milanese bourgeoisie of the inter- and post-war period represent a European architectural heritage with pronounced microclimatic qualities and specific approaches to climate control on urban scales. The microclimatic diversity in and around the buildings of architects such as Giuseppe De Finetti, Giulio Minoletti, Gio Ponti, Luigi Caccia Dominioni, Vico Magistretti and Asnago Vender arise from a "special equilibrium between the individual building and the urban space", as Giulio Bettini noted with regard to the work of Asnago Vender.[53] This equilibrium derives from the architectural play with "rules" and their deliberate transgression,[54] which in turn leads to the numerous references to the natural conditions of the city. The continuity between

the building and the city, between inside and outside, may be the reason why Milan appears, in the perception of its inhabitants, as one big "interior".[55] The reciprocal reference of buildings and urban nature means that everything comes to be inside—as in an apartment.

In this context, it is Milan's facades in particular that acquire paramount importance for an architectural renewal of microclimatic design. The interconnected "curtain" of the facades (*cortina*) of the city's buildings forms the central Milanese peculiarity. It is this that produces the "'diversity' of the street space" and its microclimates.[56] The building regulations of Milan favored not only the formal proportioning of the front facade but also other topological operations such as reductions, displacements, bulges or setbacks, which might occur to the building as a whole or in its details.

Exemplary for their play with the depth of facades are the residential and office buildings designed by Luigi Moretti at Corso Italia (1949–56). While the ground floor is occupied by a low and permeable volume, with several entrances that facilitate passage, the main functions (apartments, offices, services) are concentrated in three narrow and high volumes that occupy as little surface area as possible and are placed at carefully estimated distances from one another. This decision had a significant impact on the microclimates created in its urban context. In particular, the empty space around the buildings facilitates air circulation and provides insolation in a very dense urban context. The main volume is inserted in the center of the plot, at a distance from the adjacent buildings to gain as much air and light as possible. The building has a thin concrete structure with large areas of glass on one side, while on the other side the openings are reduced to a minimum. With no mechanical air conditioning and minimized thermal insulation, the shape of the central building facilitates cross-ventilation of the rooms in summer and solar radiation in winter.

In Asnago Vender's buildings, on the other hand, the building components are integrated into the urban fabric with strong references to Milan's urban climate. The buildings of these architects were characterized by "additions to already existing perimeter block developments, where a reference to the urban environment is unavoidable".[57] The habitable roof of a building at Via Andrea Verga (1961–65) has on the street side a multi-story attic terraced to the rear, while on the courtyard side a slight slope clad in copper suggests a hipped roof. Instead of a corner formation, an L-shaped form was selected, which was further differentiated by volumetric shifts. For example, the

→ figs. 174, 175, 176　　　　　Global Adaptations after 1945

street-facing main facade was given an oriel-like projection. Here, Vender reinterpreted the regulatory possibility of the balconies above the street into a cantilever of the whole structure and accordingly conceived of them as "closed balconies".[58] This topological work on the volumetric shape also supported a microclimatic diversity generated by facades with different orientations and different degrees of relation with the exterior space. Terraces, French balconies and simple openings in the building and different forecourts generate a variety of microclimatic conditions. The structure is developed equally from the interior and from the exterior space. "Our solution is inspired by the living green that surrounds the house from many sides and can be seen inside through the many windows on the facade; this ultimately makes the green even more alive, not only for the inhabitants of the house, but also—and, we believe, to a greater extent—for the citizens in the street, who can admire the green almost entirely as they pass by [...] So why block the view of it with an opaque structure along the street?"[59]

These kinds of differentiated references to the apartments, the public space, and the urban climate by means of openings were the result of a conscious confrontation with a Milanese set of rules that oscillated between the poles of hygienic rationalism and urban design. Loos's student Giuseppe De Finetti had brought the debate on artistic urban planning initiated by Camillo Sitte to Milan and sought applications in his buildings, among others. De Finetti built the Casa della Meridiana (1924–25) at Via Marchiondi, originally a free-standing multi-story building of luxury apartments in the Novecento style, but which is now part of a row of buildings. The terracing, with clear borrowings from the Haus Scheu of his mentor Loos, reveals a desire to design a country house in the city. It is noteworthy how the garden of the small plot of land is located, so to speak, on different terraces; it thus appears as a kind of precursor to the aforementioned structure of Asango Vender. Already in this project, an interest in staggering in all conceivable directions can be seen—a topology that is generated in particular by environmental influences.

Engaging architecturally with issues of climate control meant dealing with contrasting seasonal conditions. In the case of the Condominio XXI Aprile at Via Lanzone (1951–53), Asnago Vender designed two rectangular volumes of three and eight floors, housing offices and apartments. The functional duality is reflected in the rectangular windows of the offices and the horizontal ribbon glazing of the apartments containing winter gardens. The apartments had both

heating and air-conditioning systems, which were placed in the suspended ceilings. The apartment building Giardino d'Arcadia (1957–59) at the Corso di Porta Romana, designed by the architect Giulio Minoletti, is located in a private park. Particularly striking are its balconies, which are all oriented differently. The transition to the park is marked by planters and windows arranged at an angle. As in the case of the large-scale forest housing estate Zehlendorf in Berlin (see chapter 3), Minoletti used an articulated color register that places accents on the volumetric variety, the individuality of the apartments and the rhythm of shadows. In a watercolor drawing by Minoletti, the idea of dissolving the relationship between inside and outside becomes evident. Also in the case of the multi-functional building at Piazza San Marco (1969–71), designed by Vico Magistretti, the aesthetic potential of structuring the line of the street becomes visible. The intention of the architect to react to the urban environment is clearly articulated. The L-shaped volume was inserted at the intersection of three main streets at Piazza San Marco. The ground floor is completely permeable and hosts various public functions. The duplex apartments on the upper floor are set back, distinguished by a cross-shaped rhythm. The building has a roof garden on the first floor, which protects the longer facades from noise, pollution and provides more shade and intimacy to the apartments.

Such Milanese examples show a diversity of cooling options under ambiguous atmospheric conditions. They open up scope for architectural research and transcend the conceptual framework of scientificity. In the words of Milan-based architectural theorist Elisabetta Pero, the "continuity between building and environment"[60] creates novel demands on the building envelope and facades as transitional and mediating areas. The facade is no longer a mere separating layer between an inside and an outside, but a multi-layered and multi-scalar buffer area that mediates between different thermal zones. Pero summarized these new requirements for climate control in Milan's architecture as follows: "Two aspects are important to the exploration of aesthetics in sustainable architecture: the city and its density on the one hand, and the form of individual buildings on the other. These two aspects must be considered both discretely and jointly, as they lie at the center of far-reaching ramifications that underscore current debates on sustainability."[61]

5.1.3 Microclimatic Walks: Mobilizing the Body-Territory

For a deliberate design of urban microclimates, new collaborations must be established, rethinking "the Picturesque" as a man-made form of nature.[62] In the second half of the 20th century, landscape architects and architects aimed at reconciling the two aforementioned perspectives—the aesthetics and thermodynamics of urban microclimates—in a new understanding of microclimatic urban design. By reassessing the nature in the city, new accounts of the climate of cities emerged. The overall goals of ecology and democratic participation in urban development went along with new methodologies of how to explore and how to design the city as a green city, taking the perspectives of the residents into account.

As part of those assessments, "the relation between urban morphology, microclimate, and thermal comfort" was examined from the perspective of the user.[63] Vasilikou and Nikolopoulou speak of "the assessment of variations in the thermal sensation of users through 'microclimatic walks'", exploring the thermal qualities of the urban environment.[64] The main goal of these walks addressing "physical, psychological, and physiological"[65] aspects of microclimates was "to identify a specific change in the thermal conditions and define its quality".[66] The methodologies included the "monitoring of microclimatic conditions, field surveys with questionnaire-guided interviews with the public, statistical analysis, evaluation of the environmental performance of urban textures, comfort mapping, social study of open spaces, and design guides and proposals".[67] Overall, the perspectives of experts and citizens were amalgamated in the microclimatic walks.[68]

The pioneers, who examined the "intimate ground-level experience"[69] of the city with bodily–sensory means, are to be found in the 1960s in the fields of architecture and landscape architecture. Their ideas emerged in the context of a general empirization of the city as promoted by Kevin Lynch and Donald Appleyard (amongst others) as well as in the context of the emerging field of ecological urbanism. The systematic exploration of the urban environment led to precisely designed routes and associated notation systems, reconnecting the fragmented urban islands via predefined walks through the city. The routes, indicated in "choreographic scores", implied "time, process, and change".[70] According to the "scored' structure",[71] the participant or visitor walked "over a prescribed 'course'",[72] comparable to developments in Fluxus or in the Theatre of the Absurd—e.g. Yoko Ono's "instruction pieces" or "the rituals" in Samuel Beckett's plays.

In what follows, two of these pioneering approaches to the urban environment, aiming at exploring the environmental "archipelagos" of cities, shall be highlighted. The (garden) cities of Bath (UK) and San Francisco (USA) were the urban territories for two such multi-sensorial walks, both of which took place in 1966. Because of their event-like character, one could see these multi-sensory walks as pioneering Bernard Tschumi's "follies" in his Parc de la Villette in Paris, which opened in 1984.

Peter Smithson:

The Material Culture of Bath's Microclimates

The architect Peter Smithson dedicated a booklet to the richness of the historic city of Bath, in which the wealth of a heritage inside and outside the city is appreciated.[73] The essay *Bath: Walks within the Walls* comprised a precise description of Bath as garden city. "Bath is Rome in England: On seven hills, founded by the Romans and refounded in conscious imitation of Roman civic virtues in the 18th century, by a will as strong as that of Rome itself, it appears today like a city of palaces and gardens. Turned over to other uses. A shell of a city—as was Rome in the 18th century—with overgrown terracing and mounds, disused waterways and bridges, springs, farmhouses, cows, pigs and horses, gardens and allotments all 'within the walls'."[74]

Since Roman times, Bath has been famous for its microclimatic diversity in its hilly landscape. The spas of the city represent early examples of "thermal places" in the sense described by Lisa Heschong. The identification of thermal places (as places of particular cultural and social value) forms the starting point of any architectural approach to urban microclimates. The city was the subject of the first comprehensive empirical account of the urban climate in UK. In their study of Bath's microclimatology, conducted between October 1944 and January 1946, the meteorologists W. G. V. Balchin and Norman Pye focused on the "differences in local climate". The basis of the study was a system of measuring stations, which enabled a scientific comparison of the thermal differences and an assessment of the microclimatic diversity of the city. They speak of striking "micro-climatic differences" in the urban landscape.[75] The term "urban heat islands" was in all likelihood introduced into the

→ figs. 177, 178, 184, 185   Global Adaptations after 1945   **164**

English-speaking world within the frame of Balchin and Pye's study.[76] "The distribution of the station sites also enables an assessment of the urban effect to be made. In winter the central city area is warmer than the surrounding country areas at similar elevation by a degree or so and exhibits the characteristic heat island within a built-up area. Much is doubtless due to the increased heat generated by the city itself and the higher absorption rate permitted by the increased atmospheric pollution of winter months which also reduces outward radiation. In summer, however, the city station showed lower maxima than the surrounding country sites at similar elevation to the extent of 1–2 degrees. This was an interesting conformation of other observations that the heat island effect of winter is not necessarily continued throughout the year."[77]

In 1966, Peter Smithson provided a kind of architectural echo to these climatological insights based on four precisely designed walks through the city. His walking project reinforced the interest by other architects that had already been shown in this city earlier—for example, by Schinkel and Hilberseimer.[78] The thermal diversity demonstrated by Balchin and Pye is reflected in the diversity of the built environment as described by Smithson. Distinctive details are combined with the view of the whole city, written observations with photos and maps of the routes. Gutters in the streets, damp cave entrances and mossy rubble walls are just as important as the famous historical architecture of the curved "open-to-nature" crescents. Smithson's descriptions include typologies, materials, construction methods and urban green spaces, thus opening up an insight into the material culture of Bath's urban microclimates. As early as 1962, near the town of Bath, in the foothills of Wiltshire's famous landscape park Stourhead, Alison and Peter Smithson built their weekend home (the Upper Lawn Pavilion), which anticipated the sensorial diversity within the walls of a small plot of land, creating a microclimatology in miniature.

The movement in the space of the city played a decisive role in the description of microclimates. In his four walks through Bath, Smithson orchestrated both the *routes* and the *tempi*: "Behind Augusta Place quiet gardens and allotments with a marvellous view down the river and over to Camden Crescent high up on the opposite bank. Afterwards walk as quickly as you can up the hill and pause at the junction with Rosemount Lane. From now on the walk is real *rus in urbe* for behind the present walls and hedges are the mounds and terracing of previous occupancy now in transition."[79] Until the early 20th century, movement

was the only means of experiencing the thermal differences and thus the microclimatic diversity of (urban) landscapes. Gordon Manley referred to the work of travel writers such as Celia Fiennes, who had provided early microclimatic accounts of the region around Bath in the early 18th century. By comparing different positions in the urban landscape, Manley emphasized the critical role of the vegetation cover for microclimatic conditions. It was "evident that in England the average length of the frost-free season and hence the gardening prospects can vary to a remarkable extent with choice of location. But the minimum temperature of the air on a clear night is also markedly affected by the nature of the vegetation cover."[80] Smithson ties in with this tradition of basing microclimatic description upon movement through the city. In his walks through the city one can detect approaches to an architectural methodology of how urban microclimates can be perceived and made fruitful for architecture and its regulation.

Lawrence and Anna Halprin:

> Multi-sensory Approaches to
> San Francisco's Environment

Another city that, like Bath, is characterized by extraordinary urban microclimates, is San Francisco (in California). The city and its surroundings set the scene for a series of walks elaborated by the landscape architect Lawrence Halprin and his wife, the dancer and choreographer Anna Halprin, in 1966 and 1968. The "kinaesthetic awareness" of the movement and the everyday experience of the urban environment played a critical role in the work of the couple. Their walks "represented a curiously potent encounter between architecture and dance".[81] The three sites of the 24-day workshop "Experiments in Environment" were downtown San Francisco, woodland Kentfield and the coastal Sea Ranch.

The walks, "involving designers, dancers, musicians, visual artists, writers, teachers, and psychologists",[82] started in San Francisco by providing a "city map" and a master "score to stimulate direct interaction with the physical environment of downtown".[83] The master score "indicated the sequence in which each participant was to visit the places along the route, the time to get there, and how long to stay".[84] Each participant had different time requirements, so that a complex choreography of the group was created through the city. While in Kentfield the bodily experience was in the foreground, on the territory of Sea

→ figs. 179, 180   Microclimatic Islands   **165**

177
The Garden at Stourhead, Wiltshire (near Bath, UK), 1750–70s. Film still taken from *Barry Lyndon* by Stanley Kubrick, 1975.

178
Location of observing stations as part of "a microclimatological investigation of Bath and the surrounding district". Source: Balchin and Pye 1947.

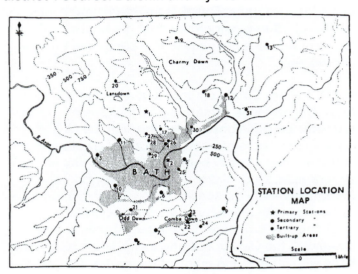

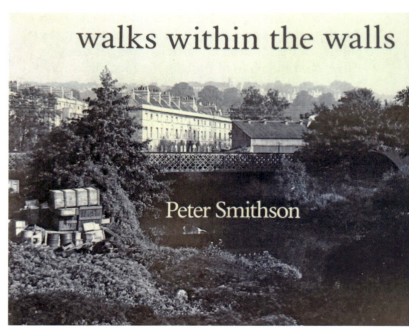

179
Peter Smithson, walk 3 through Bath (UK), 1966/68.

180
Peter Smithson, view of Bath (UK), 1966/68.

Global Adaptations after 1945

181
Anna and Lawrence Halprin, Experiments in Environment Workshop, "Driftwood Village–Community", Sea Ranch (USA), 1968.

182
Anna and Lawrence Halprin, Experiments in Environment Workshop, "Take Part Process", Kentfield (USA), 1966.

183
Anna and Lawrence Halprin, Experiments in Environment Workshop, "Market Street Walk", San Francisco (USA), 1966.

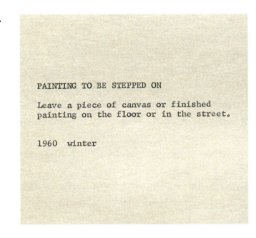

184
Yoko Ono, "Painting to be stepped on", 1960.

185
Merce Cunningham performing on Anna Halprin's dance deck, Kentfield (USA), 1957.

Microclimatic Islands

Ranch the found environment was at the center of the choreography. The participants had to observe and transform the environment: a "village" had to be built using the found driftwood. The general guidelines were designed to stimulate a multi-sensory approach to the environment: "Be as aware of the environment as you can [...]. This will include all sounds, smells, textures, tactility, spaces, confining elements, heights, relations of up and down elements. Also your own sense of movement around you, your encounters with people and the environment AND YOUR FEELINGS!"[85]

Two years before launching the workshop (1964), Lawrence Halprin had designed the layout for the 16-km-long area of second homes, the so-called "Sea Ranch". "Halprin's widely acclaimed [...] work on the Sea Ranch [...] shaped his attitude toward open space in the city."[86] The architectural quality to which the combination of *field studies, simulations in the laboratory and design work* could lead is shown by this example of early American microclimatic architecture. However, the planning process of the Sea Ranch retrospectively provides some insights on how Lawrence Halprin had imagined the architectural work *after* the walks. The deliberate exploration of the environment is the precondition for the design of a new urban environment. In numerous projects, Lawrence Halprin took "bodily movement through the city" as the basis for his designs. The walks were the basis for his attitude towards urban design.

While architectural historiography of this stretch of coastline has been quick to focus on individual buildings as stand-ins for postmodernism, its exemplary interdisciplinary character and crucial environmental scale have mostly been ignored. Located 160 km north of San Francisco, in Sonoma County, Sea Ranch forms one of the central experimental architectural sites of the 1960s that strove to incorporate the consideration of microclimates into its design. The site, exposed to the harsh northern California climate, has been conceived and built upon since 1963 by the developer "Oceanic Properties". A specific constellation led to the fact that the usually applied quality standards were clearly exceeded in this case. The responsible coordinating architect, Alfred Anton Boeke, who in his younger years had worked for Richard Neutra in Los Angeles, knew how to assemble an extraordinary talented team of consultants—architects, landscape architects and scientists. In a documentary by landscape architect Zara Muren, Boeke recalled, "There were architects talking about the work of landscape architects and landscape architects commenting on architecture and the [...] so called concept of

the Sea Ranch evolved out of the intercourse and interrelationships of the free conversation among all these people."[87]

The site and its buildings were designed with careful consideration of the territory, its ecology and microclimates. The overall concept of the Sea Ranch was shaped by three decisive approaches: 1. the cultivation or ecological further development of the landscape; 2. an arrangement of the land based on the balanced interplay of public and private area; and 3. the design of site-specific buildings. The Sea Ranch land had been overgrazed by herds of sheep and the soil was eroded. The landscape "as an ecosystem needed help",[88] as Lawrence Halprin noted. The Monterey cypress trees, systematically planted as early as 1916 to act as wind breakers to mitigate the strong northwest winds, served as a fundamental model in the planning of the site. Inside these cypress clusters, there is a calmness to the wind; Zara Muren speaks of "great wall of trees, cathedral-like within" in her documentary film.[89]

In 1964, the landscape architect Lawrence Halprin worked with the ecologist Dick Reynolds to document the site and provide the basis for the overall planning; they established the scientific basis on which the architects could draw. Halprin described the collaboration with Reynolds as follows: "We spend a lot of time just spending time there and studying the land and we did a lot of what was then unusual kind of studies. Studies of the soils, of the geology and the geography [...], wind studies, he [...] made investigations on natural winds, [...] where they came from, how they were affecting things. Wind was a very difficult problem. It was so violent [...]. It was clear from the beginning that if people would live there and enjoy being outdoors, that I had to do something about it."[90] The winds mainly generated the comprehensive regulation of the layout of the Sea Ranch's buildings. The studies of wind flows led to a rule for the arrangement of the buildings: Great importance was given to adequate proximity to the trees (the distance could not be more than ten times the height of the trees). Further reforestation was carried out and all buildings were placed in precise relation to it. The buildings were arranged in clusters so that more than half of the approximately 2,000 hectares of land could be used as common land by the residents. All the buildings of the Sea Ranch were placed in clusters in the open landscape or integrated into the existing woodlands, according to Halprin's plan.

→ figs. 181, 182, 183   Global Adaptations after 1945

Furthermore, carefully formulated building guidelines, based on an overall territorial vision and architectural prototypes, strictly regulated the variability of the buildings. Joseph Esherick designed the first six prototype single-family homes in 1965, while Charles Moore and his partners at MLTW (Moore Lyndon Turnbull Whitaker) simultaneously developed a first condominium as a model complex for ten other condominiums. The architects' greatest challenge was to develop designs that took into account the strong wind forces, which meant the winds became the form-generating motif of the project. In Esherick's words, "We had contours of wind intensity for the entire site. Those maps were very useful."[91] Both Esherick and MLTW drew on wind tunnel studies to test the design hypotheses. This resulted in arguably one of the earliest feedback loops in the history of simulation-based architectural practice. In the words of Charles Moore, "Fortunately they were building a wind tunnel at the University of California, Berkeley. So we were able to test the model of the building in the wind tunnel for times of high wind."[92] In the case of Esherick, this led to three key insights: firstly, clustering only 5–8 buildings led to the necessary calming of the winds. Isolated buildings were unbearably exposed to the winds, as the wind tunnel studies showed; acceleration occurred especially in the corner areas. Secondly, gable roofs with a pitch of 4–12° outperformed flat roofs. And thirdly, buildings slightly sunk into the terrain reduced the surface facing the winds. As a result, the first floor offered views at the level of the grassland. The MLTW-designed Condominium No. 1, on the other hand, had protected courtyards whose orientation and layout were only found after extensive wind tunnel studies; indeed, these entailed a comprehensive revision of the previously existing designs. "We changed everything and built the condominium around a motor court", says Charles Moore.[93] By placing the condominium right at the edge of the coastline, it was completely exposed to the winds, which entailed a careful arrangement of the different spaces. A few years later, the sublime interplay of topography, greening and buildings were also to be found in Charles Moore's and William Turnbull's design for Kresge College on UC Santa Cruz campus overlooking Monterey Bay, which opened in 1971. The building volumes were arranged in an L-shape along a curved main road in a redwood forest. Borrowing from vernacular cityscapes, Moore and Turnbull continued a picturesque tradition that had also been applied in the case of the Sea Ranch.

The Body-Territory

The theme of consciously creating microclimates led to a design-and-research methodology that imitated the thermal experiences of residents and integrated the two perspectives of the aesthetics and the science of microclimates. Along with history, the environment was in fact the central problem of postmodern architecture—"the problematic of ecology",[94] as Jürgen Habermas called it—something that is generally overlooked (see chapter 6). Due to its spatial complexity and the meticulous work on the microclimatic profile of the buildings, Sea Ranch has a special significance within an architectural history of appropriating the notion of microclimate. "A lot of people who influenced me, were heavily involved in the creation of this", Lisa Heschong stated. "Ed Allen, my thesis advisor, was [...] working on Sea Ranch along with Donlyn Lyndon, who was chair of the department at MIT at the time."[95] Urban gardens and urban villas were testing grounds for urban environments based on changing thermal conditions. Rather than providing stabilized thermal conditions, they conveyed the dynamics and interdependency of changing microclimates. In a nutshell, urban villas and gardens, as promoted by Ungers and others, formed a model for how to explore the thermal environment. Most critical was the agency of movement for the perception of thermal differences in cities.

In order to link the individual body with the urban environment, philosopher Tiziana Villani introduced the term "body-territory", an epistemologically novel and certainly "problematic form of bodily identity" that emerges in cities—one no longer based on the distinction between "the organic and the inorganic", but rather linking the bodies of the residents with the atmosphere of the city.[96] In the same vein, Matthew Gandy speaks of a "blurring of boundaries between the body and the city"[97] recommending the investigation of a "cyborgian sensibility",[98] which he sees emerging at the interface between infrastructure, urban environment and the human body. In this line of thought, cities are not only the result of governmental planning and development; they are also conceived by a bottom-up method as living laboratories in which the relevant approaches to urban climate from the perspective of residents emerge.

5.2 The City under a Single Roof (Closed Systems)

The longing for consciously generated and purified urban microclimates led to numerous projects that examined thermal islands, sharply delimited from the outside world, as a design principle. In this context, architectures with a large scale constituted since the early 1960s a remarkable further development of the debates on urban nature and its climates, launched in the interwar period. The impulses by engineering and infrastructures were taken up and made fruitful for extended building envelopes and underground networks. Architects experimented with closed systems that exceeded the scale of the individual building: Subterranean cities, artificial city mounds or large-scale highway covers were attempts to master precarious urban climates with design approaches. With the creation of actual underground cities, a new vision of urban planning emerged. The idea of "the building" was increasingly challenged by the idea of architecture as large-scale "environment".

The approaches of that period have to be understood as part of the new field of design research, "characterized by technological optimism and under the influence of fields of knowledge such as organization science, systems theory and cybernetics".[99] These fields of knowledge provided the underlying theory for how the technological means can be made fruitful for the management of cities.[100] Controllability was indeed the driving force and the man-made microclimate of the closed system its main asset. Juan Navarro Baldeweg's 1972 projects, such as the enclosed little tropical forest in an arctic landscape, are exemplary for the new obsession with closed systems. These projects once again gave a decidedly negative connotation to the given (outdoor) climate.

5.2.1 Underground Cities: Artificial Hills in Swiss Architecture

There are two main reasons for the erection of subterranean cities: on the one hand, unsolved organizational problems of the existing city above the ground (e.g. traffic) and on the other hand harsh climatic conditions.[101] The relationship between the above-ground and the underground city opens up new possibilities for an integration of design and urban climatology. Although the above-ground city can take advantage of solar gains and natural ventilation, harsh winter or summer conditions remain a critical concern, which is counterbalanced underground via active heating or cooling measures. The underground city forms a city within the city, creating a counterpart and an extension of the existing city and thus establishing a dual system. "Despite their structural variations, climate remains one of the most critical factors encouraging the development of these indoor environments. Hot and humid summers, and long cold winters of northern cities with their often severe conditions, such as windy and wet streets, greatly influence their usefulness and extensiveness."[102]

In the built reality, the underground city appears as a network of services and public spaces rather than an inhabited space; underground systems usually don't include residential areas (except those of homeless people). According to estimates, there are in the meantime at least 100 subterranean cities around the world. There are examples both from cold and hot climatic zones, including Southeast Asia, with many built and unbuilt projects; the underground city as controlled artificial microclimate represents in particular a solution for extreme climates. Most prominent are Montréal's RÉSO and Toronto's PATH in Canada,[103] Crysta Nagahori in Osaka (Japan), and the CityLink project in Singapore.[104] In principle, the underground city comprises a network of passages from one place to another with the purpose of avoiding cold or heat. Toronto's PATH, for instance, connects with 29 km of passageways, more than 50 office towers and buildings, 20 parking garages, 8 hotels, 1,200 stores and 6 subway stations.[105]

The Spatial City

However, since the early 1960s, resolutely *visionary design proposals* have been elaborated wherein topographic conditions combine with large-scale structures, creating specific microclimates. In this respect, the 1967 proposal for a new urban landscape developed by Merete Mattern together with her mother Herta Hammerbacher and Yoshitaka Akui as part of the *Ratingen West* urban-design competition is exemplary.[106] Their contribution aimed at an integration of architecture and landscape—and formed an important prelude for a new kind of explicitly ecological approach to the city, integrating environmental planning and considerations of participation.[107]

A generation of Swiss architects such as Justus Dahinden, Carl Fingerhuth, René Haubensack, Walter Jonas and Pierre Zoelly extensively elaborated the idea of an underground landscape (based on a man-made hilly topography) as a new urban climatic environment for Switzerland. Since the early 1960s, projects appeared that relied on a systematic, rule-based,

→ fig. 186   Global Adaptations after 1945   **170**

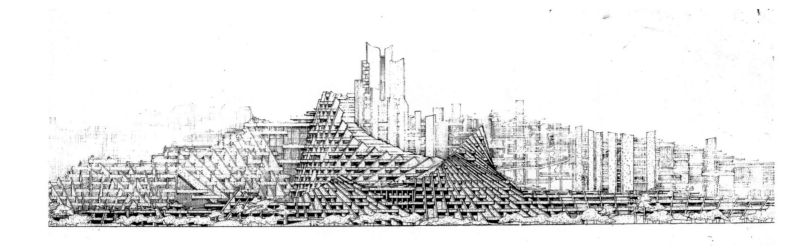

186
Merete Mattern, Herta Hammerbacher, and Yoshitaka Akui, contribution to the *Ratingen West* urban-design competition, near Düsseldorf (Germany), 1967.

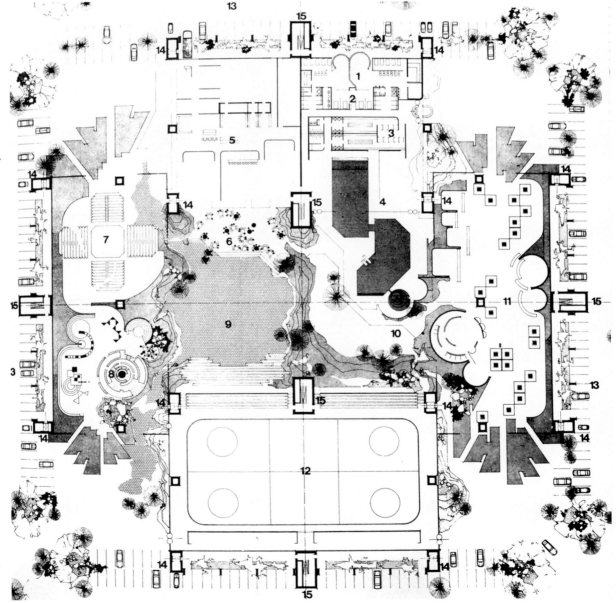

187
Justus Dahinden, layout of the Akro-Polis, 1974.

Microclimatic Islands 171

188
Justus Dahinden, section of the Akro-Polis, 1974.

Microclimatic Islands

structuralist investigation of terraced buildings that also represented artificial topographies. One can mention Haubensack's and Fingerhuth's competition entry *Werden*, in which buildings, topography and greenery become indistinguishable. The special attraction here was the combinatorial play with an artificial hilly landscape. A man-made hill was proposed by Réné Haubensak for the *Sihlraum planning* carried out in Zurich (Switzerland) in 1961. Located in a working-class district (Kreis 4), the hill was designed as a ziggurat with hanging gardens—giving the infrastructure a quasi-natural touch. In this project, a fusion of infrastructure, green space and topography occurs as well. The work of Pierre Zoelly was also devoted to the theme of the *architecture subterrané* ("underground architecture"), realized for instance in the case of the International Clock Museum in La Chaux-de-Fonds by Zoelly and Georges-Jacques Haefeli (1972–74). Walter Jonas had worked obsessively on his idea of the *Trichterhaus* ("funnel house")*,* a large-scale outwardly spiraling structure that aimed at minimizing its footprint and that can be read as an inversion of the hill motif as elaborated in particular by Justus Dahinden. With his urban-planning vision of the "Akro-Polis", Dahinden attempted to develop a new kind of *Raumstadt* ("spatial city"), connecting his approach with the ideas of Walter Schwagenscheidt, Eckhard Schulze-Fielitz and Yona Friedman.[108] "The Raumstadt is the structural, systematized, prefabricated, mountable and dismountable, growing or shrinking, adaptable, air-conditioned, multifunctional labyrinth of space."[109] Dahinden's central approach was an artificial urban topography whose profile was based on hill structures that aggregate into larger units. "Groups of urban hills result in an urban continuum: contiguous residential neighborhoods and interlocking city areas. Both areas grow parallel to each other."[110]

The Multifunctional Labyrinth

The Akro-Polis is built on a modularized approach, which includes a "long-life supporting structure" made of steel or reinforced concrete and a "short-life filling structure"[111] relying on a "plug-in principle."[112] A hill unit includes 15,000 inhabitants and has an area of 28 hectares. The radius of a single hill is 300 m; the primary structure has a height of 200 m. Mutually combinable cells are provided on green terraces, which are 9 m long and 4 and 3.6 m wide. The cells house the "air-conditioned living, sleeping and working spaces"—all of which are "windowed to the outside".[113] In their diagonal orientation, the stacked floors lead to a superposition of vertical and horizontal development.

The central hall, located within the hollow hills, is a "public area" where "urban activities are condensed in a confined space" and where a man-made climate is provided. The huge dome-like space "is intended for any type of use and can be easily adapted to changing needs. The introverted macro-container excludes all negative environmental influences."[114] The given urban climate outside is considered as contaminated, contrasted with a climatic counter-world indoors. "Traffic is murdering the human milieu. The human biosphere in the city is threatened. An urban denaturalization process manifests itself in the pollution of air, land and water. Our cities are in respiratory distress."[115] The "functionally neutral container" is "artificially air-conditioned",[116] enabling various sensory experiences that go hand in hand with the proposed "temporary event structures".[117] Uses with varying thermal conditions are provided: a multi-purpose hall for ice skating and concerts; a water-experience area with saunas, baths and relaxation; massage and fitness rooms; self-service and garden restaurants; a fairground with grandstand and an agora for meetings, events and discussions; a marketplace with boutiques; exhibitions; children's paradise and after-school care and much more.[118]

Dahinden further elaborated his Akro-Polis project as part of the Grand Prix International d'urbanisme et d'architecture. A city quarter of Kiryat Ono near Tel Aviv (Israel) was the given setting. The planning and building system was developed into a macrostructure, for which detailed plans and cost calculations were also prepared. Single- or double-sided "terraced macro-units" (from 60 to 70 m in length and 13 stories high) were to be arranged either as a ribbon or as a group. In the case of the latter, the "enclosed large space can be closed off at the top by a semi-transparent air cushion, which is both a canopy and a projection surface for cosmic light shows".[119] In this case, the man-made climate became both a thermal and a media spectacle. The electronic means (e.g. for the "formation of sound") but also mechanical elements such as "plug-in walls," "mobile grandstands," "hydraulic stages," "scenery hoists," "plastic sails," "movable air cushions," "balloons on cable bracings" and "flexible projection surfaces" were part of this artificial interior space.[120] "The economic maintenance of a uniform mesoclimate is ensured by the fact that the canopy exposed to the weather by an inhabited structure is equivalent to a static climate belt, and the heat losses to the outside or heat radiation to the inside are neutralized by cells that are also air-conditioned. The air cushion closing the public urban space at the top is divided in the middle into two chambers by a reflective metal foil. The fronts of the urban hill

→ fig. 187   Global Adaptations after 1945   174

units are shielded with heat-radiating special glass or shade-creating louvers. Special climate areas such as indoor pools or ice rinks are delineated by controllable secondary volumes; retractable plastic sails or synthetic balloons reduce the controlled volume of space."[121]

The artificial sky above the "air-conditioned common area"[122] played with experiences of a synthetic nature, which adds new facets to the concept of urban climate. In the hilly city, "electronically controlled media" and "space-changing robotic mechanisms" combine with "milieu change".[123] "The public area can be easily rearranged using room-changing robotic mechanisms. The visual, auditory and olfactory climate can be controlled."[124] This reveals the new program of an urban-climate-oriented architecture of everyday life. The main idea was to make urban climatology's concern a reality based on a sophisticated technological approach. In this respect, Dahinden also ties in with the projects of the Weimar Republic, such as the Spa Palace for Berlin (see chapter 3). The polluted urban climate has to be counterbalanced by means of an artificial urban environment. "The city's public realm has optimal bioclimatic conditions that can be conditioned as needed."[125] The public space of the hill city appeals far more to augmented sensory experiences than to sheer comfort; it thus ties in with the concept of thermal delight launched by Lisa Heschong, which links thermal perceptions with other sensorial experiences.

By making all available technical means applicable, Dahinden connects to the understanding of the "man-made climate" of the interwar period. The combination of an actively air-conditioned public interior and a landscaped and terraced hillside exterior is noteworthy. The disparity of the applied models—the simultaneous reference to elements of the Garden City movement as well as to new techno-optimist developments—place Dahinden's proposals in a peculiar light that builds on a media-based infrastructuring of the urban environment. "Akro-Polis brings people in their living environment (in the outside) into contact with nature, light, sun, and green space, and at the same time places them (in the inside) in the field of tension of overall urban activities in a confined space. The residential value of the outward-facing urban quarters with their stepping down, increasing from the bottom to the top, can be highly estimated as a result of the good sunlight and undisturbed orientation to the open nature."[126] The Akro-Polis' adaptability mirrored Cedric Price's *Fun Palace* and Archigram's *Plug-In City*, both from 1964. Dahinden speaks of "a transitional model for a limited period of time,"[127] which

was also relevant for the "renovation" and the "regeneration of the urban fabric" as a new "free-time city."[128] The large technological structure is in the service of its adaptability by users and inhabitants and of the "democratization" of planning culture.[129] The urban climate is transferred into a new synthesis; simulation via comprehensive media equipment becomes the signature of the architectural treatment of the urban climate. Dahinden was mixing nature and culture: at the base of each hill, "natural space and man-made space meet and merge organically".[130]

5.2.2 Large-scale Domes: Beyond the Interior–Exterior Dichotomy

"Bigness" had already acquired a new relevance in architecture with the construction of skyscrapers at the beginning of the 20th century[131] however, it was only with the projects of territorially conceived domes that a direct reference to entire urban communities was established. In 1931, New York architect Raymond Hood spoke of the idea of a "city under a single roof".[132] Entire neighborhoods and cities were to be brought together into an (air-conditioned) whole with the single wide-ranging gesture of the dome.[133] Here, the creation of urban climate was emphatically reformulated into a technology-driven project. The closed urban system appears as a proxy for the well-organized society, responsible for solving thermal problems.

The idea of the house saw itself replaced by experimenting architects with the idea of large-scale environments that were committed to overcoming the opposition of inside and outside. The result was spatialized environments that cultivated the dissolution of place and non-place, city and country, inside and outside. Exemplary for this are Frei Otto's project (with Kenzo Tange) in 1970–71 for a city in the Arctic with 40,000 inhabitants under a dome 2 km in diameter and Buckminster Fuller's concepts, developed from the 1940s onwards, culminating in the famous *Dome over Manhattan* of 1960. Buckminster Fuller deliberately promoted the expansion of the span length and the dissolution of the conventional building. Lisa Heschong commented in 1979 on the dome as follows, emphasizing the idea of an indistinguishability of inside and outside areas: "This climatic envelope would enable the entire city to be air conditioned, indoors and outdoors. Indeed, 'outdoors' would be a thing of the past."[134] Buckminster Fuller's tension integrity structures and Frei Otto's pneumatic structures are examples of such extended interior

→ figs. 188, 191   Microclimatic Islands   **175**

189
Juan Navarro Baldeweg, a tropical forest in an arctic landscape, 1972.

190
Juan Navarro Baldeweg, enclosed ecosystems floating in New York harbor, 1972.

191
Richard Buckminster Fuller,
Dome over Manhattan, 1960.

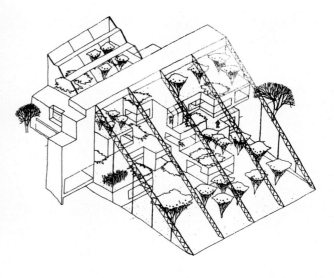

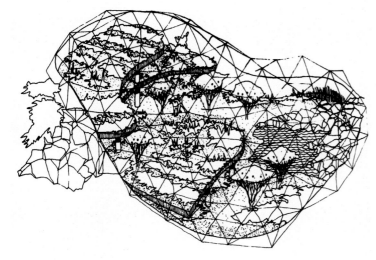

192–193
Sean Wellesley-Miller and Day Chahroudi, bioshelter, 1974.

Microclimatic Islands

spaces that are no longer buildings (in the conventional sense of the word).[135] These experiments with air and bar structures aimed to far exceed the spans of beams and conventional domes. The intended ephemerality and mobility of these structures still breathed the spirit of the war years, when the military had to rapidly build up and dismantle existing structures, which in the post-war years combined with the restless spirit of a population eager to experiment. Buckminster Fuller's concepts developed in the late 1940s—such as the *Necklace Dome* (1950), *Autonomous Living Package* (House 3) of 1952 and the *Dome over Manhattan*—deliberately pushed the expansion of spans and the disintegration of the building.

Concepts of large-scale indoor spaces such as the *Dome over Manhattan* illustrate modernity's great dream of countering the polluted and overheated city with the purified and cooled interior for the privileged. These spaces are, to paraphrase a title of the magazine *Architectural Design* from 1972, designed "for survival";[136] highlighted also in Juan Navarro Baldeweg's vision of generic ecosystems. The "pneumatic bubbles" floating in New York's harbor was only one of many projects promoting artificial urban microclimates. NASA was an important driver in stimulating design research for extraterrestrial territories, not only the most climatically extreme environments. The first steps towards the implementation of such a goal were undertaken in the Biosphere 2 program in the 1980s. The so-called "biosphere" was designed as an autonomous residential zone, equipped with a self-sufficient thermal-energetic concept. The "Bioshelters Total Energy System" by Day Chahroudi and Sean Wellesley-Miller dates from 1974; it involved autonomous living zones with air-conditioned interior areas. It used the so-called Flexahedrom Climate Envelope with a Cloud Gel from Suntek.[137]

In historical retrospect, it becomes apparent how wide-span domes (over city districts) and encapsulating bubbles remained part of a modernist frame of thought. They belonged to the large-scale technological projects of modernity, carried out in ignorance of the system dynamics of modern cities. The change in scale didn't change modernity's dream of a cultivated interior and a natural exterior. The expanded spans merely shifted the interior–exterior dichotomy from the building scale to the scale of the urban environment. Such air-conditioned interiors illustrate the desire to counter the unhygienic and air-polluted city with a purified interior. In this respect, these projects can be deciphered as allegories of the sustainable city (see chapter 4). A sense of hybridization and interaction between different thermal zones was not promoted. Instead, the idea of self-sufficiency became the new conceptual paradigm.

→ figs. 189, 190, 192, 193   Global Adaptations after 1945

6   Energy-Synergy

Energy Ecologies and Settlement Structures

Richard Buckminster Fuller firmly believed the notion of "synergy" had strong implications for the way we conceive of energy in architecture: "the words synergy (syn-ergy) and energy (en-ergy) are companions".[1] He conceived of synergy as a "generalized design science exploration", promoting cooperation among architectural elements. Starting from a superordinate view (the whole), one could explore the interplay of a system's elements. According to his famous definition, "synergy means behavior of whole systems unpredicted by the behavior of their parts taken separately".[2] Relying on Fuller's intuition, the theory of synergy leads to an understanding of energy that is shaped by the coproduction and coordination (synergy) of building groups and the urban environment as a whole.

From the early 1960s onwards there was a significant theoretical discussion around "collective form",[3] which in the 1970s would be reformulated as an explicit discourse on "energy and form".[4] Landscape architects such as Ian McHarg and architects such as Ralph Knowles explored the urban dimensions of energy. As I would like to show, the energy potential of the urban climate was of decisive importance for this. At that time, *energy* and *climate* were arguably still recognized as a promising conceptual pairing. In the words of urban-planning theorist Vladimir Matus in describing the trade-off between climate and energy, "Climatic elements, which are in fact energy carriers, transport energy by *radiation*—the process of emitting energy in the form of waves or particles, *conduction*—the use of a substance capable of transmitting heat, and *convection*—the transfer of heat by automatic circulation of fluids or gases."[5] Going beyond the scale of the individual building, new forms of architectural thinking were being put forward, which weighed natural forces and several buildings together for the *production* or *conservation* of energy.

In the late 20th century, the question of *energy synergy*, as addressed by Fuller, was pursued by two intellectual trajectories that led to noteworthy ideas in the field of climate-related urban design. With Charles Waldheim, we can speak of a *non-figurative*

and a *figurative* type of urbanism, in which ecological analysis and ecological design are closely intertwined. One school of thought aims at overcoming the urban–rural dichotomy, while the second school of thought emphasizes the great value of the existing city and its ongoing transformation. As Margaret Crawford retrospectively emphasized, energy synergy is about "a new focus and scale for saving energy, challenging the current architectural obsession with producing energy-saving buildings".[6]

6.1   Productive Energy Landscapes (Non-figurative Urbanism)

Between 1969 and 1971, the new significance of "ecology" for architecture and urban design became apparent, with energy as the pivot of the ecological rationale. In the design disciplines, these years were framed by the release of Ian McHarg's *Design with Nature* (1969) and Peter Reyner Banham's *Los Angeles: The Architecture of Four Ecologies* (1971). McHarg promoted a new understanding of "ecology of the city",[7] relying on interdisciplinary scientific knowledge. In the words of Lewis Mumford, in his introduction to McHarg's book, "The name of this effort, in so far as it draws upon science, is 'ecology,' a body of knowledge that brings together so many aspects of nature that it necessarily came late upon the scene."[8] Banham on the other hand explored the man-made ecologies of Los Angeles, referring to the notion of "urban ecology" as developed earlier in the century by the Chicago School of Sociology. In *Los Angeles: The Architecture of Four Ecologies*, he described the vast and unmanageable urban landscape of the Californian metropolitan area in terms of specific ecologies, which had developed over time in accordance to natural and infrastructural features.

The two books stand for the scope and the tension of *an urban definition of ecology*, prioritizing insights either from the natural or social sciences. As I would like to show, and following different lines of ecological thinking, urban environments become "productive" either by *integration* (relying on the existing urban fabric and energy infrastructures to a much higher degree) or by *decentralization* (harvesting the local renewable-energy sources). From an ecological point of view, the city can be conceived as the new critical place of energy production, as Dean Hawkes programmatically envisaged: "The city that produces all of the energy it needs for its buildings and the urban infrastructure is, of course, only a vision. To take the first steps towards its realization would transform the

→ fig. 195

179

agenda for research and practice in architecture more radically than any idea since the advent of the modern movement."[9]

6.1.1 Integrated Energy Infrastructures

Within one auspicious decade, several books were published that highlighted the complexity and contradiction of urban ecologies in the great American cities—Las Vegas (1966), Philadelphia (1969), Chicago (1970), Los Angeles (1971) and New York (1978)—and emphasized the critical agency of urban infrastructures for architecture and urban development. These studies provide an innovative framework for analyzing energy ecologies in an ethnographic manner. Not surprisingly, these writings were geared to an "empirical urban sociology"[10] as founded by Jane Adams, Robert Park, Ernest Burgess and John Dewey at the beginning of the 20th century in order to understand the social transformation of rapidly growing cities.[11]

Alvin Boyarsky, Peter Reyner Banham, Denise Scott Brown, Robert Venturi and Rem Koolhaas provided new insights into urban ecologies from an architectural point of view. By displaying the pictorial universe of urban landscapes, they revealed both the sublimity and the inadequacy of modern energy infrastructures. By simultaneously bringing into view anonymous urban infrastructures and the splendid high-rises of famous architects, the blueprint for a new urban narrative was established, coining in turn a new sense for *energy ecologies*. Having internalized the pictorial turn of 1960s, these architectural critics hot-wired research in urban ecologies with Kevin Lynch's "imageability" of the city.[12] Their ethnographic descriptions and phenomenological readings of the American city were full of coruscating humor, sarcasm and the desire to explore the (post-)modern city as a new cabinet of wonders.

The Urban Ethnography of Energy Ecologies

In "Chicago à la carte. The City as an Energy System," Alvin Boyarsky provided a sense for the man-made topographies of urban landscapes. "The energy system of expressways, railway, industrial and institutional land use produce characteristic residual rectangular residential pockets."[13] These urban landscapes comprise spatial layers that are located underground, on the surface and in the air. Chicago "as energy system" relies on infrastructures as the main drivers and victims of rapid modernization; Boyarsky speaks of "the world's largest railway network

and the busiest inland port facilities, delivering coal and hauling ashes from the light framed, bold and simple skyscrapers to fill in future lakeshore parks".[14] His account oscillates between historical descriptions and a contemporary viewpoint, portraying Chicago as "a masterpiece of junk culture."[15] Infrastructure as critical installations of modern society is threatened at every moment with becoming obsolete, with "its own mechanistic cycle of growth, redundancy and replacement, quite different from typical European experience which, at its best, represents an uninterrupted history in stone".[16] The modern energy system and its infrastructure form the blueprint of urban development at large.

At the beginning of the 20th century, Chicago already "possessed the basic hardware and Dionysian qualities" of the futurist visions of Marinetti, Sant'Elia and Boccioni.[17] The descriptions by Boyarsky resemble in particular Umberto Boccioni's painting *The City raises* of 1910. Looking at the architecture of Chicago in light of these visions, the city appears as one interconnected infrastructure—"the city as a building."[18] Boyarsky opens up a three-dimensional account of energy landscapes. He highlights the "complexity of Chicago's multilayered infrastructure",[19] providing the backbone of the "three dimensional tartan grid".[20] The "multi-layered section" represents, in essence, the spatial and temporal dimensions of Chicago's energy landscape. "Frayed edges and entrails reveal the anatomy of its multi-layered section which, plunging storeys deep into the earth, has been further tuned in recent decades by lower level expressways gathering up traffic and service vehicles, and a subway system, complete with accessible sewers, water supply, electric cables, communication lines, threaded through an already complex matrix, giving access to public and commercial buildings, and eventually to the commuting railway stations and massive underground parking facilities."[21]

Boyarsky's "Chicago à la carte" has been righty characterized as "a passionate history of the city based on postcards and texts, a tool to address the production of urban spaces as the complex interaction of architecture, infrastructure, engineering and labor, politics and economy".[22] The energy landscape is presented as being mediated through different types of infrastructures. Using postcards as the representing media of urban history, Boyarsky introduced an anonymous history of large technological systems as a basis for architectural history, while Banham, on the other hand, emphasized the urban palimpsest generated by infrastructure and topography. His four ecologies of Los Angeles—"Surftopia", "Foothills", "The

→ fig. 196   Global Adaptations after 1945   **180**

plains of Id" and "Autopia"—respectively depict distinct social fabrics and architectural styles related to a specific lifestyle on the coast, on the hills, on the city's plains and in conjunction with LA's vast highway network. Banham is intrigued by the infrastructures that generate architectural form.[23] Early railway lines produced the urbanization of the hills and mountains, but only the way the "views over the sea"[24] are perceived, he argues, explains the specific LA type of infrastructure. In view of Los Angeles, Banham connects the culture of "automobility" with the individual's perception of the "environment". In the double-page spread that forms part of the book's prelude, he writes "LA has beautiful (if man-made) sunsets," concluding a short series of perceptions of the city.[25] Banham introduces the notion of urban "ecology" in order to emphasize the man-made character of the sunset, accessed through the infrastructure of the Sunset Boulevard. The infrastructure of the territory is always also an infrastructure of perception; an idea widespread in popular American culture since the early 1960s.[26] The vanishing point of this narrative on urban nature and infrastructure is always the sea or the sunset, which have to be *experienced*. Nature, as Banham understands it, is not an unspoiled, virgin entity but rather fabricated by specific man-made interventions. The beauty and the ugliness of the urban landscape are inseparable.

The double reading of the urban energy landscape as *outer- and inner-world experience* has been a leading approach of these ethnographers of the postmodern metropolis. The productive character of the urban environment is mirrored in the psychological dispositions of the citizens, and vice versa. The psychology of the citizen provides an imaginative reservoir that has to be tapped by the architect interested in new forms of urban nature. In the words of Boyarsky on the urban development of Chicago, "Psychologically embraced by the overtones and associations of an aggressively assembled environment of simultaneously colliding elements, deformed and multiplied images, restrained by superimposed geometrical networks, buildings like giant machines, containing within themselves the architectonics of the environment around them, are perceived in successive state by primitives of a new sensibility that has been completely overhauled."[27] In Rem Koolhaas's *Delirious New York*, the artificial character of urban landscapes becomes a psychogram of the bachelor visiting the "Downtown Athletic Club". His inwardness is reflected in a new type of urban nature located in the interior of the city's towers. The Downtown Athletic Club is the cipher of a new "synthetic" urban nature that is coproduced by architecture and infrastructure. "Eating

oysters with boxing gloves, naked, on the nth floor—such is the 'plot' of the ninth story, or, the 20th century in action. In a further escalation, the tenth floor is devoted to preventive medicine. On one side of a lavish dressing lounge an array of body-manipulation facilities is arranged around a Turkish bath: sections for massage and rubbing, an eight-bed station for artificial sunbathing, a ten-bed resting area. [...] On the 12th floor a swimming pool occupies the full rectangle; the elevators lead almost directly into the water. [...] Of all the floors, the interior golf course—on the seventh—is the most extreme undertaking: the transplantation of an 'English' landscape of hills and valleys, a narrow river that curls across the rectangle, green grass, trees, a bridge [...]. The interior golf course is at the same time obliteration and preservation: having been extirpated by the Metropolis, nature is now resurrected inside the Skyscraper as merely one of its infinite layers, a technical service that sustains and refreshes the Metropolitanites in their exhausting existence. The Skyscraper has transformed Nature into Super-Nature. [...] In the Downtown Athletic Club each 'plan' is an abstract composition of activities that describes, on each of the synthetic platforms, a different 'performance' that is only a fragment of the larger spectacle of the Metropolis."[28] Explorations of energy ecologies reveal the interplay between technical infrastructures and urbanization, between the urban nature and the man-made institutions. In the narratives and images of Boyarsky, Banham and Koolhaas, urban ecologies appear as visual entities co-fabricated through infrastructures and imagination. For these authors, the modern energy landscape has the fundamental potential to generate a new type of urban nature, going beyond the limited conception of the public park and the urban garden. Specific indoor microclimates contribute to the experience of such a "synthetic" nature in the city.

However, the most radical insights into the interactions between nature and the city derive from the more explicitly political perspective of ecology. Like Ludwig Hilberseimer 25 years earlier, Boyarsky in his account on urban ecologies highlights the distribution of urban poverty depending on race. He refers to "Chicago's black community," which comprises "approximately one third" of the population, concentrated in two areas, the South side and the Near West side. "The lines of energy and communication define the neighbourhoods which contain the isolated symbols of religion, education and civic authority."[29] Boyarsky refers to the hygienic and climatic conditions of the housing blocks: "The majority of the rooms receive their light and ventilation from narrow wells between properties."[30] "Isolated, crowded, and

→ figs. 194, 197, 198   Energy-Synergy

181

194
In Los Angeles. *The Architecture of Four Ecologies*, Peter Reyner Banham connects the culture of automobility with the individual's perception of the urban environment.

195
Table of contents foregrounding the four ecologies of Los Angeles. Source: Banham 1971.

Views of Los Angeles   1

1. In the Rear-view Mirror   3
2. Ecology I: Surfurbia   19
3. Architecture I: Exotic Pioneers   39
4. The Transportation Palimpsest   57
5. Ecology II: Foothills   77
6. Architecture II: Fantastic   93
7. The Art of the Enclave   119
8. Ecology III: The Plains of Id   143
9. Architecture III: The Exiles   161
10. A Note on Downtown . . .   183
11. Ecology IV: Autopia   195
12. Architecture IV: The Style that Nearly . . .   205
13. An Ecology for Architecture   217

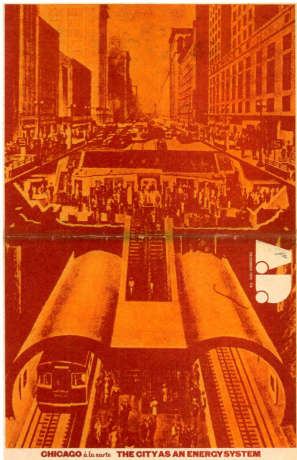

196
AD special issue on "Chicago à la carte. The City as an Energy System", edited by Alvin Boyarsky, December 1970.

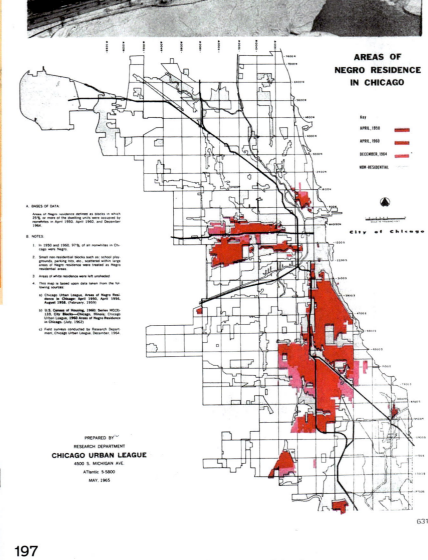

197
Plan, showing the ghettos of Chicago's black community. Source: Boyarsky 1970.

198
Urban poverty depending on race: "The majority of the rooms receive their light and ventilation from narrow wells between properties." Source: Boyarsky 1970.

Energy-Synergy 183

199
SITE, "Highrise of Homes" proposal for urban environments, ink and wash rendering by James Wines, 1981.

200
Superstudio, proposal for a forest cube crossed by the Golden Gate Bridge in San Francisco (USA), 1972.

201
SITE, "Highrise of Homes" proposal for suburban contexts, watercolor rendering by James Wines, 1981.

Global Adaptations after 1945

202
Angela Danadjieva (Lawrence Halprin & Associates), opening day of the Lovejoy Fountain, city park in downtown Portland, Oregon (USA), June 23, 1970.

203
Early initiative in decentralizing the energy supply by using urban climatic forces: Solar collectors and a windmill installed by the residents, Lower East Side, New York, 1976.

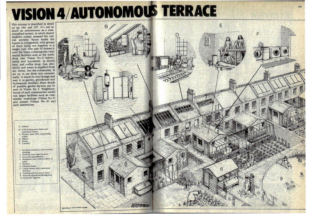

204
Visions of "radical technology". Integrated energy infrastructures for urban environments. Source: Harper and Boyle 1976.

Energy-Synergy

with an extremely young population, its service requirements" exceed by far the current situation.[31] Boyarsky addresses infrastructure as public goods, which depend on democratic accessibility, and he criticizes in this regard the uneven distribution of such infrastructures in Chicago.

Almost at the same time, Ian McHarg correlated social and economic data with data on health and hygiene. For *Design with Nature*, McHarg (and his students) mapped social and physical factors of Philadelphia in parallel.[32] "It was decided that we would collect all available statistics on these three categories of health and, in addition, that we would compile information on economic parameters, ethnicity, housing quality, air pollution and density. All data were divided into three categories, highest, intermediate and lowest incidence. All data were mapped on transparencies so that concurrence of factors could be observed. It was believed that the mapping of these data would reveal the environment of pathology and that the absence of this would reveal the areas of relative health."[33] McHarg questions "the pitfalls of causality";[34] he speaks rather of the "concurrence" of environmental and social factors. "By superimposition and photography, it is possible to add the factors concerned and obtain a summation that is as accurate as the components."[35] In the superimposition of these data, patterns of urban ecologies are revealed that could be summarized as the *political ecology* of the city. "It is possible to begin to investigate the factors of the social and physical environment that are identified with these polarities and with the intervening stages. At the moment, it seems clear that crowding, social pressure and pathology do correlate sufficiently to justify more serious investigation. In this vein, it is reasonable to ask why the poor should live on the most expensive of central urban land, which requires the highest density."[36]

Energy ecologies do not simply follow evolved structures of the city; rather, they often require fundamental conceptual re-orderings of the city. In this respect, they are to be understood as a radical contribution to a new type of energy governance of the city. Following McHarg's *Design with Nature*, research into energy ecologies began to generate insights into the manifold intersections of political, infrastructural, social and natural conditions of cities. By assessing the geographical distribution of potential energy sources of a city (e.g. geothermal energy), new epistemic entities were generated that left the traditional political and cultural order of the city behind. There was a fundamental superimposition of active and passive climate-control systems in cities, relying on a complex and interconnected energy supply, which in turn took into account various lifestyles and changing comfort requirements. Urban ethnography contributed to the deciphering of these superimpositions.

Integrated Urban Energy Landscapes

In Europe, already in the early 1950s one finds a decided shift in the discursive structure of urban climatology towards the integration of energy. Instead of the "thermal effect of radiation", there was increasingly talk of the "energy content of radiation".[37] Urban climatology explicitly entered into a relationship with energy supply, and economic efficiency became a function of the city-wide energy balance. What in 1937 was still awkwardly called "fulfilment of the metropolitan space by economic man"[38] with regard to the so-called *heat economy*, could in 1956 be elegantly summarized by the new term *urban energies* (*Stadtenergien*). For the second edition of his book, *Das Stadtklima* ("The Urban Climate"), Albert Kratzer introduced this term, with which he was able to succinctly address numerous problems of urban climatology. "An agglomeration of people also results in an agglomeration of energies, which must have an effect on the climate."[39] The artificial, externally introduced energies of the city become themselves a constituent element of the (urban) climate. The enormous accumulation of carbon-based energy resources in the city changes the form of the city, and at the same time shapes the urban climate.[40] The "energies are used for the most part within the large cities and introduce masses of heat and impurities into the city air. In this form, they then appear as a climatic element."[41]

Kratzer compares the heat supply from coal with that from the sun. Drawing on Wilhem Schmidt's pre-war calculations, the sun is seen as the city's natural heat source, competing with artificial ones. With such considerations, Kratzer opens up an interesting dual perspective on the energies of a city. Besides increasingly important energy infrastructures, it is influencing factors such as the sun and wind that make up the energy resources of a city. In the case of the Berlin of 1931, 3,600,000 tonnes of black coal and 2,100,000 tonnes of lignite were combusted, resulting in "a heat supply to the territory of Berlin of about 20 × 12 kcal/year.[12] Since the area of buildings and roads in Berlin is about 30'000 ha, this equates to a heat flow to each square centimeter of 7 kcal per year or roughly 20 cal per day. [...] For Vienna W. Schmidt calculates 8.1 (12.2) kcal/cm²/year) compared to 52.3 kcal/cm²/year) of solar radiation. According to him, in winter 22.1 (33.3) cal/cm²/day are provided by combustion, in November 29, December 15, January 23 and

February 52 cal/cm²/day by the sun. So, in winter the artificial heat almost balances the heat given by the sun."[42] There are basically two approaches to urban energies, providing access to the natural resources of the city as well as the resources brought to the city. Bringing these elements together is the actual work of an ecologically oriented architecture and urban design. If Kratzer is to be followed, climatic factors such as the sun must be included in the heating of flats, and made usable for possible energy savings.

In the same vein, and one decade earlier in *The New City*, Ludwig Hilberseimer criticized "the haze and smoke which hangs over our cities," impeding direct sunlight. Hilberseimer mentioned studies by the National Conference Board of Sanitation that referred to such a loss of solar radiation. "In one year a city like New York loses 35 per cent of the sunlight that should be available. On certain days as much as 73 per cent is cut off by smoke and fog, and during a certain period the loss is 50 per cent."[43] According to Hilberseimer's proto-ecological argumentation, solar radiation cannot be considered separately from this plague smoke or from the wind, just as the air conditioning of the buildings cannot be considered separately from the layout of urban districts. Urban nature equally provides energy resources and is subject to contamination and transformation by the artificially introduced ones.

Starting with public art projects, New York-based architecture and environmental collective SITE (Alison Sky, Patricia Phillips, James Wines) developed a novel interdisciplinary attitude towards urban environments. Their most famous projects were those for the retailer BEST Products Company, for which they developed a series of large-scale warehouses. The relationship between inside and outside was at the center of many projects. Opened facades and integrated natural elements started to characterize the projects from the second half of the 1970s onwards, exemplary in the case of the BEST Forest Building in Richmond (VA, USA) in 1980. The cutaway of one section of the facade allowed "the existing forest to penetrate the building as an 'inside-outside' experience."[44] Beyond these place-specific interventions, SITE developed a sense for the climatic conditions in cities, pioneering indoor and outdoor environmental strategies. The cooling effect of walls of water, for instance, played an important role both in outdoor and indoor projects. Alongside water, the agency of vegetation for air conditioning was considered. SITE developed a model for how to enrich high-rise buildings with green spaces. With their "Highrise of Homes", they gave the idea of vertically stacked individual sto-

ries a decidedly urban-ecological dimension.[45] In SITE's designs, buildings consist of platforms of greened urban landscapes. Expanding upon Koolhaas's *Delirious New York*, for SITE a high-rise building was no longer a homogeneous interior but a framework for story-specific interventions, each with its own particular microclimate. Conceptually, their high-rise buildings resembled the cartoon published in *Life Magazine* in 1909 (and made famous by Rem Koolhaas), which conceived of the high-rise typology as a stacking of individualized stories and microclimates. The residential high-rise appears, as Koolhaas writes, as a "utopian device for the production of unlimited numbers of virgin sites on a single metropolitan location".[46] However, while Koolhaas emphasized the programmatic aspects, SITE foregrounded the surplus potential for creating urban climates.

Playing with the quality of "bigness", the Italian architecture group Superstudio made a casual contribution to the reinterpretation of a bridge, which must be considered a remarkable contribution to urban-climatic thinking in architecture. The so-called "Cubo di Foresta" was introduced in 1972 on a postcard depicting San Francisco Bay. The forest cube was crossed by the Golden Gate Bridge by exactly placing the cube in the middle section of the bridge.[47] With its rigid geometry, the artificial forest oscillates between architecture and landscape architecture. Conceptually, Superstudio's work is reminiscent of Marcel Duchamp's 1919 work *L.H.O.O.Q.*, in which the artist added a moustache to a postcard reproduction of Leonardo da Vinci's Mona Lisa. On second glance, however, the Dadaist gesture proves to be a precise urban-climatic architectural intervention, which makes use of the exceptional microclimatic conditions found on site. In the interplay with the city's summer fog, the forest goes beyond the mere gesture of megastructure. The pronounced fog formation in San Francisco Bay turns the forest cube into a kind of rainforest at low temperatures, which would lead to an enhanced experience of this bioclimatic phenomenon. It would be architecturally transformed under urban climatic premises and the bridge would be transformed into a way of accessing this microclimatic sensation.

Within only two decades, a paradigm shift had taken place in the scientific understanding of the urban climate: away from hygiene towards an ecology of urban energies. From then on, the urban climate was in a reciprocal relationship with urban energies, and its energy balances were examined in the form of interactions between the interior and exterior. After the oil crisis, research on urban energy became more urgent

→ figs. 199, 200, 201   Energy-Synergy

still, accompanied by new ecological thinking in architecture and urban design. Such research explored the relationship between the density of settlement structures and energy infrastructures. The overall goal was, as Niels Schulz and others emphasized only recently, "to identify energy saving potentials through system integration at the level of entire urban settlements, with the aim of at least halving the energy intensity of cities".[48] The notion of "energy landscape" turned out to be of fundamental significance for the development of such an ecological understanding of architecture and urban development. The landscape architect Michael Hough elaborated this notion, pioneering "an ecological view that encompasses the total urban landscape".[49] In Hough's analysis, buildings are in a systematic relationship with the urban environment. He focused on energy savings through a more adequately designed urban environment; the perspective of the individual building was abandoned in favor of a more cooperative view of the city. The notion of energy landscape implies the total urban landscape, and thus a continuity and exchange between inside and outside conditions. The deliberately designed urban space appears as the principal means through which an ecological way of energy production becomes possible.

In their 1976 book *Radical Technology*, Peter Harper and Godfrey Boyle point to the multi-scale specificity of an energy supply for the city, critically questioning the ecological concepts of self-sufficiency and decentralization that were in vogue at the time as alternative technologies to the carbon-based and conventionally planned city. The system of global energy infrastructures had not yet conceived integration of its various scales and networks as a critical asset. "Autonomous units [...] tend to be impractical in large cities. It is almost certainly better to concentrate on making better use of existing services, possibly supplemented by autonomous systems but chiefly by clever ideas and making do. It is necessary to find a rational mixture of mains, communal and household scale."[50] Harper and Boyle address the integration of energy infrastructures, while also considering the city's natural energies. In 1979 and in the same vein, Karl Ganser and Wolfgang Bahr emphasized the "need for coordination between concepts and measures of energy saving and spatial planning," which potentially "exist at all planning stages".[51] In the context of Germany in the 1970s, their reflections on "heat supply systems and urban design" clearly considered such systems to be relevant for energy conservation. These investigations scrutinized energy within larger energy landscapes. "More complex coordination processes arise at the level of building

blocks and neighborhoods. These are accumulations of buildings with different structures, different kinds of heat supplies and different periods for renewal. In addition, at least in the more densely built-up quarters, there are performance-based energy supply systems of different ages with specific capacities and renewal needs. Finally, different willingness to invest, investment power and organizational capacity of home owners, energy supply companies and the public sector must be taken into account. The various demands should be balanced on the basis of 'local care concepts'. So far, there are no binding coordination rules at this level."[52]

6.1.2 Decentralized Energy Commons

While the concept of energy infrastructures is historically intertwined with the use of carbon-based energy sources, energy commons are centered around the use of the natural forces (renewable energies) present in cities. Whereas over the 20th century, "the oil and gas industry" generated "a landscape of lines, axes, nodes, points, blocks, and flows", the city addressed by energy commons highlights primarily the natural forces and the buildings found in cities.[53] Rather than purely technological entities, energy commons are novel forms of infra-architecture, relying on an "architecturalization" of vegetation and topography, as well as renewable energies, the sun and the wind.[54] The extension of the modern notion of energy infrastructure was theorized by Vladimir Matus, when he made the observation that "the thermal environment within and surrounding a house" has to be brought into "equilibrium" with "a steady input of energy".[55] The city based on energy commons harnesses the natural forces in the city via appropriate design. As Matus pointed out, the sun, winds and the shading of buildings are all potential energy commons. "Relatively recently, it was discovered that it is possible to tap, tame, and store the wild and fluctuating energies around a building; most important, it was found that these energies can replace the more expensive conventional energies from a central distribution grid."[56] In contradistinction to such terms as energy infrastructures, the notion of energy commons proved to be accessible to the insights of urban climatology. In this respect, Ian McHarg emphasized the "public good" character of natural forces and buildings.[57]

→ fig. 204   Global Adaptations after 1945   **188**

Energy Commons

The theory of *common-pool resources* offers alternatives to the infrastructural thinking of modernity by distinguishing a natural "typology of goods (private, club, public, common)";[58] it is linked to the the concept of the commons—the forest for firewood and construction wood—which had been widespread in Europe for centuries, forming forerunners of new types of energy commons. As late as the 1970s, the research literature on this topic was still sparse, as Ian Laurie has argued.[59] Laurie traced the development of the European commons as follows: "The commons which now exist in the towns and cities were formerly land 'held in common' (i.e. privately owned land with common rights for the community) and were located on the countryside beyond the urban boundaries. With the spread of towns a few commons were [...] kept and encompassed by the towns within their sprawling environs. Although still grazed, they were also recreational parks long before the nineteenth century town park movement in Europe and America. Furthermore, their natural landscape character was entirely different from the formal ornamental country parkland of the Renaissance design tradition which had influenced the new designs for the town parks. Essentially, the urban common represents 'rus in urbe'—country in towns."[60] In the course of democratization and modernization, the commons of European cities have increasingly become public parks and recreational areas. Their enclosed and insular character makes them important urban spaces where urban nature unfolds.

Large-scale green spaces of a city such as the commons have, over the course of the 20th century, become significant as microclimatic islands and as thermal reservoirs counterbalancing urban heat islands; furthermore, they secure the biodiversity of urban habitats, which is also reflected in thermal diversity.[61] Since cities always store ambient energies, the urban commons also play an important role in the adequate transfer of these energies. Ian Laurie describes urban commons in Wimbledon (London), Southampton and Clifton (Bristol) as "representative of many of the best qualities that these forms of open space create. All are situated well within the boundaries of large cities and are intensively used and enjoyed by the townspeople. They are all 'lowland' commons less than 100 m above sea-level, and therefore in comparatively sheltered climates within large city 'heat envelope', and they are not seriously affected by industrial pollution."[62] The climatic elements of a city can serve as common-pool resources that are subject to adequate management. However, climate control is now conceived as a societal rather than a technological mechanism.

The notion of energy commons relies on the historical development of the European commons, accompanied by a fundamental shift in the perception of urban green spaces. McHarg distinguishes "the ecological method" from conventional approaches to the formulation of public spaces in the city. "The ecological method would suggest that the lands reserved for open space in the metropolitan region be derived from natural-process lands, intrinsically suitable for 'green' purposes: that is the place of nature in the metropolis."[63] According to McHarg "the problem of determining the form of metropolitan growth and open space" is critical. "The hypothesis [...] is that the distribution of open space must respond to natural process."[64] In a similar vein, Michael Hough has criticized the lack of productivity of the city's green spaces that has resulted from a one-dimensional focus on recreation—"Leisure has become the prime function of urban parks, while other environmental and productive functions that the city's land resources must serve have largely been forgotten."[65] While Anne Whiston Spirn has particularly emphasized the epistemic deficits in the treatment of urban nature: "Nature has been seen as a superficial embellishment, as a luxury, rather than as essential force that permeates the city. Even those who have sought to introduce nature to the city in the form of parks and gardens have frequently viewed the city as something foreign to nature, have seen themselves as bringing a piece of nature to the city."[66] Spirn calls for an actual cultivation of the nature in the city, which leaves far behind the mere greening of the inner cities. Instead of control or defense, she introduces the idea of *activating natural processes*. Conceived in this way, natural processes are also always energy processes that need to be understood and applied. "The city must be recognized as part of nature and designed accordingly. The city, the suburbs, and the countryside must be viewed as a single, evolving system within nature, as must every individual park and building within that larger whole. The social value of nature must be recognized and its power harnessed, rather than resisted. Nature in the city must be cultivated, like a garden, rather than ignored or subdued."[67]

Spirn and Hough both analyzed the role of various natural parameters such as climate, soils, water, plants, wildlife and agricultural management in the city. Their books *The Granite Garden: Urban Nature and Human Design* and *City Form and Natural Process*, both published in 1984, provide a comprehensive reception (and extremely rare synthetization) of the contemporary urban climate research from an *ecological design* point of view; both publications ultimately aim at developing new methodological approaches to the design process. A central concern of theirs is to use the energy resources provided by nature for the city. By allowing the physical–chemical–biological play of natural forces to permeate every corner of a city, their aim is to reclaim the relevance of nature in the conception of the city. By subjecting everything to the ecological process, the physical side of everything becomes all the more apparent.[68] "Nature in the city is an evening breeze, a corkscrew eddy swirling down the face of a building, the sun and the sky. Nature in the city is dogs and cats, rats in the basement, pigeons on the sidewalks, raccoons in culverts, and falcons crouched on skyscrapers. It is the consequence of a complex interaction between the multiple purposes and activities of human beings and other living creatures and of the natural processes that govern the transfer of energy, the movement of air, the erosion of the earth, and the hydrologic cycle. The city is part of nature."[69] The idea of the wholeness of the system prevails in both publications. They thus explicitly follow on from the preliminary work by McHarg, stating that "nature is a single interacting system and that changes to any part will affect the operation of the whole".[70] The comparison of the contents of Spirn's and Hough's books reveals the striking parallels in terms of approaches, subjects and order:

Michael Hough: *City Form and Natural Process*	Anne Whiston Spirn: *The Granite Garden. Urban Nature, and Human Design*
Urban Ecology	City and Nature
Climate	Air—Dirt and Discomfort Air—Improving Air Quality, Enhancing Comfort and Conserving Energy Earth—Shifting Ground and Squandered Resources Earth—Finding Firm Ground and Exploiting Resources
Water	Water—Floods, Droughts and Poisoned Water Water—Controlling and Restoring the Waters
Plants	Life—Urban Plants: Struggle for Survival Life—Nurturing the Urban Biome Life—Pets and Pests
Wildlife	Life—Designing Wildlife Habitats
City Farming	
Making Connections	The Urban Ecosystem—The City as an Infernal Machine The Urban Ecosystem—Designing the Urban Ecosystem Epilogue—Visions of the Future

205
Striking parallels in terms of approaches, subjects, and order:
The contents of Anne Whiston Spirn's and Michael Hough's books of 1984.

→ fig. 205   Global Adaptations after 1945   **190**

Attempts that conceive of the comprehensive meaning of urban commons have aimed at making public spaces the central intervention sites of an ecological urbanism. The "Portland Open Space Sequence" (Portland, Oregon), designed and built in 1965–70 by the American landscape architecture firm Lawrence Halprin & Associates, can be conceived as an example of this. The intervention by Halprin & Associates foregrounds the ecological character of city form. In this case, natural processes generated a new urban form. By combining various installations of flowing water, a striking cooling of the microclimates was achieved. In addition to Ira Keller's Fountain, the Lovejoy Fountain, which was accompanied by a pavilion designed by Charles Moore (1966–68), was also included. The Bulgarian landscape architect Angela Danadjieva, who worked for Halprin & Associates at that time, designed the Ira Keller Fountain in Portland (Oregon) as well as, later, the Freeway Park in Seattle (Washington). Danadjieva, who trained as an architect, started her career as a set designer for the Bulgarian film industry. The jagged and backdrop-like urban topography points to the staged nature of this type of urban design.

One early initiative in decentralizing the energy supply was the East Village Solar Collectors and the Windmill, built in 1976 in the Lower East Side of New York (519 East 11th Street). The set-up of these devices represented a new form of bottom-up initiative towards *decentralizing* the heat-and-electricity network of a big city.[71] As much as the sun can be understood as a generally accessible good (common-pool resource), the technical infrastructures required for its efficient use to supply hot water and heat to households are accessible only to a few, who thus form a "club". The concern for decentralization of energy networks, as has emerged since the 1960s, was shaped by technological restrictions and the individualism of those times.[72] The possibility of decentralization empowered individuals to find their own solutions for their own energy supply in the city; however, it has also released them from thinking about the greater whole. In the field of the use of renewable energies, limited access still prevails.[73]

Energy Hinterlands
(Non-Stop Cities and Supersurfaces)

As they developed over the course of the 20th century, cities came to represent patterns of enormous densification in terms of energy while at the same time depending on the resources of a globalized "energy hinterland."[74] Today, cities use "more than two-thirds [...] of the technical primary energy [...]

while covering less than 2 percent of the terrestrial planetary surface."[75] The distribution of energy in the modern city was closely intertwined with the development of infrastructures. Already at the beginning of the 20th century, the architects Tony Garnier and Antonio Sant'Elia envisaged the architecture and urban layout of the infrastructure-dependent city. In Garnier's "cité industrielle", the interplay between the city and its energy hinterland is depicted in the form of dams and power stations. Infrastructures were often the main promotors of "explosive rates of urbanization"[76] tightly linked to the consuming areas of transportation and architecture. The city of the 20th century with its culture of auto-mobility and mechanical air conditioning was, generally speaking, the outcome of a "vast oil wealth". The notion of energy infrastructures refers mainly to the "oil city." "Much of what is modern in the modern city is [...] the by-product of oil."[77]

Architecture and energy are always part of a superordinate hierarchy of city and hinterland, city and world. The energy hinterland as the basis of an oil infrastructure of the modern city comprises "refineries, gas plants, petrochemical plants, export terminals"; "oil-producing regions" (such as Baku, Cabinda, Fort MacMurray, Houston, Kirkuk, Luanda, Murmansk, Odessa, Port Harcourt, Warri); "hubs of corporate power" (such as Irving, Texas); and cities that are the "products of vast oil wealth" (such as Dubai).[78] Energy infrastructures have dimensions that are global (transnational), regional (nation-wide), local (city-wide) and micro-scale (buildings). As Kiel Moe has pointed out, buildings form transit points of "global material geographies" and "sensations of the human body." "Buildings—and their operations of energy—must be understood as physically connected through the cascade and recurrence of energy that courses through buildings, cities, and the world. To place a building in the nested cascade of its energy hierarchy opens the formation of architecture to new questions and new opportunities."[79]

However, energy hinterlands are also the territory of other—namely, renewable—resources, as Ian McHarg has emphasized. "The hinterland of cities" is not only the area where fossil fuels are extracted and processed but also the area where clean air is to be found; a heat-supply system can be based on both technological and natural resources. By highlighting natural forces as sources of energy his research aimed at closing a fundamental research gap that had already been identified earlier in the century (see chapter 3). McHarg sees three ways to treat hinterlands, depending on their capacities: "the hinterland

→ figs. 202, 203, 209, 210   Energy-Synergy

206
Superstudio, "Supersurfaces" as globalized infrastructures, 1972.

207
Archizoom, perspective of the *No-Stop City*, 1970.

Global Adaptations after 1945

of a city is the source of the clean air that replaces the pollutants discharged by the city. The rural hinterland also contributes to a more temperate summer climate. Can we use this information to discriminate between lands that should remain in their natural condition, lands that are permissive to certain uses but not to others and those lands that are most tolerant to urbanization—free from danger, undamaging to other values?"[80] Thus, regional planning plays a decisive role in connecting the city with its hinterland. Biologist and sociologist Patrick Geddes was groundbreaking in reconnecting the different (interacting) scales of architecture. With the first publication of his so-called "valley sections" in 1909, he introduced a central "graphic tool for planning"[81] proto-ecological urban landscapes. The valley section placed the city in a systematic relationship to its hinterland, implying an "interdependence of biogeography, geomorphology and social systems" and introducing a "new method for grasping larger interrelationships" in the field of modern regional planning.[82]

The thermal loads of a city depend comprehensively on its hinterland. "Airsheds," for instance, have a fundamental impact on the thermal and environmental loads of a given city. "The satisfaction of these two requirements, in the creation of urban airsheds as responses to atmospheric pollution control and microclimatic control, would create fingers of open space penetrating from the rural hinterland, radially into the city. This is perhaps the broadest conception of natural process in urban growth and metropolitan open-space distribution. Clearly, this proposal directs growth into the interstices between the airshed corridors and suggest that metropolitan open space exist within them."[83] McHarg proposes the establishment of airsheds 16–24 km long in order to discharge the polluted and heated air. Urban energy production begins with the conception of the "open spaces," as the work of McHarg clearly demonstrates. "Under the heading of atmosphere the subject of climate and microclimate was raised. In the study area, the major problem is summer heat and humidity. Relief of these conditions responds to wind movements. Thus, a hinterland with more equable temperatures, particularly a lower summer temperature, is of importance to climate amelioration for the city. As we have seen, areas that are in vegetative cover, notably forests, are distinctively cooler than cities in summer. [...] We can then say that the areas selected as urban airsheds are likely to be those selected as appropriate for amelioration of the urban microclimate."[84]

208
Archizoom, perspective of the *Wind Town*, 1973.

Energy-Synergy

193

209
Hermann Sörgel, map showing the energy network and reclaimed land around the Mediterranean Sea, 1932.

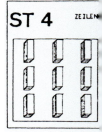
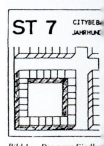

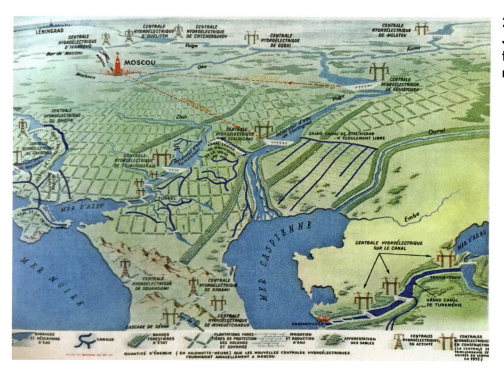

210
Joseph Stalin, great plan for the transformation of nature, 1948.

Global Adaptations after 1945

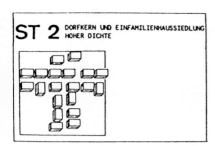
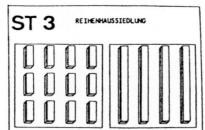
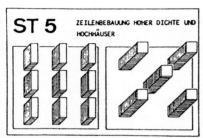
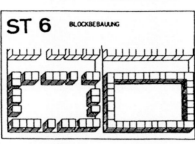
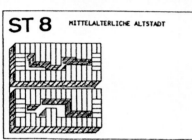
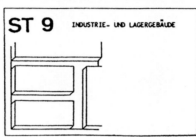

211
Implications of the settlement structure for the heat-supply system (and vice versa). Source: Roth and Häubi 1980.

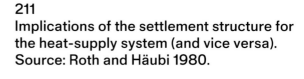
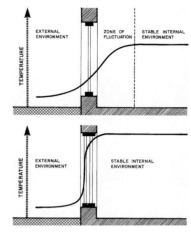

212
The highly insulated building envelope eliminates the zone of thermal fluctuation. Source: Matus 1988.

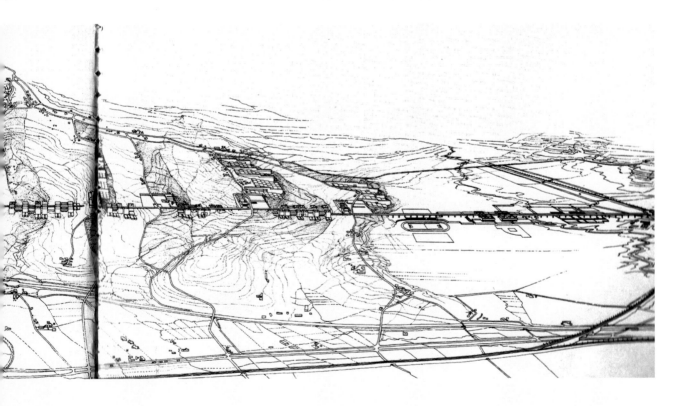
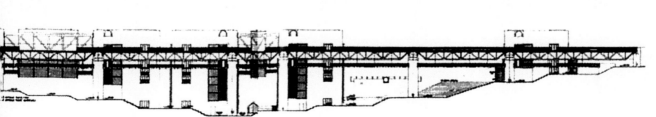

213
Vittorio Gregotti, University of Calabria (Italy), 1978–80.

In the discussion of *environmental building*, the figure of the bridge formed an architectural leitmotif. In view of the semantic indeterminacy of peri-urbanized environments, the potential of infrastructure was considered. By integrating structure, infrastructure and territory, strategies for new energy landscapes emerged. As a structure, a bridge floats above the territory or connects to it only selectively. The public baths complex in Bellinzona (Switzerland) designed 1967–70 by Aurelio Galfetti, Flora Ruchat-Roncati and Ivo Trümpy is captivating in its fusion of a public pathway and the furnishings of the bath with changing rooms. The six-meter-high passageway cuts through the valley, creating a new kind of connection while at the same time providing access to the baths. In the valley, which was still largely undeveloped, the bridge formed a new connection between the Ticino River and the city of Bellinzona, also spanning the new highway. The course of the river and the green belt were tied together with the bridge-like infrastructure. The wooden changing rooms of the baths are suspended from the concrete passageway. The water basins and the sprawling lawn are connected to the bridge structure.[85] The convergence of structural and infrastructural requirements is taken to the extreme in the projects of Italian architect Vittorio Gregotti. In a housing project for Cefalù near Palermo from 1976, eight double buildings span a valley—housing not only the dwellings but also other facilities for the neighborhood, including parks and car-parking spaces that were planned on the roofs. In the case of his project for the University of Calabria (near Cosenza), built 1978–80, a bridge road literally forms the infrastructural backbone for the provision of the public buildings that adjoin it, in which teaching and research rooms are housed, along with halls of residence.

In particular, the urban projects of *architectural groups* (such as Superstudio and Archizoom), and *environmental collectives* (such as SITE and Ocean Earth) explored the possibilities of decentralizing and integrating urban energy.[86] This type of urbanism clearly went beyond the conceptual basis of the garden city movement. The main goal was to transcend the city/countryside dichotomy and instead promote the productive dimension of cities.[87] According to Andrea Branzi, "the failure of the urbanistic development of suburbs, and hence the management of territorial organization, was an undeniable fact".[88] In the idea of the "non-stop city", proclaimed by the Italian group of architects Archizoom, the idea of an interconnection between countryside and city returned but was expanded into a new horizontal development of the city in general. With Charles Waldheim, one can speak of "an urbanism of continuous mobility, fluidity,

and flux";[89] of "an urbanism without qualities, a representation of the 'degree-zero' conditions for urbanization."[90] "As a deliberately 'nonfigurative' urbanism, 'Non-Stop City' renewed and disrupted a longstanding traditional nonfigurative projection as socialist critique. In this regard, Branzi's 'Non-Stop City' draws on the urban planning projects and theories of Ludwig Hilberseimer, particularly Hilberseimer's 'New Regional Pattern' and the project's illustration of a proto-ecological urbanism. Not coincidentally, both Branzi and Hilberseimer chose to illustrate the city as a continuous system of objects."[91] The other important group of architects of an *Architettura radicale*, the neo-avant-garde group Superstudio—with their supranational infrastructures, the "supersurfaces"— also created a new approach to the city, again with urban-climatological implications. Based on the analysis of new technological possibilities of communication and an "extension of senses", an infrastructural thinking was launched that also had implications for "the creation of shelters and microclimates".[92]

Synthesizing Urban Nature

Beginning in the 1960s, an awareness of large-scale energy landscapes emerged as an attempt to make the goals of urbanism and climatology more compatible. A synthesis between nature and the city was the central theoretical problem of an urban-design practice that was newly oriented towards ecological forms of modernization that took the urban climate into account. Anne Whiston Spirn's and Michael Hough's theoretical approaches consisted of an all-encompassing concept of urban nature. Even more than Ian McHarg, who referred to an *urban* nature (among other kinds),[93] Hough and Spirn struggled for a theoretical standpoint that did justice to the hybrid status of the city's nature, between natural and man-made dimensions, nature and artifact. City and nature no longer formed a conceptual opposition, rather the city was understood as part of nature. "We fail [...] to see nature as an integrated connecting system that operates in one way or another regardless of locality."[94] This conceptual incorporation and re-hierarchizing was intended to take account of the paramount importance of scientific knowledge. "Nature is a continuum, with wilderness at one pole and the city at the other. The same natural processes operate in the wilderness and in the city. Air, however contaminated, is always a mixture of gasses and suspended particles. Paving and building stone are composed of rock, and they affect heat gain and water runoff just as exposed rock surfaces do anywhere. Plants, whether exotic or native, invariably seek a combination of light, water, and air to survive. The city is neither wholly natural or

→ figs. 206, 207, 208, 213　　　Global Adaptations after 1945　　　**196** wholly contrived. It is not 'unnatural' but, rather, a transformation of 'wild' nature by humankind to serve its own needs".[95] Referring to the writings of Anne Whiston Spirn, the historian of technology Thomas Hughes pointed to the "overlapping natural and human-built systems found in cities".[96]

In the perspective of an urban political ecology, one could speak of a new "re-naturing [of] urban theory",[97] that was launched from late 1960s onwards in the works of McHarg, Hough and Spirn as part of an emerging ecological urbanism. "While late-nineteenth-century urban perspectives were acutely sensitive to the ecological imperatives of urbanization, these considerations disappeared almost completely in the decades that followed (with the exception of a thoroughly 'de-natured' Chicago school of urban social ecology)."[98] Matthew Gandy emphasized the fact that "the design, use, and meaning of urban space involve the transformation of nature into a new synthesis".[99] In terms of a history of theory, however, the re-naturalization of the city has also led to an urbanization of the concept of nature and thus to a transformation of what is understood by "nature," as the publications of Boyarsky, Banham, and Koolhaas show. "The Skyscraper has transformed nature into super-nature."[100]

6.2 Energy-conserving Urban Environments (Figurative Urbanism)

In the 20th century, "the availability of cheap energy has been the overriding determinant of urban form" with the discipline of architecture playing a substantial role as the unwitting promoter of the carbon city.[101] As Michael Hough noted, visions of modernity were explicitly or implicitly related to the use of oil. The modern city, with its culture of personal mobility and mechanical air conditioning, was the outcome of an unprecedented abundance of (carbon) energy. Historically however, "groupings of houses around greens and courtyards were arranged on the basis of functional necessity to conserve heat, minimize winds and provide access to sunlight and space".[102] In the same vein, Vladimir Matus speaks of "the manipulation of ambient energy", which is to be performed in the interplay of urban buildings.[103] "In high-density urban areas, every building plays two important roles. First, it provides a protected internal environment—a stable thermal milieu within the narrow limits of the human comfort zone. Second, it is responsible, to a great extent, for the creation of the urban microclimate."[104]

While traditional solar-house research hasn't been concerned with the neighboring built environment, in the case of dense urban structures the built context requires a profound analysis. Urban architecture, which refers to climatic factors such as the sun or the winds, entails the examination of its contexts. Among other things, theoretical efforts had to be made to reflect the architectural epistemology of groups of buildings and, from a technological point of view, their energy performance. The notion of "collective form" prepared, as I would like to show, the ground for those architects examining *energy synergy* within groups of buildings. "The aesthetics of the collective form necessitate new definitions of scale and proportion of buildings."[105]

6.2.1 Energy in Collective Form

The notion of collective form set a new "relational" order, beyond architecture's traditional focus on objects.[106] According to the definition by architectural theorist Fumihiko Maki, "collective form represents groups of buildings and quasi-buildings—the segment of our cities. Collective form is [...] not a collection of unrelated, separate buildings, but of buildings that have reasons to be together."[107] No less than the fundamentals of architectural theory were questioned by this definition, relating architecture to urban environments. According to Maki, it is the modern city that challenges the scope and scale of traditional architectural theory. At the beginning of the 1960s, there was a fundamental lack in reflecting the growth in scale. "The theory of architecture has evolved through one issue as to how one can create perfect single buildings whatever they are. A striking fact against this phenomenon is that there is almost a complete absence of any coherent theory beyond the one of single buildings. We have so long accustomed ourselves to conceiving of buildings as separate entities that, today we suffer from an inadequacy of spatial language to make meaningful environment."[108] The key problem of conceptualizing architectural theory lies in fact in transcending the scale of the individual building and conceiving architecture as part of larger urban aggregates. "Investigation of the collective form is important because it forces us to reexamine the entire theory and vocabulary of architecture, the one of single buildings."[109] If Charles Waldheim conceives of "infrastructure and ecology as nonfigurative drivers of urban form",[110] the collective form, as outlined by Maki, can be seen as a figurative driver of urban infrastructure and ecology.

→ fig. 212 Energy-Synergy

The Urban Ethnography of Collective Forms

Introducing the notion of collective form, Maki formalized the comparative urban analysis of collective housing, as it had been developed by the Modern Movement in the context of CIAM 2 and 4; the CIAM programs had already anticipated the growth in scale of analysis, from the individual building to the collective urban form.[111] However, as the architectural historian Manfred Speidel pointed out, Japanese architectural ethnography played a significant role for a broader understanding of the notion of collective form.[112] Maki's research was part of a larger development in Japanese architecture analyzing traditional village structures. In a significant way, the modern metropolis (in Japan) evolved out of vernacular village structures: There was a thorough interest in the everyday behavior of people and the methodological implications of deducing design strategies from an ethnographical description. In the case of his design for the University Seminar Houses in Hachioji (Tokyo) in 1964, Takamasa Yoshizaka speaks of "group organization and physical structure".[113] Here, vernacular group forms were taken as a starting point. Yoshizaka called his approach "method with discoveries"—that is, first, exploring the behavior of people and their place, and, second, translating it into a contemporary design proposal.

Maki shared his interest in the *relational logic* of urban growth with other architects and theorists of the first half of the 1960s, conceiving of architecture as part of larger wholes: think of Peter and Alison Smithson's notion of "conglomerate ordering", Kevin Lynch's notion of "city form", Richard Buckminster Fuller's notion of "synergy" or Aldo Rossi idea of "typology". Kevin Lynch's *The Image of the City* (1960) with his contribution to large-scale design theory helped to establish the term "urban design" in contrast to urban planning, promoting foremost "urban form" and thus the formal aspects of designing the city.[114] Buckminster Fuller explored the implications of synergy as a design method to further develop the architecture–energy relationship. There is a surplus given by the collective form that is not achieved by simply assembling the elements without considering the whole. Buildings should be conceived as elements of larger urban systems. According to Peter Smithson, "it is useless to consider the house except as a part of a community owing to the inter-action of these on each other".[115] The Smithsons used the *valley section* of Patrick Geddes to show the interdependency of the house with its habitat and environment. In the middle of the 20th century, this led to the development of the concept of "habitat" in architecture by Team X and others.

Aldo Rossi reacted to advancing scientification of the design disciplines with his design methodology that conceived the architectural space of the city as a space of "collective memory" (see chapter 4). The "design" and the "analysis" were closely related, which culminated in the notion of "typology."[116] He speaks, for instance, of "the typology of the courtyard and the portico house as composite elements of the collective residential building".[117] According to Miroslav Šik, "coherent ground plan photographs" of old towns and city districts formed the "basis for analysis and design."[118] In this context, the research carried out in Ticino (Switzerland) in the 1970s is noteworthy, translating the theoretical approaches into a kind of architectural ethnography of Ticino.[119] Among other things, typological building surveys of 18 villages were made, also bringing to light the thermal qualities of aggregated units. Rossi's "typologies" (relying on collective memory) appear as a mirror of "collective form" as developed by Maki.

Although Maki was only vaguely concerned with energy, he was an important forerunner on energy issues in urban design. According to Maki, "society" must be conceived of as an "urban society",[120] requiring new heuristic instruments; urban communities are "heterogeneous" and under continuous transformation.[121] Maki conceived of the city as a "dynamic equilibrium" that is constantly threatened with dissolution and which needs new ways of conceptual integration.[122] Flows of energy—or, more generally speaking, "circulation dynamics"—become the generator of urban form. Maki focused on the "forces" as they were described by sociologists, economists, and writers. He speaks of "the specific formal result of forces in the city".[123] His definition of the city as "a dynamic field of interrelated forces"[124] already strongly resembles the definitions of ecology, as highlighted a few years later by McHarg and others. Although not yet contributing to the project of what Lewis Mumford called "ecological design,"[125] he was already reflecting the city shaped by the collective. For Maki, architecture is the central agency for the mediation of the urban whole. "We lack an adequate visual language to cope with the superhuman scale of modern highway systems and with views from airplanes. The visual and physical concepts at our disposal have to do with single buildings, and with closed compositional means for organizing them."[126]

Designed Collective Forms

Maki raised the problem of *intentionally designed collective forms* in contradistinction to historically emerged forms.[127] According to Maki's approach,

→ figs. 214, 220   Global Adaptations after 1945   **198** there are three fundamental manifestations of the collective form, representing the city and its growth: the "compositional form", the "megaform" and the "group form." The character of the collective form is defined by its intrinsic "linkages". The rules of growth and the rules of the elements of a collective form are reciprocal.[128] As Maki's understanding of the "megaform" shows, energy can be a critical linkage for a collective form.

Mega(structure)forms

More explicitly than in the other two types of collective form, energy plays a central role in megaforms. It is because of the infrastructural nature of the mega(-structure)forms that energy supply is part of their conception. The mega(structure)form is both a load-bearing structure and a building-services technology. The collaboration of structural and environmental engineers opens new possibilities: "Large scale climatic control will be studied further. A new type of physical structure, environmental building, will emerge."[129] The appropriate mixture of the different subsystems results in what Maki calls "an environmental control system."[130]

Maki was not alone in developing a climatic approach. With his project *The Agricultural City* (1960), Kisho Kurokawa launched a mode of the city in which natural and built elements were fundamentally intertwined. Against the background of the replacement of the Japanese agricultural city of Aichi, which was destroyed by the Ise Bay Typhoon in 1959, a year later Kurokawa developed his design for a city on stilts, the so-called *Agricultural City*. This was intended to provide security against future flooding by lifting the city off the ground. Based on a basic 500 m square grid, a megastructure was developed to accommodate the new city; it comprised 25 (100 m square) blocks, each accommodating 200 people. The new city was characterized by its arrangement four meters above the ground, where all public buildings and all infrastructures for water, electricity and public and private transport (such as the railway lines) were accommodated. The agricultural area on the other hand was located below the new city, which was therefore characterized by simple and short access routes. The microclimates above ground level were shaped by agriculture and recurrent flooding. The residential parts of the city consisted of individual units in the form of mushrooms with 1 to 3 stories, punctually penetrating the new grid. The infrastructural core of the mushrooms made of reinforced concrete, to which the apartments were attached, provided water, electricity and gas; accordingly, the serving spaces such as

kitchens and bathrooms were directly adjacent to them. The mega(structure)form relied on the superordinate grid that bound everything together.

With this double coding of the city on two levels with two ecologies, Kurokawa introduced a totalizing understanding of productivity in the city. Kurokawa was concerned with transcending the widespread urban–rural antagonism in favor of a new type of urbanization; he spoke of "rural communities" that "are cities whose means of production is agriculture." "It seems to me that there exists a city versus village concept with an emphasis toward cities. We say 'the flow of agricultural population into cities' or 'dispersion of urban population.' I am of the opinion that rural communities are cities whose means of production is agriculture. Agricultural cities, industrial cities, consumption cities and recreation cities should each form an integral part of a compact community."[131] The new city combined town and countryside, urban and rural ways of life—leading to an "agri-industrial way of life," using this term coined by Percival and Paul Goodman.[132] More explicitly than in the other two types of collective form, mega(structure)forms rely on different speeds of renewal. Kenzo Tange highlighted a number of different cycles, anchored in modern life. He spoke of "short-lived items" and "large-scale operations" in order to illustrate the different scales and speeds implied by mega(structure)forms.[133] Concepts of transformability are required for this. Maki emphasized the idea "of several independent systems that can expand and contract with the least disturbance."[134]

Another exemplary mega(structure)form is the *Bridge City across the English Channel*, elaborated by Eckhard Schulze-Fielitz and Yona Friedman in 1963. This bridge city exemplified the *Infra-Structure* envisioned by Friedman and Schulze-Fielitz, which was open to in-fills of various kinds. The gigantic spatial structure of the bridge city had eight floors, where the different uses—roadways for car and train traffic, storage rooms, offices, pipelines, tourist and the actual residential facilities—were accommodated. The bridge city was not only a transit space and common port for the two metropolises of London and Paris; it was also intended as an urban extension, housing 30,000 people. The water below the city can be conceived of great significance for the microclimatic conditions. The concept of the mega(structure)form can also be seen as an attempt to introduce a broader understanding of climate control on the urban scale, which can create different microclimates. The mega(-structure)form represents a framework that ensures the basic supply of a city. The territorial scale of the

→ figs. 215, 216   Energy-Synergy   **199**

215
"Collective Form" according to Fumihiko Maki (from left to right): "Compositional Form", "Mega-Structure Form", and "Group-Form", 1964.

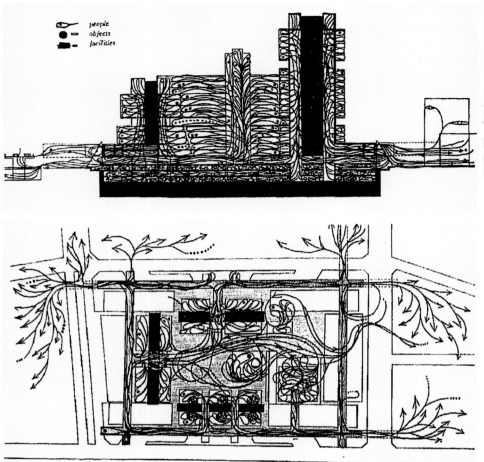

214
Fumihiko Maki, section and footprint of Dojima Redevelopment Project, Osaka (Japan), 1962. The movement diagrams are drawn by graphic designer Kiyoshi Awazu.

216
Kisho Kurokawa, section and perspective view of the *Agricultural City*, 1960.

Global Adaptations after 1945 — 200

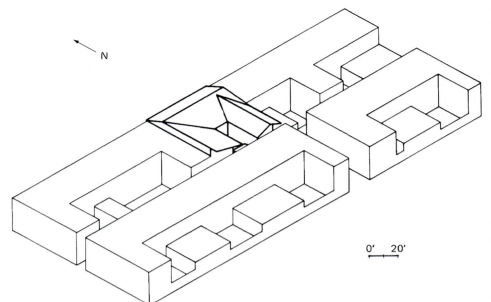

217
Isometric of a block of the Greek city of Olynthus, taking advantage of the south orientation of the court. Source: Knowles 1974.

218
"The relationship between shadows and insolation at street level in different street grid layouts." Source: Matus 1988.

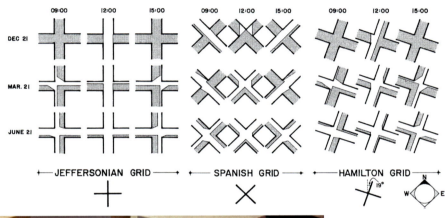

219
Ralph Knowles, design study on the equal solar access of buildings in Southpark, Los Angeles. USC School of Architecture, Solar Studio, 1981.

Energy-Synergy

220
Aldo Rossi et al., village structures and their implications for (passive) climate control, Ticino (Switzerland).

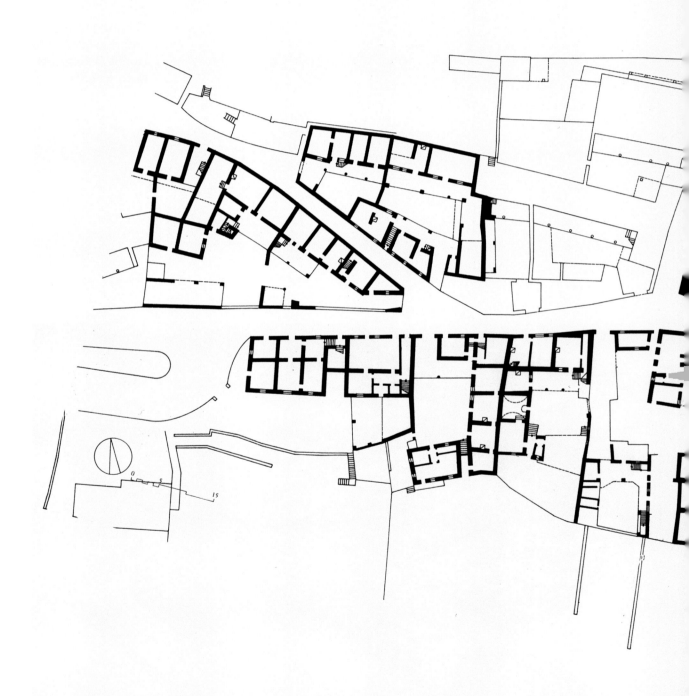

megaform is one of its main features. This is the reason why Maki considered the mega(structure)form as a feature of the territory and thus as an artificial landscape. "The megastructure is a large frame in which all the functions of a city or part of a city are housed. […] In a sense, it is a man-made feature of the landscape. It is like the great hill on which Italian towns were built."[135]

Group and Compositional Forms

Only by considering architecture as part of collective forms can urban climates (and their energy-related potential) be examined properly. In this sense, Ralph Knowles in the 1970s systematically examined the *geometry of the city* regarding its potentials for solar access and for improved energy performance, also considering more broadly the political aspects of passive climate control in the city. "The act of making linkage" was the main challenge with regard to the use of urban energy.[136] Knowles placed energy (as a "linkage") at the center of the collective form; he spoke of "building forms and their interactions".[137] "If we are willing to organize the increments, rates and directions of that change around a clear purpose, cities like Los Angeles and New York could, within a decade, be transformed into energy-conserving systems. Our cities are being transformed anyway. Can we get more benefit from that transformation than higher land values and more office space?"[138]

Referring to Greek settlements and American pueblos as models of ecological urban growth (see chapter 2), Knowles launched a comprehensive reflection on the energy-related implications of Maki's notion of collective form. Based on an understanding of natural "forces" shaping sites and buildings, Knowles investigated larger urban ecologies and the dependency of the settlements on these natural and man-made environments. "The purpose of this case study was to develop ecological frameworks within which buildings could be formed and located to minimize the energy cost of maintaining their equilibrium in a natural environment of cyclic forces."[139] Among other things, Knowles focused on the "growth" of the settlements and highlighted in particular the need to coordinate "energy and form:"

"Buildings in modern cities have tremendous dimensions when compared to the Greek house and the Acoma pueblo. They also interact; and because of their size and proximity the interaction occurs over long periods of time. But neither form nor location are planned for their considerable potential as responses to the natural environment. Building forms and their

interactions may be simple or complex but the purpose for their forms and their interactions is generally set from the developer's point of view. The form of a modern building is most often limited by the commercial value of the land, by the limits of the site, and by any code restrictions having to do with how much the site may be covered or possible restrictions governing setback. The way buildings interact with each other is usually not planned and remains a matter of circumstance that results from a rapid rate of change based on short-term profit. But form and the interactions resulting from location, seen in the pueblo studies, can be used to mitigate the effects of daily and seasonal insolation. In modern building, they could replace some of the need for mechanical support systems."[140]

Knowles's concept of the "solar envelope" included (like zoning plans) specifications for sustainable growth as well as the aesthetic potential for architecture and urban design.[141] What was "graphed" is "the impact of sunlight on form".[142] "The solar envelope […] regulates development within imaginary boundaries derived from the sun's relative motion. Buildings within this container will not overshadow their surroundings during critical periods of solar access for passive and low-energy architecture. If generally applied as an instrument of zoning, the solar envelope will not only provide for sustainable growth but will open new aesthetic possibilities for architecture and urban design."[143] The solar heat gains have to be transformed into an architecture of the collective form that avoids excessive overheating and stores the usable heat. Or, in the words of Vladimir Matus, "The major force behind the weather engine and its climatic elements is the sun": "it is very important in shaping the urban form and urban spaces (urban masses and voids)."[144] The theories of ecological urban growth emphasized methodological aspects rather than case studies and aesthetic scrutiny. The "solar envelope" as elaborated by Knowles derived from (visual) form rather than (tactile) materiality. Although materialization was a key issue for the architects' thermal concepts, Knowles's approaches had a clear focus on geometry that seems to disregard material aspects. Although he emphasized the aesthetic dimensions of his comprehensive investigations, they clearly foreground rigorous methodology and, associated with it, new regulation.[145] Heliomorphic design introduced the scientific rationale of the *method* in design processes relying on the new practice of *design research* (see chapter 3).

→ figs. 217, 218   Global Adaptations after 1945

Already in the 1920s and 30s in New York, Henry Nicolls Wright (1878–1936) had anticipated important considerations on heliomorphic design as later elaborated by Knowles. Using proto-empirical methods in the laboratory, Wright further developed the geometrical studies of William Atkinson and anticipated the scientific approach of simulation. He took the speculative "subdivision" of land as a starting point for re-thinking the design of larger planning units. "Land, road, utilities, grading, house were all one—must be conceived of as an integrated unit to serve living, not selling."[146] Influenced by Ebenezer Howard's *Garden Cities of Tomorrow* (1902) and Raymond Unwin's *Town Planning in Practice* (1909), Wright was aware that "the slow, unrelated process of land subdivision and speculation, followed by scattered building of houses, was inexcusably wasteful and antagonistic to the attainment of desirable housing."[147] His approach explained the systematic interest in climatic aspects of urbanization, which he pursued especially toward the end of his life—such as the planning of Sunnyside Gardens in Queens, New York (1924–29), Radburn in New Jersey, and Chatham Village. Wright investigated how solar radiation in dwellings was related to summer cooling and winter radiation.[148] He worked in the early 1930s on a project called "Hillside Group Housing", which opened up new approaches to modern terrace construction.[149] Philip Steadman speaks of "heliothermic site planning by Henry Nicolls Wright".[150]

6.2.2 Regulating Urban Climates (as Energy Sources)

From the beginning, the central goal of urban climatology was to develop scientific methods for understanding the thermal conditions of cities. In doing so, the methods were in a field of tension between the actual and the target state: It was, after all, the central insight of urban climatology that the microclimates of the city are to a large extent man-made and that the climate should therefore be described after any architectural intervention.[151] In this respect, the statement by the British geographer Charles Ernest Brooks "that not only the existing climate is to be considered, but what it will be after building is completed"[152] expresses the central epistemological prerequisite behind the notion of city climate from a design point of view. In particular, the tool of urban climatic maps has the great disadvantage of merely depicting the current state without providing the architect with a tool indicating the microclimates following the design processes.[153] *Mapping* (measurements) has therefore to

be complemented by another fundamental skill: *simulation*. Only simulation would be able to scientifically fathom the effects of architectural–urbanistic interventions.

Simulating Urban Environments

In the second half of the 1960s, a new form of climate-related, empirical research emerged at American universities that transcended previous work with the solarometer. This new form of empiricism had an urban, even territorial, focus; it took place in the field as well as in the laboratory. While the Olgyay brothers produced primarily passive strategies for individual buildings, the ecologically oriented microclimatology has led to novel passive concepts for urban energy landscapes of the post-industrial society. For this purpose, urban laboratories were established.

In these new laboratories, the effects of urban constellations on urban climatic conditions were examined. MIT's Laboratory in Microclimate Studies was not the only laboratory specializing in an urban scale. Rather, it must be seen in the context of other examples such as the Berkeley Environmental Simulation Laboratory (founded in the late 1960s and initially led by Donald Appleyard) and the USC Natural Forces Laboratory in Los Angeles, founded and led since 1967 by Ralph Knowles.[154] Thus, there was an actual tendency at that time towards empirical–simulative investigation of the natural play of forces for the purpose of an ecologically rational conception of urban landscapes, as well as an exchange between authoritative American institutions.[155] It is important to emphasize that these laboratories were run by architects and landscape architects, among others, and not only by physicists and engineers, which led to a possible transfer of knowledge between natural science and architectural production. Ralph Knowles represented this transfer in an exemplary manner.

Lisa Heschong, in particular, embodied the conceptual developments by means of a biographical inter-disciplinarity. She was employed as an assistant at MIT's Laboratory in Microclimate Studies in the first half of the 1970s, while still a graduate student; there, she was part of an interdisciplinary team of landscape architects, architects and engineers who used new empirical methods to support the planning of large-scale territories. Because of her hybrid educational background, Heschong was predisposed to subject the scientific concepts of microclimate to a cultural–architectural transformation. Under the guidance of landscape architect Terry Schnadelbach and architect Richard Brittain, Heschong collaborated on an

→ figs. 222, 223, 224   Energy-Synergy

205

early ecological housing project in *Pequannock Watershed* in New Jersey, among other projects. Brittain was responsible for studying the microclimates and Schnadelbach for examining geomorphology and plant populations, while Heschong herself conducted the "ecological survey"—i.e. her task was "to describe the biological environment that they were going to put this housing in".[156] While the laboratory primarily developed empirical planning principles for large-scale housing projects, in addition undergraduate students at MIT were introduced to new simulation techniques and new forms of ecological study of the territory (wind-tunneling and water tables); visualized fluid dynamics were used to study the effects of individual structural measures on the microclimates of a landscape.[157]

The establishment of such laboratories at major institutions "responded to two important trends that emerged in the late 1960s: the growing importance of citizen participation and an early wave of concern for the environment [...]. Both developments led to a regulatory framework, the National Environmental Policy Act of 1969." The mandate for the new labs, therefore, "was to develop methods to measure and communicate the effects of urbanization on cities and landscapes".[158] Thus, the driver for the emergence of university laboratories from the 1960s onwards was the requirement to translate everyday experiences into planning-communication processes. Users were to be involved in planning processes through simulations. Scientific and governmental interest in the urban climate thus led to an experimental space of experience, which was enriched by a civil-society perspective. The methodologization and technologization of the urban climate was thus oriented towards the sensory experience of the users. With some justification, such an enriched experiential space can be circumscribed by the notion of the *living lab*. The idea was to combine scientific findings with their direct application in real-life situations through appropriate feedback.

With the construction of wind tunnels and, later, with the development of simulation software for computers, the basic problem of simulating urban climatic conditions became apparent. The abstraction of the model raised doubts about whether it had any predictive power regarding actual conditions. Both the physical city model in the wind tunnel and the virtual 3D model involved dramatic simplifications, which often hardly exceed the foreseeable. Nonetheless, simulation remained the only method that allowed designers of urban microclimates an evidence-based check on their design assumptions. The new laboratories, as they emerged, especially in the USA, were

situated in the field of tension between technological autonomy and reality-based validity. The university-based research approach made its integration into the design processes of architecture and landscape architecture a great challenge. The use of wind models was expensive and time-consuming, and difficult to reconcile with the rather spontaneous approach in design processes. In addition, there was a clear focus on geometry; materiality, on the other hand, and thus materialization-related thermal conditions were much more difficult to fathom in simulation. However, in the history of urban climatological knowledge, cities and wind tunnels are epistemically interdependent. Not only have urban designers used wind tunnels since the 1960s to simulate urban climatological phenomena and anticipate structural solutions; the establishment of these laboratories consolidated an idea of the city as a perfectly air-conditioned interior space.[159] The city itself thus became the experimental counterpart to the model-based simulation, allowing conclusions to be drawn about the epistemic nature of the laboratory-based wind tunnels. The urban environment itself became a 1:1 model laboratory, which could also be used for testing and establishing desired thermal regimes via new spatial constellations.

Governing Solar Allocation

According to Ian McHarg, no new science was needed for applying ecological design. However, the new methodology of simulation played a critical role in assessing the permissive or prohibitive character of urban environments. Simulations were of fundamental significance not only for finding appropriate design solutions but also for a new form of thermal governance. McHarg had already recognized the need for the regulation of land use, should the natural forces of the city be moderated and their energies extracted. "There is a need for simple regulations, which ensure that society protects the values of natural processes and is itself protected."[160] The form of growth is dependent on a "shared resource management".[161] The usage of energy commons must be regulated and stipulated. "We can initially describe the major natural processes and their interactions and thereafter establish the degree to which these are permissive or prohibitive to certain land uses."[162] McHarg addressed "natural processes" as "public goods" and the "responsibility" of the "landowner" and "developer" for the transformation of natural processes in the city. "It seems clear that laws pertaining to land use and development need to be elaborated to reflect the public costs and consequences of private action. Present land-use regulations neither recognize natural processes—the public good in terms of flood,

→ fig. 221 Global Adaptations after 1945

drought, water quality, agriculture, amenity or recreational potential—nor allocate responsibility to the acts of landowner or developer."[163]

The protagonists of alternative energy resources recognized the need to develop legal solutions for the dissemination and use of solar energy. For this, the legalization of so-called "solar access" was critical. The sun and associated technologies "are framed as common pool resources".[164] With his "concept of solar envelop zoning,"[165] Ralph Knowles tried to introduce "solar access as a zoning criterion".[166] Knowles's request "to govern solar allocation"[167] was an expression of society's need for coordination in order to generate the best performance of buildings and the highest comfort of outdoor spaces: "To use the sun, you must have access to it. To have access there must be a public recognition of the right of access—a solar policy. Access can be achieved by private design or development initiatives, but a commitment expressed as public policy best assures access on an equitable, comprehensive, and long-term basis. The process of developing a public policy also provides a forum for public expression of the value of urban solar access. A solar policy may be based on a modification of traditional zoning principles, resulting in the solar envelope [...]. Since given conditions of climate and land use vary from place to place, the specific solar policy selected for particular locations will also vary. To assure solar access without misunderstanding and conflict among neighbors, the access policy must be fair, unambiguous, and applied evenhandedly. A balance must be struck between the right to develop property and the right of access to the sun."[168]

The politicization of the collective form was a prerequisite for potentially overcoming the dominance of private property as the basis of urban development. In this sense, Norman Coplan wrote in *Progressive Architecture* in 1981, "The development of solar energy as an alternative source of energy may be dependent in part upon the resolution of legal problems which relate to solar access."[169] The decisive step was to be taken where the classical means of bioclimatic urban design were combined with the modern "instruments of zoning".[170] "Great benefit can be derived from an understanding of the principles of location and form, which are the keys to solar design methodology. Although the orientation of future design to assure solar access will be a significant contribution to development of the use of solar energy as an alternative source, the problem will nevertheless remain of how solar access may be promoted or guaranteed, under our prevailing legal framework, in connection with ex-

isting structures as well as in respect to future development."[171]

The collective form needs coordination between the different agents of urban development. Although initiated by design, form as collective form is ultimately political and regulated by legislation. While every landlord, directly or indirectly, tries to control shading from neighboring buildings, an urban community has to guarantee a superordinate perspective on the interplay of buildings and to ensure the synergies emerging out of this interplay.[172] Although ambient energy can be conceived of as the central link of the collective form, higher-level attempts at integration are necessary—encompassing the spatial, social, as well as the legal dimensions of energy-related processes.[173] The real significance of the debates around 1980 lies in the systematization of legal and parametric questions of architecture, which in turn referred back to the heliomorphic-design tradition (see chapter 3).

Climate and Energy

The theorists of man-made climate of the interwar period opened up a space of thought that would be further differentiated into *ecological urbanism* in the second half of the 20th century.[174] Two aspects, however, distinguished the approaches of the 1970s and early 1980s from the previous attempts at heliomorphic design carried out in the German-speaking countries in the interwar period: on the one hand, they aimed at an explicit legal framework of solar access (incorporating a theory of the commons), and on the other hand, they aimed at a quantitative description of the energy performance in and between buildings. The main achievement of heliomorphic design in the late 20th century was to have recognized and outlined these two objectives without, however, bringing urban societies closer to them.[175] Simulations neither made it possible to quantitatively assess the impact of design decisions on performance nor succeeded in providing urban energies within an appropriate regulatory framework. The question of how a conscious approach to the urban climate can contribute substantially to energy savings remains unresolved to this day. While investigations into the "collective form" led to the epistemological embedding of architecture in larger groups of buildings, the investigation of the *nature of the city* promoted the ecological embedding of the urban climate in superordinate energy systems. The problem of the production, transmission and storage of energy appeared differently in the light of viewing architecture as a result of the developmental processes of collective forms;

→ fig. 219　　　　　Energy-Synergy

221
Berms lift the South wind over houses. Wind tunnel tests in the laboratory, MIT.

222
Environmental Simulation Laboratory, UC Berkeley (USA).

223
Smoke simulations in the wind tunnel, Colorado State University (USA).

224
Exploring the wind profile of downtown San Francisco (USA). Environmental Simulation Laboratory, UC Berkeley.

Energy-Synergy

architecture had to be rethought in light of urbanization processes and urban growth. In the words of Michael Hough in 1984, "As pressures for energy conservation [...] become more urgent, we must look for environmentally sounder ways of manipulating the climate of cities than the present total reliance on technological systems." The central question was "how the exterior environment can usefully [...] conserve the city's energy".[176] A city's climate and territory, and all the buildings found within it, are thus conceived as energy-influencing factors. It was recognized that there are alternatives to the predominant culture of mechanical air conditioning. "The process of keeping cool in summer inside increases temperatures outside—a non-productive transfer of heat from one place to another. There are cheaper and more effective ways of achieving similar results. Since open space comprises a large part of the city environment, it can contribute to the modification of climate."[177] Urban nature appeared as a provider of energy sources and a generator of urban form; the conservation of energy on an urban scale through the integration of all urban systems was a new goal.

The dialectic between urban *climate and energy* became a critical subject of an ecologically oriented urban-design practice of the late 20th century. In his 1981 review of Lisa Heschong's book *Thermal Delight in Architecture*, Doug Kelbaugh interpreted Heschong's "thermal concerns" as mere "metaphor for energy concerns which in turn are the cornerstone of an ecological consciousness".[178] The central issue was the conceptualization of the *time dimension* of the urban climate (related to modern life), and thus of a practice of climate control that systematically takes time into account; Ralph Knowles conceived of climate control in the city as a "space-time construct."[179] He spoke of "an allegory of rhythm as a design strategy for improving the quality of urban life".[180] However, so far there have been only rudimentary experiences with such a collective approach to climate control. Great efforts are still needed to make fruitful this approach relying on a design perspective on urban energies.

Global Adaptations after 1945

Epilogue

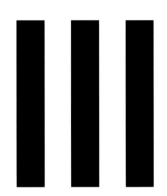

Singapore as a Model?

7   Singapore as a Model?

Urban Climate Control in Practice

Contrary to a widespread historiographical commonplace, 20th-century architecture did not dissolve the boundaries between inside and outside but rather established them in the form of its thinking on comfort and systematically anchored them in the built environment. Due to its endemic overheating, the early glass architecture of modernity was a driver for the spread of air conditioning in the residential sector and in office buildings, and thus an impetus for a dichotomizing way of thinking about the inside and the outside. Previously, buildings had always been thermally connected to their surrounding environment; microclimates of varying degrees formed fluid transitions between inside and outside. Thermal continuity and diversity, rather than the dichotomy of an air-conditioned interior (culture) and a non-controlled exterior (nature), formed the normal case in human history until the middle of the 20th century. By focusing entirely on stabilizing and homogenizing thermal conditions inside buildings, modern comfort thinking failed to further perceive and cultivate the potential richness of architectural experiences of different microclimates. Lewis Mumford very vividly described the aforementioned different historical normality in *The City in History*: "That the medieval house was cold in winter, hardly less in the south than in the north, perhaps accounts for the development of inner rooms, insulated from the outer walls by air, as it surely does for the development of the alcove for the bed, or the curtaining around the bed, to make the enclosed heat of the bodies warm the stale air."[1]

7.1   Climate Change in Southeast Asia

There are various factors which suggest that the greater region of Southeast Asia should be considered as a singular entity. In addition to today's tendencies towards political and economic integration, for a long time it was *climate* that promoted a unifying reading of the region. The climate of Southeast Asia, as the English historian John Villiers noted at the be-

ginning of the 1960s, is "remarkably uniform and characterized by uniform temperatures with high relative humidity, heavy rainfall and regular returns of the monsoon winds".[2] The climate of the region, colloquially known as a "monsoon" climate, is characterized by the rhythmic alternation of a wet summer season (with heavy rainfall from June onwards) and a rather dry winter season (with little rainfall from December onwards), which are separated by inter-monsoon periods. The word monsoon comes from the Arabic *mausim*, which means "season." Arab traders, who dominated the trade between China and Europe in the (European) Middle Ages, linked their economic activities to the course of the monsoon winds, which, depending on the seasons, brought them and their ships from the West to the East and back again. Today, the term *Southeast Asia* encompasses those land masses and island groups that are formed by the countries of Brunei, Cambodia, Indonesia, Laos, Malaysia, Myanmar, the Philippines, Singapore, Thailand, Timor-Leste and Vietnam.[3]

The distinctive feature of the Southeast Asian monsoon, up until the second half of the 20th century, was that it emerged from a territory that was structured by islands and surrounded by water masses, and which almost entirely without exception was covered by rainforest. The monsoon cycles[4] were based on different thermal characteristics of the water and land masses, as well as the tropical vegetation that stabilized the strong winds and absorbed moisture from the extremely humid air. Rainforests of the Malaysian Archipelago were the main thermal and hydrological regulators on land; they provided the natural resources of the indigenous building cultures of the nomads and those who were settled. Apart from the ubiquitous bamboo and rattan, it was teak, in particular, which formed the basic structural material for the numerous ethnic groups of Southeast Asia. In order to counteract the warm humid climate, a culture of passive climate control—now commonly referred to as *natural ventilation*—developed throughout Southeast Asia.

7.1.1   The Agency of Cities

Reconstructions of the forms of ventilation practiced by various Southeast Asian ethnic groups depend today on an analysis of natural-historical and ethnographic records, as exemplified by the legacy of the English naturalist Alfred Russel Wallace (1823–1913)[5] and the American architect Dorothy Pelzer (1915–1972).[6] In the notes made by Wallace and Pelzer, one finds the "classic episteme," as Michel Foucault called them, relating to climate—which are distinguished by the close links between bodies, races, cultures and climate.[7] The climate—and, indirectly, the tropical ecology—is understood as an external "force," which has a more or less deterministic effect on cultural manifestations in general and on domestic construction in particular.[8] Indigenous and, more generally speaking, vernacular construction techniques can be conceived, in the sense of the evolution of construction, as being the result of a precise, albeit pre-scientific observation of the natural environment and the climate.[9] The climate was a transcendent force, to which people were subordinate inasmuch as they adapted to it as exactly as possible by means of their architecture, using the available materials, energetic and human resources as sustainably as possible.

Vernacular Construction

With the help of Pelzer's records, one can still gain an idea of what was once everyday culture in the *kampongs* (villages) and *hamlets* of this region, a central part of which was techniques of natural ventilation. This everyday culture rested upon the interplay of monsoon winds, tropical ecology, vernacular domestic construction and thermally relevant behaviors of the inhabitants, which were oriented both towards the time of year and the time of day. Natural ventilation was a cultural practice that combined *material culture* with the *immaterial knowledge* of a group. Without the relevant knowledge, both nature and artifacts themselves risk becoming meaningless, as Pelzer writes: "My project was a book on traditional house types of Southeast Asia. The most interesting of these houses everywhere were fast becoming lost—built as they were in perishable wood, bamboo, and thatch, in a physical climate taking heavy toll of such materials, and in a mental climate fast abandoning old forms in the rush for imported 'progress'."[10] There are numerous observations in Pelzer's notes and photographs that provide clues as to how the climate was addressed through construction before architectural modernism spread to the region. The construction materials were primarily wood, bamboo and natural fibers, and these had to be selected and prepared with regard to the tropical conditions.[11] The warm humid conditions led, for reasons that are easily explained by building physics, to filigree—i.e. air-permeable and elevated—building structures with comparatively high ceilings and wide projecting roof structures.[12] *Pile construction* and *permeability* formed the two central constructional approaches relating to ventilation in the indigenous house of

Southeast Asia. The ventilation of the houses combined natural and cultural, material (tangible) and immaterial (intangible), object-based and practical aspects into a complex whole; it was based on knowledge that involved both the builders and the users of the houses, and which had needed to be learned, practiced and passed on by them.

While indigenous Southeast Asian domestic construction in the depths of the forested hinterland was almost entirely based on filigree construction (made of long, thin elements),[13] in the coastal cities an urban culture of natural ventilation developed from the 11th century onwards, which was shaped by Chinese and later by European settlers. While the urban culture of massive construction in the case of the "shophouse" of the Chinese settlers produced an almost generic Southeast Asian building type, in the Public Work Departments of the European colonial regimes a climate-adapted architecture developed, which came to be incorporated into architectural historiography as *Tropical Architecture*.[14] These non-mechanized forms of ventilation in buildings influenced the urban development of Southeast Asia particularly between 1860 and 1960. Shophouses are *terraced courtyard houses* in warm humid climates: with their thermodynamic effect, the courtyards of the shophouses form a central point of contact for today's urban culture of natural ventilation in the region.[15] Hassan Fathy distinguished two basic thermodynamic principles for the natural ventilation of courtyard houses; today, both principles are also highly relevant for the natural ventilation of high-rise buildings: "The architectural design can ensure such natural air movement through two principles. In the first, differences in wind velocity produce a pressure differential which results in air flowing from the higher to the lower air pressure region. In the second, air is warmed, causing convection, with the warm air rising and being replaced by cooler air."[16]

The Urban Question

The inclusion of the monsoon climate in the vernacular domestic construction of Southeast Asia has produced a built heritage that, following the anthropologist Roxana Waterson, can be called the *Architecture of Southeast Asia*.[17] In 1983, the Indonesian architect Topane-petra Pandean noted in a dissertation written at the University of Hannover (Germany) that the numerous ethnic groups of Indonesia developed "different house forms"; but these had "relatively identical characteristics with regard to natural ventilation", which was attributable to the "identical climatic conditions" in many regions of the country.[18]

This trans-ethnic finding can still be understood as a starting point and a connecting point for transnational, passive-oriented research into ventilation in Southeast Asia. Natural ventilation as a traditional cooling system is part of a regional heritage, which embodies centuries-old tradition (from pre-colonial to post-colonial times) and culturally connects the different countries of the region. In recent decades, however, as a result of the progressive urbanization of Southeast Asia, this heritage has rapidly lost its relevance.

Accordingly, today the central task for rethinking natural ventilation in architecture is to bring *the critical agency of cities* to the fore. As long as comprehensive urbanization as the dominant social phenomenon of Southeast Asia is not properly taken into account, the formula of "learning from the vernacular,"[19] which is all too often carelessly used, will remain an expression of a Eurocentric projection.[20] The demand for sustainable construction and ways of living is certainly undisputed; passive climate control, however, only becomes an architectural and technical challenge in the context of contemporary large-scale housing projects, in which the need for climatic permeability must be reconciled with numerous other (and contradictory) demands of modern life, including privacy, tranquility, individuality and comfort. A reconnection with forms of natural ventilation must be based on the contemporary urban realities of the region—that is, in settlements which, in European terms, are characterized by extremely high densities as well as by both informal and formal types of urbanization. Contemporary architectural concepts of natural ventilation, in other words, will not become widespread if just used in villas but rather must be employed in horizontally and vertically oriented mass housing.[21] It is in this context that attempts to realize an updated notion of natural ventilation in architecture and urban development have to be measured today, for which the two cities of Medan (Sumatra, Indonesia) and Singapore serve as representatives for many other regions of Southeast Asia.

Although today only a ninety-minute flight apart, the two metropolises show complementary patterns of urbanization, characterized by a strongly divergent economic development; an average annual income is twenty-seven times higher in the *finance metropolis* of Singapore[22] than in the *plantation city* of Medan.[23] In contrast to the poorly regulated housing sector of Medan, Singapore imposes strong political control on its housing sector; Singapore's state housing agency, the "Housing Development Board" (HDB), is responsible for the construction of 85 percent of all housing

units, while around 60 percent of all buildings in Medan are built outside of state regulations. Medan's housing infrastructure is built both informally and by profit-oriented developers, while Singapore's housing program is built by the state, which has far more influence on the forms of cooling used in housing projects. Although they have comparable numbers of residents (Greater Medan: 4.1 million, Singapore: 5.5 million), the two cities differ fundamentally in their densities and urban structure, which are characterized by horizontality in the case of Medan and verticality in the case of Singapore. While Medan's one- to three-story urban structure extends far into the plantations (density: 1,500 inhabitants per km²), the growing island state of Singapore is compelled to build residential towers as tall as 50 stories, characterized by so-called "void decks" and surrounded by open green spaces (density: 8,141 inhabitants per km²).[24] The forms of urbanization of Medan and Singapore, which differ greatly in their political and spatial articulation, lead to divergent challenges for the natural ventilation of interiors.

7.1.2   The Haze as a Symptom

Today, the climate of Southeast Asia represents an increasingly less timeless, unifying framework of historical development; instead, it has become a dynamic entity that has neither a stable rhythm nor a clearly socially defined field of influence.[25] While climate change creates new ecologies with new territorial profiles heralding floods and droughts, the transformation of the tropical ecosystem through deforestation favors new local weather profiles with new wind conditions and levels of precipitation.[26] This is accompanied by a transnationally effective air pollution, which entails a transformation of the once "natural" outdoor climate into an "anthropogenic" urban climate. Concentrations of pollutants consort with city-specific distributions of heat and moisture, from which ever-widening social strata try to protect themselves with face masks when they are outdoors and with new technologies when they are indoors.

A good example of a transnational form of air pollution is the so-called "Haze," which in recent times has begun to cover and paralyze sections of Southeast Asia between May and October. Thousands of fires are ignited by the slash-and-burn agriculture on the plantations of the Indonesian and Malaysian islands Sumatra and Borneo, which, in dry monsoon weather lead to exceptional concentrations of pollutants in urban agglomerations such as Singapore and Kuala

→ figs. 225, 226   Singapore as a Model?   215

Lumpur. While the historical origin of this fire clearing technique—which, meanwhile, is practiced throughout eastern Sumatra—is to be found around Medan, the headquarters of the palm-oil conglomerates ultimately responsible for the fires are located in Singapore and Kuala Lumpur. In both 2013 and 2015, the concentrations of pollutants in the air exceeded, by more than a factor of three, the highest concentrations deemed safe for human health; for months on end, the inhabitants of Singapore and other parts of the region were enveloped in a murky haze—often resembling the romantic fogs common in Europe in November.[27]

As the phenomenon of Haze in Southeast Asia shows, a *political–economic integration* of an old type is increasingly being overlaid by a new type of *political–ecological integration* that short-circuits cities such as Medan and Singapore in a manner that was, until recently, completely unthinkable. The forms of relations between the countries of Southeast Asia, institutionalized today in ASEAN, find a new form of transnational integration in anthropogenic environmental change, the consequences of which are largely unexplored for an urban culture of natural ventilation. The system dynamics on which the new urban climates are based contradict the idea of a human-controlled climate, as suggested by Buckminster Fuller's famous 1960 proposal for a *Dome over Manhattan*, which was typical of such visions of modernity. Rather, cities such as Medan and Singapore experience urban climates in which buildings, green spaces, traffic, industry, aerosols, local weather conditions and global climate phenomena are combined into a hypercomplex system: a new kind of urban interior that has so far been anything but predictable.

Paradoxically, people are forced to ventilate the interior of their homes with the very same outside air from which they want to protect themselves. The unwelcome arrival of the Haze into the urban environments of Southeast Asia is an awkward contradiction of the government of Singapore's vision of creating, by mechanical means, *a stable thermal regime* for its entire population. As in any other city in Southeast Asia, the economic upswing in Singapore since it gained independence has been accompanied by the spread of mechanical air conditioning. On the one hand, the relevance of climate for the design of buildings and their natural ventilation continues to diminish throughout Southeast Asia with the spread of (mechanical) air conditioning; on the other hand, however, a "return of the climate"[28] can be discerned, which accompanies a new "semantics of climate."[29] Rather than an abstract, global, geophysical system, the

climate appears as an "intimate ground-level experience";[30] as a "fulcrum of social action",[31] which manifests itself in urban- and microclimatically relevant strata, and which again becomes more relevant in times of rapidly growing energy consumption.

In the following passage, some of the challenges involved in the architectural implementation of natural ventilation in the cities of Southeast Asia will be discussed: firstly, the emergence of new urban climates, due to global climate change and the transformation of urban ecologies; secondly, the increasing overlapping of passive and active cooling systems, due to the non-regulated use of air conditioning; and, thirdly, the general transformation of construction, due to the fundamental shift from filigree to massive building methods.

7.2 Cooling the Urban Heat

In Southeast Asia, current debates on climate control generally make a strict distinction between natural ventilation and air conditioning. While natural ventilation is typically associated with a lack of comfort but also the value of culture and sustainability, air conditioning is associated with a waste of energy but a surplus of modernity and comfort. However, as I want to show, these are not two separate systems of climate control but only an entangled hybrid one; natural ventilation and air conditioning are superimposed on each other. The simultaneity of different types of ventilation is the empirical reality that planners and architects have to address in the elaboration of their approaches to cooling. The re-appropriation of forms of natural ventilation has to be able to cope with this reality and to find answers as to how natural ventilation can be strengthened without disregarding prevailing lifestyles. As a result of new lifestyles, purely passive forms of ventilation are increasingly difficult to advocate—as are purely active forms of ventilation, due to the accompanying rapid growth in energy consumption.

7.2.1 Singapore's Entangled Ventilation System

In Singapore, the state plays a key role as "provider" and "social engineer" in the field of social housing. By combining all public-housing efforts under one single authority—the so-called Housing Development Board (HDB)—the greatest power over all dimensions of the social housing sector was achieved. Historically, so-

Epilogue

216

cial housing was "the prime mover in the formulation of a national identity,"[32] which can be easily seen by traveling across the Singaporean territory, which is laced with HDB settlements. The visual appearance of this nation is formed by its mass-housing program. Science-fiction author William Gibson remarked that in Singapore, "somehow it's all infrastructure"—which is to be understood against the backdrop of these New Town facilities' sheer presence.[33] According to Gibson, an actual amalgamation of architecture and infrastructure has occurred in Singapore. According to the HDB's *Annual Report 2013*, 83 percent of the resident population lives in HDB apartments and around 95 percent own the apartments they occupy.[34] Gibson has variously pointed to the outstanding importance of graphic novels such as *The Long Tomorrow* for his own imaginative writing about the future city.[35]

Natural Ventilation

Even in the early 1990s, mechanical air conditioning was hardly to be found in Singapore's public-housing sector; natural forms of ventilation were the prevailing means of cooling and dehumidifying dwellings.[36] Accordingly, architects could expect residents to know how to make use of architectural opportunities for cross-ventilation. Over a period of five decades, a Singapore-specific culture of passive climate control developed in its large-scale housing projects, which inherited and perpetuated vernacular forms of ventilation on the Malay Peninsula and in China.

Indeed, popular practices of ventilation and state-driven measures of bioclimatic design in public housing in Singapore have been two sides of the same coin. The vast majority of the approximately 11,000 public-housing buildings are organized into 22 so-called "New Towns." Ventilation was certainly not the only rationale determining the layouts of the settlements and apartments of these new HDB satellite cities. But ventilation was certainly a crucial parameter for the design of the buildings and neighborhoods since the tropical climate of Singapore is rarely bearable in a badly ventilated apartment.[37] Even now HDB apartments are still officially "designed to be naturally ventilated".[38] This basic assumption has far-reaching impacts on the conception of the architectural envelopes as well as the current level of energy consumption in Singapore. Two aspects are crucial for naturally ventilated blocks: 1) their climate-related typologies, 2) the thermal practices of their residents.

On the one hand, urban and architectural modifiers of indoor and outdoor microclimates can be identified in the dwellings. At an urban scale, notable modifiers include the deliberate cultivation of tropical green spaces between the buildings, the two-sided orientation of the buildings, their specific height and orientation, the distances between the buildings and between the void decks—and on an architectural scale, the facade design, the building materials used, the size of windows and solar shading, the volumetric development of buildings, and the floor-plan arrangement of the apartments and their effect on cross-ventilation. However, New Towns from different eras possess distinguishable characteristics in the arrangement of their blocks, which in part relates to different bioclimatic design strategies. It is important to note that there has not been a linear development from sufficiently to insufficiently or from unsatisfactorily to satisfactorily ventilated HDB blocks and apartments. From the 1960s to the 1980s "the nation was built almost only in the East-West direction". 1960s estates were "composed only by linear slab blocks (corridor style) in [the] most common height of 10 stories."[39] Estates of the 1970s were dominated by "big slab blocks in most common heights" of 12 to 16 stories and "point blocks of 20 and 25 stories".[40] Estates of the 1980s were "composed by slab blocks in [the] most common heights" of 10 to 16 stories, "plus 25-story point blocks."[41] "Compared with previous decades, 1980s block corners were bent to give [...] a sense of enclosure."[42] Estates since the 1990s reflect the growing demand for privacy and individuality. There are no corridor-facing units built later than 2004. The new blocks "tend to be around 40-storey[s] high."[43] "Today, HDB blocks amalgamate point- and slab-block designs, featuring taller blocks."[44]

On the other hand, there are behaviors and objects in the vicinity of the dwellings whose microclimatic effects depend on how they are handled by the inhabitants. Though these elements are partly provided by the HDB and designed by architects, the thermal effects broadly depend on way the residents use them. The entrance corridor, for example, is a widespread bioclimatic element that creates shade for the apartments that adjoin it. In many housing projects, it serves not only as a transition space but also as a comparatively cooler outdoor space. Plants along the access corridors, introduced by the inhabitants, absorb the warmed air and provide additional cooling. Metal-grating doors at the entrance to the apartments allow requirements for comfort and safety to be combined. With an apartment door left open, the apartment benefits from cross-ventilation effects.

Singapore as a Model? **217**

Adjustable louvres allow the ventilation of the apartments without having to suffer an "undue loss of privacy."[45] As such, these are used particularly along the main access corridors. Clothes-drying poles attached to the facade are a characteristic feature of Singapore's large-scale residential buildings. The garments hung on them shade the exposed facades and, through evaporation of moisture, contribute to a cooling along the facades.

The task of ventilating large-scale blocks of urban mass housing by natural means is not an easy one and fundamentally different to that of rural settlements. The high-density population of mass housing leads, design-wise, to recurrent trade-offs between privacy and cross-ventilation requirements.[46] Many HDB apartments have difficulties "to cope with noise problems, privacy necessities and thermal comfort for different activities".[47] For example, the access to units along single-sided buildings results in reduced privacy as windows open to the common corridor. Bedrooms, particularly, lack cross-ventilation since doors tend to be closed for privacy during the night. Social change has led to a degree of individualization that is increasingly incompatible with traditional forms of natural ventilation.

Residential Air Conditioning

HDB apartments are, even now, still officially cooled by means of *natural ventilation.* This state-sanctioned basic assumption is increasingly contradicted by the fact that more and more apartments are equipped with mechanical air-conditioning systems. The 30 years from 1978 to 2008 facilitated the necessary conditions for a comprehensive social penetration by the required technology, allowing the proportion of HDB apartments in Singapore with air conditioning to increase by more than 70 percent. Residential air conditioning is thus a very young phenomenon in Singapore. As air conditioning has increasingly infiltrated not only workspaces in factories and offices but also private dwellings in the form of a kind of "domestic furniture",[48] in recent years an all-encompassing regime of "man-made weather" (Willis Carrier) has taken the place of the naturally given (and experienced) climate. Three aspects lie at the center of today's debates concerning residential air conditioning: 1) the productivity of the population, 2) its growing energy consumption and 3) the transformation of the Singaporean everyday culture.

The paradigm of the *control* of indoor climate has supplanted the traditional *adaptation* to outdoor climate, not only in the *working environment.* According

225
Satellite Map of the Southeast Asian Haze, June 2013, showing the fire hotspots in Riau province on the east coast of Sumatra (Indonesia).

Epilogue

Singapore as a Model?

to journalist Cherian George, Singapore had become an "air conditioned nation".[49] For Lee Kuan Yew, its first prime minister, air conditioning was the most important technical invention of the 20th century; like many other commentators, he thought that productivity was directly dependent on climatic conditions. "The humble air-conditioner has changed the lives of people in the tropical regions. [...] Before air-con, mental concentration and with it the quality of work deteriorated as the day got hotter and more humid... Historically, advanced civilisations have flourished in the cooler climates. Now, lifestyles have become comparable to those in temperate zones and civilisation in the tropical zones need no longer lag behind."[50] According to Reyner Banham, "air-conditioning was a way of losing less, or making more, money," inasmuch as the productivity of human labor was increased.[51] "To govern the absolute water content of a body of air"[52] is much more than just a technical problem: it is the cipher of a biopolitics that exposes things and people to uniform temperatures and low humidity levels in order to increase productivity.

In Singapore, energy is mainly derived from fossil fuels. Electricity is produced by co-generation plants using natural gas, as well as imported from Malaysia (relying on similar sources).[53] Although Singapore has one of the highest per capita energy consumptions worldwide, household energy consumption remains moderate in a global comparison; its energy consumption is high because of industry and aviation. Residential energy use comprises electricity (from natural gas) and the direct use of natural gas. In 2012, total residential energy consumption was based on 89 percent electricity with one third of household electricity consumption used by air conditioning.[54] "Over the last years the high demand in the use of air-conditioning [...] in the residential sector has contributed to the increase in energy consumption levels. This is basically due to the low costs of electricity and household A/C systems allied to a social lifestyle change, demanding better comfort levels. Sales of A/C equipment have increased considerably over the past few years [...] especially small packaged A/C systems that can be easily installed by homeowners."[55] Since air conditioning in private dwellings was not planned, Singapore now faces a systematic challenge due to the peak load that occurs when people return from work, threatening to overwhelm the electric grid. However, by entering the privacy of people's apartments, air conditioning transformed the public discourse once more "from efficient production to human comfort."[56] With the proliferation and affordability of residential air conditioning units, the technology changed its character from an amenity to a necessity.

The proliferation of residential air conditioning has had extensive impacts on the everyday culture of Singapore. As per a statement by the historian Raymond Arsenault, "in varying degrees virtually all" its residents "have been affected, directly or indirectly, by the technology of climate control. Air conditioning has changed ... [their] way of life, influencing everything from architecture to sleeping habits."[57] Arsenault speaks of an "air-cooled privatism," a phenomenon with far-reaching consequences for public urban spaces.[58] "Over the past [...] years, extraordinarily varied methods of living with heat, of calibrating clothing, of adjusting social and seasonal rhythms and fine tuning the built environment have been eroded and increasingly replaced by a resource-intensive [...] culture of mechanical cooling."[59] Together with man-made outdoor microclimates, the increasing spread of air conditioning throughout Southeast Asia favors forms of living in the interior of buildings and leads to a depreciation of outdoor space.

Hybrid Ventilation Systems

Natural ventilation and air conditioning presuppose two completely different design strategies: in theory, air-conditioned buildings require airtight envelopes whereas naturally ventilated buildings require permeable ones. A passively ventilated building relies on as much atmospheric exchange with its surrounding environment as possible, while the mechanically cooled building works by creating a temperature difference between the inside and the outside. Maintaining this temperature difference requires either insulation or a constant input of energy.[60]

However, natural ventilation and air conditioning are becoming increasingly co-present in HDB apartments. It would be necessary to reconsider the building envelope, both architecturally and in terms of building physics, as it nowadays presents itself increasingly as a semi-permeable compromise. The current architectural answer to the superimposition of the two climate-control systems is, as I would put it, *the semi-permeable envelope*. Architecturally, the semi-permeable envelope reflects the contradictions between the two climate-control systems; regulation-wise, it is stipulated by approving high U-values.[61] HDB buildings are thought of as naturally ventilated despite the fact that 80 percent of households in Singapore have access to air conditioning. In Singapore, there are no regulations for thermal transmittance in buildings that are officially passively ventilated. The only specification for non-air-conditioned buildings is that walls should not exceed 3.5 W/m²K.[62] This is a regulatory level completely unheard of in Europe. The

→ fig. 227   Epilogue

220

majority of regulatory values in Europe range between 0.25 and 0.40. The current *U-value regulation* co-explains and co-legitimizes the high energy consumption of HDB apartments. It is remarkable that the buildings of the HDB are, even today, supposed to be naturally ventilated while in fact they are widely air conditioned. In terms of *housing policies*, natural ventilation continues to be the state's official concept and air conditioning the ventilation system of the individual owner. It is up to the owner to install their air-conditioning units under guidelines provided by the HDB. Only very recently did HDB started to provide ledges for the window units to sit on when mounted—a tacit acknowledgment of the reality.

Providing sustainable cooling systems in Singapore's mass housing is a typical example of a "wicked problem" for planners and architects.[63] In HDBs, the two entangled climate-control systems are full of contradictions, which cannot be resolved by means of conventional bioclimatic design alone. Instead, approaches from the social sciences and cultural studies are needed to shed light on current ventilation practices. There's a "scholarly neglect"[64] in terms of ethnographic description and architectural analysis of the entangled ventilation system of today's HDBs. Such a description and analysis of the current state of ventilation would be the precondition for future architectural concepts and new building regulations. Still outstanding are the innovative concepts that would determine how naturally and artificially ventilated blocks could be realized in a more sustainable combination. Or, to cite Horst Rittel and Melvin Webber, "To find the problem is thus the same thing as finding the solution; the problem can't be defined until the solution has been found. The formulation of a wicked problem is the problem itself! The process of formulating the problem and of conceiving a solution (or re-solution) are identical, since every specification of the problem is a specification of the direction in which a treatment is considered."[65]

7.2.2 Medan's Material Change

The modernist paradigm of comfort—the controlled separation of *inside and outside*, cultural and natural spheres—is questioned throughout Southeast Asia, but especially in Medan. As a result of social change, the climate loses its relevance for the design of buildings; on the other hand, disregard for the climate (and other natural parameters) is reflected in widespread thermal deficiencies of the dwellings of broad swathes of the population. Unlike Singapore, Medan

lacks both comprehensive state-driven climatic measures in the residential sector and ubiquitous air conditioning in the workplace. While the Haze afflicts the Malaysian Peninsula and Singapore only during certain seasons, Medan is affected by high air pollution all year round. According to comparative measurements, Medan is one of the cities in Southeast Asia with the highest concentration of fine particulates in the air. Together with the aforementioned slash-and-burn agriculture, industrially managed plantations and traffic are the main causes of the high concentrations. For a large proportion of Medan's population, thermal sensitivity is manifested as an interaction between high temperatures, humidity, precipitation, dust and aerosols in the air, which are linked to economic distress and inadequate housing infrastructure. In the heat of tropical nights, residents in the outskirts of the city move their mattresses outside because they cannot stand it any longer in the humid interior of their houses.

The Plantation City

In the mid-19th century, the east coast of Sumatra was still "an unknown and inhospitable jungle of no economic significance".[66] Fifty years later, the so-called Plantation Belt along the east coast had become a global center of colonial extraction of raw materials; "rubber, tobacco, oil palm, tea, and fiber, became the five most important plantation crops in East Sumatra both in regard to export value and acreage".[67] "Over the past century North Sumatra has been the site of one of the most intensive and successful pursuits of foreign agricultural enterprise in the Third World", writes the anthropologist Ann Laura Stoler.[68] "Under Dutch rule, the plantations [...] were virtual laboratories for technical and social experimentation."[69]

Since the first tobacco plantations in the mid-1860s, the landscape of East Sumatra underwent a complete transformation. The territory possesses, as Stoler says, a specific "colonial imprint" that is shaped by "the juxtaposition of trees and factories" and the settlements on the edges of plantations.[70] "Subsistence farming and wage labor are part of a single economic system;"[71] centuries-old agricultural technologies (such as "shifting cultivation, swidden agriculture, or slash-and-burn cultivation")[72] exist alongside the intensive industrial use of the land, or are even integrated within it. In the last 30 years, Sumatra has lost "over 50% of its natural forest" through deforestation.[73] The rainforests now are mere remnants between settlements, transport infrastructures and plantations.

→ figs. 228, 229, 230   Singapore as a Model?   **221**

226
Singapore, Changi Airport, glass-domed 10-storey mall extension (2019) with 40-metre-high waterfall.

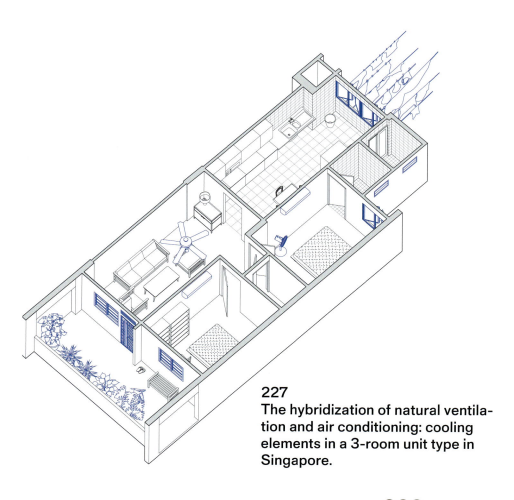

227
The hybridization of natural ventilation and air conditioning: cooling elements in a 3-room unit type in Singapore.

Epilogue 222

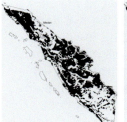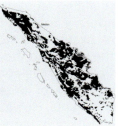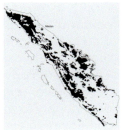

228
Cleared forest and drainage canal under construction at a tobacco plantation. Sumatra (Indonesia), around 1900.

229
Natural forest cover change in Sumatra 1985–2009.

230
Dutch school plate: a tobacco plantation in Deli, Sumatra (Indonesia), around 1945.

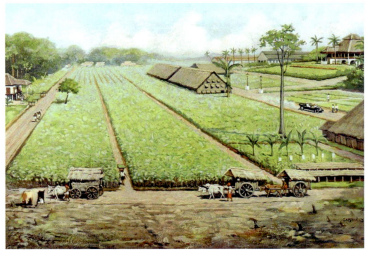

Singapore as a Model? 223

The general shift in construction from wood and bamboo to brick and reinforced concrete has had a profound influence on the *microclimates* in the buildings and the perceived sense of comfort of the inhabitants. Due to the deforestation of the rainforests, the traditional building material, timber, is hardly available and has been replaced by other materials. "Proper kinds of timber having almost disappeared from the forests or become too expensive, they [buildings] are mostly built of concrete and thinly covered with wood or simply painted the colour of wood."[74] Today, brick is not only far cheaper than wood; the large-scale use of high-quality timber is also prohibited so as to protect the remaining forests. Summarizing this development in Indonesia, anthropologist Christian Pelras emphasizes that "the main architectural change occurring nowadays [...] is not evolution but technical change."[75] Pelras situates a real gap in the adaptation of housing to climatic requirements (addressing also the missing agency of the architect): "Suffice it to say that [today's houses] often have features that are much less well suited to the climate than those of the former wooden houses. For instance, instead of having elevated floors, they are built at ground level, and often have no crawl space and may even lack a foundation. Because they are frequently located in areas susceptible to flooding, many of them are flooded every year during the rainy season. Likewise, instead of having natural ventilation, they are hermetically sealed by windows that are seldom opened, so that the air inside is often stifling. Unfortunately, little if any effort has been made to adapt this new kind of architecture to local conditions and lifestyles, or to come up with satisfactory solutions by adapting time-honoured techniques to the new situation."[76]

The middle and upper-middle classes of Medan compensate for these shortcomings by the use of air conditioning, despite the fact that Medan is afflicted by regular power cuts. The Indonesian term *biarpet* describes the recurring switching on and off of the electricity networks that often takes place daily and lasts up to six hours. Medan thus experiences the peak-load effect that Singapore tries to avoid at all costs. Buildings that rely solely on cooling by air conditioning are particularly affected by this phenomenon. As buildings are no longer erected that are designed for cross-ventilation, or fitted with convection openings, those without their own diesel generators (for the constant operation of the air-conditioning system) become, in effect, microclimatic traps.

The territorial transformation of East Sumatra is exemplary of how intensive production and modernization require adjustments of traditional concepts of the natural.[77] Climate, in particular, is greatly affected by this. It no longer suffices to conceive of the climate as an external influencing force. Rather, it represents a hybrid and relational object: part of a system made up of natural and social agents. Of particular interest here is "the reordering of the natural environment"[78] through the plantation economy, deforestation and modernization in general—and the implications of this reformation of the territory for a contemporary architectural epistemology of climate. Through social change, the climate loses its relevance to the design of buildings, while disregard for the climate (and other natural parameters) is reflected in the widespread inadequacies of the dwellings of large parts of the population.

Towards Sustainable Practices of Cooling

The climate of Southeast Asia is changing, and with it the implications for architectural applicability. Obviously, the transformation of natural territories leaves an epistemic void in its wake: the traditional interplay between climate and building is no longer in force, without a new sustainable regime clearly taking the place of the old. The various living conditions in Southeast Asia—from rural to urban, from poor to wealthy—require case-specific answers that take the thermal needs of the groups concerned into account. In the case of Singapore, two suggestions seem to be obvious.

Firstly, in terms of building regulations, at the very center of a contemporary analysis of the two entangled climate-control systems has to be the conceptual gap between the state and the resident–owner. Through the privatization of the apartments, HDB handed over responsibility for large parts of the energy supply to the residents. However, practices of climatization and energy consumption in HDB apartments can no longer be a privatized issue.[79] Sustainability is a public endeavor that needs political and public decisions by means of suitable housing policies and building regulations, a fact that is not only true for Singapore. But especially in Singapore, it would be the government's role to rethink a sustainable ventilation system for urban mass housing.

Secondly, in terms of architecture, it is time to question the concept of semi-permeable envelope. Passively and actively ventilated spatial areas have to be disconnected to a much higher degree than has happened so far. Disentangling the entangled climate-control systems by a clear separation of active and passive ventilation systems speaks for a centralization of the air-conditioning system in HDB buildings

Epilogue

and, above all, a conscious mixed-mode use of active and passive ventilation systems. This implies an onion-skin arrangement of rooms, with cooled chambers in the center and more permeable spaces. The semi-permeable envelope contains permeable and nested spaces, vertical arrangements of spaces and courtyards.[80] A contemporary urban notion of ventilation has to anchor itself within the different ventilation traditions in Southeast Asia, with air conditioning as part of the thermal heritage.

Novel problems require an architectural epistemology of climate that transcends the aforementioned classical (bioclimatic) *episteme*, irrespective of whether they are vernacular or modern. Neither the vernacular notion of *adaptation* nor the modern notion of *control* on their own form strategies that would bring the tropical winds back into a sustainable relationship with today's domestic construction in Southeast Asia. Relying on new cooling concepts and sensory patterns, the global, carbon-based standard of the 20th century has to be attenuated. The traditional adaptation of indigenous architecture is not suited to provide comprehensive solutions for how the living spaces of large groups of urban populations could be better adjusted to tropical conditions.

It is Bruno Latour's concept of a "political ecology" that appears to be the most appropriate approach to this dilemma.[81] Political ecology is both an empirical survey of the novel interplay between natural and social agencies and the disciplinary project of an epistemic and normative reorganization of the "collective." As Latour puts it, it must be about consciously disentangling things that are at present increasingly superimposed, so that we might assemble them in a new way. For architecture and urban design, this means rethinking the (natural) ventilation of buildings as part of a political ecology of the city. The varied kinds of coexistence between forms of passive and active ventilation suggest the need for new types of experimentation, bringing these two basic strategies together, questioning the homogeneous interior atmosphere.

7.3 The Dialectic of Division

Along with the "mechanization" of architecture, a modern dichotomous metaphysical thinking of inside and outside has taken "command",[82] which nowadays shapes our most advanced concepts of sustainability in architecture. Peter Sloterdijk speaks of the "explication of human residence […] in an interior through

the housing machine, climate design, and environment planning",[83] referring to the transformational effect of modern air-conditioned living on the general understanding of what is worthy of protection as one's own (i.e. the interior). Never before has climate been considered so crucial to the experience of a spatial interiority. Sloterdijk has avowedly based his analysis on the fundamental considerations of the philosopher Gaston Bachelard. In his seminal book *The Poetics of Space* (1958), Bachelard scrutinized the fundamental significance of the terms "inside" and "outside" for our thinking. He speaks of a "dialectic of division" that "outside and inside form"[84] and he criticizes "the obvious geometry which blinds us as soon as we bring it into play in metaphorical domains. It has the sharpness of dialectics of yes and no, which decides everything."[85] Bachelard himself referred to an idea of the philosopher Jean Hyppolite[86] on the true ambiguity of this division. Hyppolite distrusted the real-world relevance of the divide, and he recommended we search for what might lie beyond this dichotomy. "You feel the full significance of this myth of outside and inside in alienation, which is founded on these two terms. Beyond what is expressed in their formal opposition lie alienation and hostility between the two".[87]

Bachelard's reflections on the "dialectic of division" are a timely account of today's alienated situation in the Anthropocene—and of particular relevance for a critical evaluation of modern thermal concepts, which presuppose a sharp division of outside and inside conditions.[88] To conceive of thermal environments through this dichotomy is a widespread but meanwhile contested narrative of 20th-century architecture. While the epistemic division between cultural and natural spheres becomes more and more blurred, a new consciousness of the mutual dependence of inside and outside conditions is growing. Since we are neither fully inside nor outside, we aim to redefine our disordered relationship to climate. The dichotomy of inside and outside, which has risen to the status of an attitude towards life has, I have argued in this publication, lost its conceptual relevance in the 21st century. It is not two spheres thermally separated from each other but the continuities and interactions between the microclimates of the city that need to be conceptualized. A new architectural understanding of the dialectics between inside and outside is urgently needed. If, as the Indian historian Dipesh Chakrabarty writes in his 2009 essay *The Climate of History*, "globalization" and "global warming" emerged from interrelated processes, then the question arises as to how we are to bring these two developments together in our understanding of climate in architecture today.[89]

Following Chakrabarty, thinking about climate means transcending both Eurocentric and anthropocentric constrictions. In times of climate change and man-made urban climate phenomena, architectural theory's focus on the climate of the building interior—exemplary in Sigfried Giedion's *Mechanization Takes Command* (1948) and Peter Reyner Banham's *The Architecture of the Well-tempered Environment* (1969)—is no longer sufficient. This is the reason, I have argued, why the practice of climate control has to be reconsidered in light of urban climate phenomena. As a cultural technique of coping with environmental conditions, its main strength is the ability to *connect*—and not to separate—inside and outside conditions. Climate control must be conceived of as a collective practice helping to readjust our alienated relationship to climate today.

As I have tried to show, only an architectural theory that is both scientifically informed and imaginatively tuned can make the architectural implications of 20th-century urban climatological knowledge fruitful for today's urban development. The approaches to urban climate that I have presented form components of a genealogy of architecture related to urban climate, which makes clear the limited horizon of today's debates on "heat precautions" in cities. It is *climate control on urban scales* that is the still-unacknowledged legacy of urban climatology for architecture, as evidenced by the notion of *man-made climate*. Instead of fighting the geographer's battle as an architect—i.e. fighting heat islands—it is necessary to radically pose the design question (whenever there is talk of urban climate) and cultivate the variations of thermal offerings in the space and time of the city, as the foundations of new forms of democratic urban development. Climate control should no longer be conceived of as a problem of the single building but as a problem of groups of buildings: a collective notion of *climate control* as the fundamental legacy of 20th-century urban climatology. Such an understanding of architecture as aggregated architecture would fundamentally change design in cities. The central approach to reducing the thermal load in cities would be to integrate interior and exterior perspectives, a task which has to be done foremost by the architect.

Appendices

Endnotes
References
Index
About the Author
Acknowledgments
Illustration Credits

Endnotes

I Urban Studies of Man-made Climates, *An Introduction*

1 See Denzer, Anthony. 2013. *The Solar House. Pioneering sustainable design.* New York: Rizzoli; Barber, Daniel. 2016. *A house in the sun. Modern architecture and solar energy in the Cold War.* New York: Oxford University Press.

2 Brezina, Ernst and Wilhelm Schmidt. 1937. *Das künstliche Klima in der Umgebung des Menschen.* Stuttgart: Ferdinand Enke Verlag, p. 184. Translation by the author.

3 See Fumihiko Maki's concept of "collective form" or the Cambridge School with its focus on "the city." Maki, Fumihiko. 1964. *Investigations in Collective Form.* St. Louis: The School of Architecture, Washington University; Hawkes, Dean. 1996. "The Cambridge School and the environmental tradition." In *The Environmental Tradition. Studies in the Architecture of Environment,* edited by Dean Hawkes. London: Routledge, p. 112. "A recurrent concern in the Cambridge School's environmental work has been to move from the study of the individual building to that of the city."

4 The term "man-made climate" originally appeared in the second half of the 19th century in the context of the military (where it was about clothing as a way to provide the body with a man-made climate) and in medicine (where it was about creating a climate conducive to health). Since the second half of the 20th century, the notion of man-made climate (or artificial climate) came to be exclusively associated with mechanical air-conditioning systems.

5 Schmauss, August. 1914. "Meteorologische Grundsätze im Haus- und Städtebau." *Bayerisches Industrie- u. Gewerbeblatt 19:* 181.

6 Kratzer, Albert. 1937. *Das Stadtklima.* Braunschweig: Vieweg & Sohn, p. 4. Translation by the author.

7 Ibid., 4. Translation by the author.

8 Moore, Jason W. 2015. *Capitalism in the Web of Life. Ecology and the Accumulation of Capital.* London: Verso, p. 4.

9 "The notion of 'metabolism' is the central metaphor for Marx's approach to analyzing the dynamic internal relationships between humans and nature that produces socio-natural entanglements." Kaika, Maria and Erik Swyngedouw. 2006. "Urban political ecology, policing the production of urban natures." *In In the Nature of Cities. Urban Political Ecology and the Politics of Urban Metabolism,* edited by Nik Heynen et al. London, New York: Routledge, p. 7.

10 Ibid., 11.

11 "The political meanings of climate and its social implications are due to the fact that the concept itself refers to a hybrid realm comprising land, water, air, living beings, people, and cultural institutions. Climate, in this sense, is paradigmatic in its binding of culture and nature that represents civilization as a result of materiality, contingency, and particularity of place." Fleming, James R. and Vladimir Jankovic. 2011. "Revisiting Klima." In *Klima,* Osiris Vol. 26, p. 10.

12 Important insights are, however, provided by: Jankovic, Vladimir. 2013. "A historical review of urban climatology and the atmospheres of the industrialized world." *WIREs Climate Change /* Horn, Eva. 2016. "Air Conditioning: Taming the Climate as a Dream of Civilization." In *Climates: Architecture and the Planetary Imaginary,* edited by James Graham et al. Columbia Books on Architecture and the City. Zurich: Lars Müller Publishers. / Gissen, David. 2006. "Thermopolis. Conceptualizing Environmental Technologies in the Urban Sphere." *Journal of Architectural Education* 60 (1): 43–53.

13 In this respect, see: Hebbert, Michael, Vladimir Jankovic and Brian Webb (eds.). 2011. *City Weathers. Meteorology and Urban Design 1950–2010.* Manchester: University of Manchester Press.

14 In this sense "energy-conscious planning and building" in cities does not only mean promoting the usage of renewable energy sources, such as wind, but also and in particular analyzing "the characteristic ability of the wind to significantly influence the heating requirements of interior spaces through the transfer [...] of heat". Accessed on October 16, 2018. http://www.staedtebauliche-klimafibel.de/?p=26&p2=3.4 Translation by the author.

15 See the statement of the physician Carl Flügge from 1916: "The terms 'air and light' are unfortunately united as a catchword, which is often misused and characterizes existing hygienic deficiencies quite inadequately." Flügge, Carl. 1916. *Grossstadtwohnungen und Kleinhaussiedlungen in ihrer Einwirkung auf die Volksgesundheit. Eine kritische Erörterung für Ärzte, Verwaltungsbeamte und Baumeister.* Jena: Fischer, p. 75.

16 Hebbert, Michael and Brian Webb. 2012. "Towards a Liveable Urban Climate: Lessons from Stuttgart." *Isocarp* 07: 122.

17 Ludwig Hilberseimer had already criticized the unresolved thermal significance of the sun in modern planning. While "the therapeutic value of sunshine is, of course, undisputed," the thermal effects were much more difficult to evaluate. See Hilberseimer, Ludwig. 1944. *The New City. Principles of Planning.* Chicago: P. Theobald, p. 77.

18 Markus Kilian criticizes the rudimentary reception of Hilberseimer to date and aptly emphasizes the fact "that he published art criticism and art theory on a scale as a text author and that from this position of operating critic he embarked on a specific path in the planning and design of architecture and urban planning. The combination of art theory, architecture, and urban planning united in the person of Hilberseimer, which is unique in this intensity, has so far been insufficiently considered." In addition, Kilian emphasizes Hilberseimer's importance as a planning methodologist: "The contents, which are valuable from a planning and methodological point of view, have so far been obscured by an aesthetic and architectural-historical reception." Kilian, Markus. 2002. *Großstadtarchitektur und New City. Eine planungsmethodische Untersuchung der Stadtplanungsmodelle Ludwig Hilberseimers.* Dissertation, Technische Hochschule Karlsruhe, p. 13 as well as 15. Translation by the author. See also Pommer, Richard et al. 1988. *In the shadow of Mies: Ludwig Hilberseimer, architect, educator, and urban planner.* Chicago: Art Institute of Chicago in association with Rizzoli International Publications.

19 Givoni, Baruch. 1969. *Man, Climate, and Architecture.* Amsterdam: Elsevier / Givoni, Baruch. 1998. *Climate Considerations in Building and Urban Design.* New York: Wiley.

20 Exemplary among the correspondences between meteorologists and urban planners is the 1937 publication by Brezina and Schmidt, which reproduces plans of European tenements originally found in Hermann Josef Stübben's canonical *Der Städtebau.*

21 Graham, James. 2016. "Climatic Imaginaries." In *Climates: Architecture and the Planetary Imaginary,* edited by James Graham et al. Baden: Lars Müller, p. 11.

22 Kähler, Gert. 1985. *Wohnung und Stadt. Hamburg Frankfurt Wien. Modelle sozialen Wohnens in den zwanziger Jahren.* Braunschweig / Wiesbaden: Springer, p. 22. Translation by the author.

23 Lange, Torsten. 2019. "Forschungsgeleitetes Entwerfen." In *Architekturpädagogiken. Ein Glossar.* Edited by Heike Biechteler and Johannes Käferstein. Zurich: Park Books, p. 57. Translation by the author.

24 With their argumentation of "causal connections" between climate and society, the historians of science James Fleming and Vladimir Jankovic renewed an idea that was widespread in the first half of the 20th century. Fleming and Jankovic speak of "climate as a framing device in that the verities of life such as food, wars, housing, economy, social movement, or local identity change synchronously with climate." Fleming and Jankovic (2011, 9).

25 Glacken, Clarence J. 1967. *Traces on the Rhodian Shore. Nature and Culture in Western Thought from Ancient Times to the End of the Eighteenth Century.* Berkeley, CA: University of California Press, p. xi.

26 Ibid., xiii.

27 In the words of Glacken, what is pending are "comparative studies with Indian, Chinese, and Muslim thought" on climatic determinism. Ibid., xii.

1 Thermal Geographies of the European City

1 Kähler (1985, 24).

2 Eve Blau and Ivan Rupnik have pointed to this in the context of (carbon-based) energy production. Blau, Eve with Ivan Rupnik. 2018. *Baku. Oil and Urbanism.* Zurich: Park Books, p. 27.

3 Thomsen, Christian W. 1985. "Die Hure Babylon und ihre Töchter." *Du, Städtephantasien: Architekturutopien in der Literatur* 45 (2): 22.

4 Esser, Katja. 2011. "City Weathers in Aachen during the 1800s," in *City Weathers. Meteorology and Urban Design 1950–2010,* edited by Michael Hebbert, Vladimir Jankovic and Brian Webb. Manchester: University of Manchester Press, p. 99.

5 See, for instance, English Parliament's Report (1844), Friedrich Engels' report on the "Situation of the Working Class in England" (1844/45), Viennese physician Eugen von Philippovich (1894), Truxa's "Aus dem Armenleben Wiens" (1905),

Bruno Schwan's "Wohnungsnot und Wohnungselend in Deutschland" (1929). Data according to Kähler (1985, 24).

6 Giedion, Sigfried. 1928. *Bauen in Frankreich, Bauen in Eisen, Bauen in Eisenbeton,* Leipzig, Berlin: Klinkhardt & Biermann. Giedion, Sigfried. 1941. *Space, Time and Architecture. The Growth of a New Tradition.* Cambridge, MA: The Harvard University Press.

7 Riemann, Gottfried. 1971. "Frühe englische Ingenieurbauten in der Sicht Karl Friedrich Schinkels. Zu einigen Skizzen und Zeichnungen der englischen Reise von 1816." In *Forschungen und Berichte.* Staatliche Museen zu Berlin, Vol. 13, Berlin: Akademie-Verlag, p. 75.

8 July 4, 1826. Schinkel, Karl Friedrich. 1993. *The English Journey. Journal of a Visit to France and Britain in 1826,* edited by David Bindmann and Gottfried Riemann. New Haven: Yale University Press, p. 150.

9 The Derbyshire General Infirmary on London Road, Derby. See ibid., 134.

10 Ibid., 135.

11 June 23, 1826. Ibid., 136.

12 Data according to: Usemann, Klaus W. 1993. *Entwicklung von Heizungs- und Lüftungstechnik zur Wissenschaft. Hermann Rietschel. Leben und Werk.* Munich, Vienna: Springer, p. 24

13 June 10, 1826. Schinkel (1993, 110).

14 "Window glass is permeable to sunlight and heat (but not UV radiation). Objects behind glass heat up almost as much as when they are exposed to direct sunlight. They absorb the heat, the molecules tremble more and the object expands. Finally, they themselves become heat emitters in the infrared range. However, the glass does not allow these long-wave rays to pass through. They are reflected on the inside of the glass. This is how the heat trap is created, which is called the greenhouse effect after its most conspicuous manifestation. The principle applies to all glazing." Althaus, Dirk. 2005. *Bauen heute – Bauen morgen. Architektur an der Schwelle zur postfossilen Zeit.* Berlin: Bauwerk, p. 91. Translation by the author.

15 Accessed on February 9, 2022. http://solarcooking.org/saussure.htm

16 Atkinson studied the temperatures at different levels of sunlight. The two boxes, with one glazed side, were constructed to be as impermeable as possible by inserting insulation material ("flax fiber quilted between two thicknesses of building paper") and a layer of air; inside, a thermometer was installed that could be read from the outside. "A sun box is essentially

a box or chamber of non-heat-conducting material, having on one side a window or light of glass, sealed tight to prevent air leakage. Such a box, when the window is turned toward the sun, will accumulate heat much faster than it is lost by radiation." Atkinson, William. 1912. *The Orientation of Buildings or Planning for Sunlight.* New York: J. Wiley & sons, p. 63.

17 Ibid., 78.

18 Riemann (1971, 77). Translation by the author.

19 June 20, 1826. Schinkel (1993, 128).

20 Ibid., 130.

21 July 17, 1826. Ibid., 175.

22 June 20, 1826. Ibid., 128.

23 The inversion weather creates, as the geographer Gordon Manley writes, an anthropogenic interior beneath the fog. "If smoke is rising from a number of chimneys below the inversion, it cannot escape upward; it will merely spread out horizontally, forming the unpleasant gloomy 'lid' we know so well in our cities, and giving the fog its acrid quality and brownish color." Manley, Gordon, 1949. "Microclimatology – Local Variations of Climate likely to affect the Design and Siting of Buildings", *The Journal of the Royal Institute of British Architects* 56 (7) May 1949 (read at meeting of RIBA on April 12, 1949): 320.

24 June 3, 1826. Schinkel (1993, 89).

25 June 5, 1826. Ibid., 91.

26 The term "black rain" was coined by Jean Paul Grosley in 1765. See Pevsner, Nikolaus. 1956. "The Geography of Art," in *The Englishness of English Art.* London: The Architectural Press, p. 14.

27 Schinkel (1993, 76).

28 Thomsen (1985, 22). Translation by the author.

29 Ibid., 22. Translation by the author.

30 Pevsner (1956, 13).

31 Ibid., 14.

32 "Sheltered valley sites inland may appear attractive but they are particularly subject to fog, and in many districts, not merely in towns, foggy weather is now made much more unpleasant by smoke." Manley (1949, 318).

33 Charles Ernest Brooks commenting on Manley presentation. Ibid., 322.

Endnotes

229

34 "In the planning of industrial areas likely to produce smoke and other forms of pollution during quiet weather or at night careful attention should be paid to the local topography. If possible the emission of noxious smoke should be confined to periods when the air is turbulent and convection is active." Ibid., 320.

35 Kähler (1985, 25). Translation by the author.

36 "In 1900, the German Association for Public Health Care conducted a survey. The result: among all German cities with more than 15,000 inhabitants, a quarter to a fifth suffer from the plague of smoke. In the larger cities, smoke from thousands of chimneys now determines and changes the climate. Soot, constantly trickling down, and fog, vaulting over everything like a dense bell, are part of the fixed experiences of urban life." Spelsberg, Gerd. 1988 (1984). *Rauchplage. Zur Geschichte der Luftverschmutzung.* Cologne: Kölner Volksblatt Verlag, p. 75. Translation by the author.

37 Fassbinder, Horant. 1975. *Berliner Arbeiterviertel 1800–1918. Analysen zum Planen und Bauen 2.* West-Berlin: Verlag für das Studium der Arbeiterbewegung, p. 66. Translation by the author.

38 Fear of the class struggle prevented a comprehensive hygienic examination of the residential and working quarters. See Posener, Julius. 1979–1985. "Vorlesungen zur Geschichte der Neuen Architektur." *arch+ Sonderhefte.* Cited according to Maciéczyk, Alex and Marc Syfrig. 1985. *Das steinerne Berlin, das grüne Berlin.* accompanying publication for the seminar week of Prof. Dolf Schnebli, summer semester, ETH Zurich, p. 10.

39 Hilberseimer, Ludwig. 1927. *Großstadtarchitektur.* Stuttgart: Julius Hoffmann, p. 2. Translation by the author.

40 Fassbinder (1975, 66). Translation by the author.

41 Ibid. Translation by the author.

42 The financing of actual transportation infrastructures for the workforce only became an urgent necessity towards the end of the 19th century in order not to inhibit economic development. Ibid., 62. Translation by the author.

43 See Spelsberg (1988 [1984]) / Stolberg, Michael. 1994. *Ein Recht auf saubere Luft? Umweltkonflikte am Beginn des Industriezeitalters.* Erlangen: Harald Fischer Verlag / Strauss, Sarah and Ben Orlove (eds.). 2003. *Weather, Climate, Culture.* Oxford, New York: Routledge / Uekötter, Frank. 2003. *Von der Rauchplage zur ökologischen Revolution. Eine Geschichte der Luftverschmutzung in Deutschland und den USA 1880–1970.* Essen: Klartext Verlag.

44 Uekötter (2003, 202). Translation by the author.

45 "That the hygienic factory space and all that goes with it, Volkswagen and sports palace, bluntly liquidates metaphysics, would still be indifferent, but that in the social whole they themselves become metaphysics, the ideological process behind which real calamity is contracted, is not indifferent. This is the starting point of our fragments." Horkheimer, Max and Theodor Adorno. 1988 (1947). *Dialektik der Aufklärung. Philosophische Fragmente.* Frankfurt am Main: S. Fischer Verlag, p. 5. Translation by the author.

46 "The quality of a yarn or fabric depends largely on the humidity of the air." Usemann (1993, 30). Translation by the author.

47 "The moist, equable climate of Lancashire provides favorable conditions, and this helps to account for the early development of the cotton industry there." Brooks, Charles Ernest. 1951. Climate in Everyday Life. New York: Van Nostrand, p. 32.

48 Usemann (1993, 230). Translation by the author.

49 "But there is another important thing that the land owes to the nearby sea: the humid sea breezes that condense into precipitation on the hills. After all, the humidity level of the air under the heights is on average only 10% below saturation. This humidity should later make it possible to spin the building wool here to a fineness which, on the other hand, is impossible or can only be achieved at great extra cost. How much this climatic advantage comes into consideration is shown by the fact that the spinning mill increasingly seeks out the leeward side of those hills where the rainfall is greatest, so in particular Oldham instead of Manchester." *Von Schulze-Gävernitz, Gerhard. 1892. Der Großbetrieb, ein wirtschaftlicher und sozialer Fortschritt. Eine Studie auf dem Gebiete der Baumwollindustrie,* Leipzig. Cited in Grundke, Günter. 1955. *Die Bedeutung des Klimas für den industriellen Standort. Eine Studie auf dem Gebiete der Technischen Geographie.* Gotha: VEB Geographisch-kartographische Anstalt, p. 108. Translation by the author.

50 Grundke (1955, 1). Translation by the author.

51 Brooks mentions the "katathermometer" as "suitable for comparing conditions in different parts of a factory". Brooks (1951, 27).

52 Grundke (1955, 1). Translation by the author.

53 Ibid. Translation by the author.

54 On the subject of industrial buildings, see: Architectural Record. 1957. *Buildings for Industry.* New York: F. W. Dodge Corp. / Ellsässer, Karl and Horst Ossenberg. 1954. *Bauten der Lebensmittelindustrie. Anlagen und Arbeitsablauf erläutert an 112 Beispielen des In- und Auslands aus allen Gewerbezweigen.* Stuttgart: Hoffmann / Mills, Edward. 1951. *The Modern Factory.* London: Architectural Press / Salzmann, Heinrich. 1911. *Industrielle und gewerbliche Bauten (Speicher, Lagerhäuser und Fabriken).* Vol. 1–3, Leipzig: De Gruyter.

55 Hilberseimer (1944, 76).

56 Brooks (1951, 21).

57 Ibid., 32.

58 Ibid.

59 Missenard, André. 1949 (1940). *Klima und Lebensrhythmus.* Meisenheim am Glan: Westkulturverlag Anton Hain, p. 154. Translation by the author.

60 "In the early and mid-1900s, building ventilation standards called for approximately 0.5 cubic meter per minute (cmm) of outside air for each building occupant, primarily to dilute and remove body odors. As a result of the 1973 oil embargo, however, national energy conservation measures called for a reduction in the amount of outdoor air provided for ventilation to 0.15 cmm per occupant. In many cases, these reduced outdoor air ventilation rates were found to be inadequate to maintain the health and comfort of building occupants (US EPA, 1991)." Kallipoliti, Lydia. 2018. *The Architecture of Closed Worlds or What is the Power of Shit?* Baden: Lars Müller Publishers, p. 20.

61 Brooks (1951, 30).

62 Salzmann (1911, 30).

63 Heinze, W. 1935. "Welches Klima hat ihr Wohnzimmer? Jede Stadt hat ihr eigenes Klima. – Rivieraluft im Zimmer. – Je nach Wunsch: Arktis oder Tropenhitze." In *Die Berner Woche in Wort und Bild: ein Blatt für heimatliche Art und Kunst* 25 (1): 7. Translation by the author.

64 Brooks (1951, 32).

65 Ibid., 31.

66 Concerning the historical reception, see also the 1932 dissertation of Hans Plessner: *The Sonnenbaulehre of Dr. med. Bernhard Christoph Faust.* Plessner's work appeared simultaneously with the modernist debates on air, light and sun

Appendices

67 Esser (2011, 100).

68 Ibid., 101.

69 Ibid.

70 Faust was, among other things, the author of a popular health guidebook ("Health Catechism") for the population, in which the basic hygienic requirements of his architectural considerations are also recognizable. "Education for health" represented the basic concern of Enlightenment medicine. Data according to Prinz, Regina. 2012. "'Freyes Licht, Freye Luft, Freyes leben von Pol zu Pol' – Die Idee einer 'Sonnenstadt' von Bernard Christoph Faust und Gustav Vorherr." In *L'architecture engagée–Manifeste zur Veränderung der Gesellschaft,* edited by Winfried Nerdinger. Munich: Edition Detail, p. 53. Translation by the author.

71 Ibid., 53.Translation by the author.

72 Faust had been presenting the subject to friends and building experts since 1807. "All documents, writings, drawings as well as a model passed into the archive of the Berlin Architects' Association as 'Der Sonnen- und Sternenbau des Dr. Faust'." Prinz (2012, 53). Translation by the author.

73 Faust, Bernard Christoph. 1802 (1792). *Entwurf zu einem Gesundheitskatechismus.* Stuttgart: Franz Steiner Verlag. Cited in Prinz (2012, 53). Translation by the author.

74 Ibid., 55. Translation by the author.

75 Ibid. Translation by the author.

76 Ibid. Translation by the author.

77 Ibid., 58. Translation by the author.

78 "About 95% of Berlin housing during the Industrial Revolution was built as rental housing by design, and the proportion of housing occupied by homeowners fell from about 9% to about 4.5% between 1843 and 1875." Fassbinder (1975, 66).

79 Hilberseimer (1927, 21). Translation by the author.

80 Flügge (1916, 69). Translation by the author.

81 Ibid., 128. Translation by the author.

82 In the portrait photography of Martin Munkacsi and in films such as Walter Ruttman's "Berlin: The Symphony of the Great City" (1927), Albrecht Viktor Blum's "In the Shadow of the Cosmopolitan City" (1928) and Slatan Dudov's "How the Worker Lives" (1930), the poor living conditions of the working class were revealed, and with them empirical evidence of the need for hygienic and climatic reforms.

83 Until around 1880, in Europe false winds were made responsible for different forms of diseases as part of the so-called "miasma theory." Miasma is an old Greek term for "pollution." "The air is a threatening broth in which everything is mixed: smoke, sulfur, watery, volatile, oily and salty vapors that rise from the earth, even, if necessary, the fiery matter that our soil spews out, the vapors coming from swamps, as well as tiny insects, their eggs, all kinds of infusion animals and, even worse, the contagious miasmas of decaying bodies". Corbin, Alain. 1984. *Pesthauch und Blütenduft, eine Geschichte des Geruchs.* Berlin: Wagenbach, p. 23. Cited in Prinz (2012, 61). Translation by the author.

84 "Two diagrams, illustrative of the progress of cholera, and of the meteorological phenomena on every day of the year 1849, are appended to the Report. The second plate shows the temperature, the fall of rain, the direction of the wind, and the height of the barometer on every day at Greenwich. The meteorological phenomena admit therefore of strict comparison only with the line of cholera in London, Plate III.; but the general character of the weather all over the country is indicated by this series of observations. Thus, the weather was cold in the first half of January; and the line fluctuates, but does not rise in January, February, March, April, while the cholera line falls; the hues of temperature and of cholera rise in May, June, and July; in August the temperature remained high, then cholera rose much higher; in September the temperature begins to descend, and the cholera line shoots up its highest points, attaining the greatest elevation when the curve of temperature is descending. The form of the cholera curve for all England is very remarkable: the successive terraces and pinnacles of the Plate resemble sections of the primitive mountain formations, surmounted by spires and aiguilles of granite; or recall the lines of a strange Gothic architecture. The circular diagrams are of a new form, which may serve very well to illustrate periodic phenomena. From the observations of 79 years (1771–1849), Mr. Glaisher had represented the mean temperature of each day of the year by a curve laid down in the usual way on a straight line; and it appeared natural to join the two ends of the line by substituting the radii of a circle for the ordinates, and the angular divisions of the circle for the abscissas. By this arrangement a diagram of a compact form is obtained. It is a modification of the simple dial arrangement in general use. The circle struck round the radii representing the averages-with the different coloring of the radii "extending beyond and falling short of the circle-makes the diagram represent the facts in a striking miner to the eye." Farr, William. 1852. *Report on the Mortality of Cholera in England, 1848–1849*. London: W. Clowes and Sons, p. xlvi.

85 For the scholarly literature on the metropolitan slum (in its Berlin guise), see: Hegemann, Werner. 1930. *Das steinerne Berlin. Geschichte der größten Mietskasernenstadt der Welt.* Berlin: Vieweg / Brüggemeier, Franz and Lutz Niethammer. 1976. "Wie wohnten Arbeiter im Kaiserreich". In *Archiv für Sozialgeschichte,* Jg. 16. / Baier, Rosemarie. 1982. "Leben in der Mietskaserne. Zum Alltag Berliner Unterschichtsfamilien in den Jahren 1900 bis 1920." In *Hinterhof, Keller und Mansarde. Einblicke in Berliner Wohnungselend 1901–1920,* edited by Gesine Asmus. Reinbek: Rowohlt Taschenbuch Verlag / Geist, Johann Friedrich and Klaus Kürvers. 1984. *Das Berliner Mietshaus,* Vol. 2. Munich: Prestel / Vögele, Jörg. 2001. *Sozialgeschichte städtischer Gesundheitsverhältnisse während der Urbanisierung.* Berlin: Duncker & Humblot / Thies, Ralf. 2006. *Ethnograph des dunklen Berlin. Hans Ostwald und die "Großstadt-Dokumente".* Köln: Böhlau Verlag / Volker, Ullrich. 2007. *Die nervöse Großmacht. Aufstieg und Untergang des deutschen Kaiserreichs 1871–1918.* Frankfurt am Main: Fischer Taschenbuch Verlag.

86 Posener, Julius. 1982. "Vorlesungen zur Geschichte der Neuen Architektur IV. Soziale und bautechnische Entwicklungen im 19. Jahrhundert." *arch+ Sonderhefte* 63/64, Oct. 1981: 36.

87 Genzmer, E. 1911. *Sonder-Katalog für die Gruppe Städtebau der wissenschaftlichen Abteilung der Internationalen Hygiene-Ausstellung.* Berlin: Rudolf Mosse.

88 Esser (2011, 101).

89 Flügge (1916, 6). Translation by the author.

90 Royal physician and chief state medical officer L. Fromey stated in 1796, "In general, the miserable dwellings that the common man has in Berlin contribute a great deal to the illnesses of this industrious class of our fellow citizens, and the many buildings in Berlin are a real misfortune for them. [...] the poor man can hardly find shelter for himself and his family. He therefore restricts himself more and more and helps himself with a single room, in which he not only does his craft, but also lives and sleeps with his entire household. With the high price of firewood, it now carefully blocks all access to the outside air in winter, and so these people live in an atmosphere that threatens to suffocate

Endnotes

any stranger on entering such a room. It would therefore be desirable that in the frequent royal buildings more consideration be given to this estimable class of our fellow citizens and that the often fatal consequences of their wretched and small dwellings be remedied." Quoted from: Fassbinder (1975, 58). Translation by the author.

91 Flügge (1916, 83): "all the known factors accompanying 'pauperism,' especially malnutrition, lack of outdoor exercise, unhealthy occupation, alcoholism, etc."

92 Kassner, Carl. 1908. *Das Wetter und seine Bedeutung für das praktische Leben.* Leipzig: Quelle & Meyer, p. 127. Translation by the author.

93 Ibid., 138. Translation by the author.

94 Brezina and Schmidt (1937, 190). Translation by the author.

95 Ibid., 188. Translation by the author.

96 Ibid., 193. Translation by the author.

97 Data according to: Posener (1982, 43).

98 Flügge (1916, 131).

99 Brezina and Schmidt (1937, 187). Translation by the author.

100 Stübben, Josef. 1890. "Der Stadtebau." In *Handbuch der Architektur,* vierter Theil, 9. Halb-Band. Darmstadt: Stuttgart, p. 14–15.

101 Brezina and Schmidt (1937, 184). Translation by the author.

102 Posener, Julius. 1981. "Vorlesungen zur Geschichte der Neuen Architektur III. Das Zeitalter Wilhelms II." In *arch+ Sonderhefte* 59 Oct. 1981: 72. Translation by the author.

103 Hegemann, Werner. 1911. *Der Städtebau nach dern Ergebnisse der allgemeinen Städtebauausstellung in Berlin nebst einem Anhang: Die internationale Städtebau-Ausstellung in Düsseldorf,* Erster Teil. Berlin: Verlag Ernst Wasmuth, p. 7.

104 Flügge (1916, 127). Translation by the author.

105 Usemann (1993, 3). Translation by the author.

106 Ibid., 27. Translation by the author.

107 Ibid., 26. Translation by the author.

108 Kassner, Carl. 1910. *Die meteorologischen Grundlagen des Städtebaues,* Urban-planning lectures from the Seminar for Urban Planning at the Royal Technical University of Berlin, Vol. III, Issue VI, edited by Joseph Brix and Felix Genzmer. Berlin: Ernst & Sohn, p. 15..

109 Flügge (1916, 138). Translation by the author.

110 Flügge (1916, 69). Translation by the author.

111 While the paradigm of as-much-as-possible-sun was established among modern architects, a differentiated attitude was present among physicians.

112 Flügge (1916, 84). Translation by the author. The original can be found in: Carrington, Thomas Spees. 1912. *Fresh air and how to use it.* New York: Nat. Assoc. for the Study and Prevention of Tuberculosis.

113 Flügge (1916, 127). Translation by the author.

114 Ibid., 70. Translation by the author.

115 Ibid. Translation by the author.

116 Ibid. Translation by the author.

117 Ibid., 72. Translation by the author.

118 Ibid., 71. Translation by the author.

119 Ibid. Translation by the author..

120 Ibid., 72. Translation by the author.

121 Thomsen (1985, 22). Translation by the author..

122 See the overview of European cities in: Geist, Johann Friedrich. 1969. *Passagen, ein Bautyp des 19. Jahrhunderts.* Munich: Prestel, pp. 123–128.

123 Ibid., 9. Translation by the author.

124 Passages spread "concentrically" around the Palais-Royal in Paris; the earliest examples appeared in the Paris of the Napoleonic period – 1799, 1800, 1808, 1811 – especially in the "triangle" of the new boulevards, Rue St. Honoré and Rue St. Martin. Data according to: Ibid., 256. Translation by the author.

125 Ibid., 11. Translation by the author.

126 Ibid., 12. Translation by the author.

127 Ibid. Translation by the author. "It is not until the 1920s that the full, continuous glass roof becomes common." Ibid., 18. Translation by the author.

128 Potvin, André. 2004. "Intermediate environments," in Steemers, Koen and Mary Ann Steane (eds.). *Environmental Diversity in Architecture.* London / New York: Routledge, p. 121.

129 Geist (1969, 9). Translation by the author.

130 Ibid., 12. Translation by the author.

131 Ibid. Translation by the author.

132 Ibid., 131. Translation by the author.

133 Ibid., 18. Translation by the author.

134 Kracauer, Siegfried. 1930. "Abschied von der Lindenpassage." In Mulder-Bach, Inka and Ingrid Belke (eds.). 2011. *Kracauer, Siegfried: Werke,* Vol. 5.3. Berlin: Suhrkamp. p. 393–400. Cited in Geist (1969, 39). Translation by the author.

135 Kracauer (1930). Cited in ibid., 140. Translation by the author.

136 Geist (1969, 20). Translation by the author.

137 Ibid., 10. Translation by the author.

138 Ibid., 253. Translation by the author.

139 Grämiger, Gregory. 2018. *Baugesetze formen. Architektur und Raumplanung in der Schweiz.* Zurich: gta Verlag.

140 De Decker, Kris. 2012. "The solar envelope: how to heat and cool cities without fossil fuels." Accessed on February 9, 2022. http://www.lowtechmagazine.com/2012/03/solar-oriented-cities-1-the-solar-envelope.html

141 Kassner (1908, 127). Translation by the author.

142 Brezina and Schmidt (1937, 196). Translation by the author.

143 Brezina and Schmidt emphasize how difficult (if not impossible) it is to correlate a city's mortality rates with its climate. "One must, in order to obtain useful statistics, compare populations in dwellings which are markedly different climatically, without any difference in the economic conditions of the inhabitants, but the statistical material is difficult to find." Ibid., 194. Translation by the author.

144 Flügge (1916, 73). Translation by the author.

145 Data according to De Decker (2012).

146 Flügge (1916, 72). Translation by the author.

147 Posener (1982, 40).

Appendices

2 Man-made Climate by Design

1 "There had been extensive research on German and Austrian urban climates in the early part of the century." Grimmond, Sue. 2011. "London's urban climate: historical and contemporary perspectives." In *City Weathers. Meteorology and Urban Design 1950–2010*, edited by Michael Hebbert, Vladimir Jankovic and Brian Webb. Manchester: University of Manchester Press, p. 67.

2 Kratzer (1937, 63). Translation by the author.

3 Ibid. Translation by the author.

4 Myra Warhaftig et al. postulate a transition from hygienic to climate-adapted building in the work of Alexander Klein ("From Hygienic to Climate-Adapted Building"). See Warhaftig, Myra and Susanne Rexroth, Philipp Oswalt. 1994. "Gebäudeklimatische Studien von A. Klein." In *Wohltemperierte Architektur. Neue Techniken des energiesparenden Bauens*, edited by Philipp Oswalt together with Susanne Rexroth. Heidelberg: Verlag C. F. Müller, p. 53.

5 Brezina and Schmidt (1937, VII). Translation by the author.

6 Geiger, Rudolf. 1941 (1927). *Das Klima der bodennahen Luftschicht, ein Lehrbuch der Mikroklimatologie*. Braunschweig: Vieweg & Sohn, second edition, p. 352. Translation by the author.

7 Missenard, André. 1938 (1937). *Der Mensch und seine klimatische Umwelt*. Stuttgart, Berlin: Deutsche Verlags-Anstalt, p. 15. Translation by the author.

8 Missenard (1949 [1940], 9). The preface was written in May 1939. Translation by the author.

9 Rodenstein, Marianne. 1988. *"Mehr Licht, mehr Luft". Gesundheitskonzepte im Städtebau seit 1750*. Frankfurt / New York: Campus / Reulecke, Jürgen and Adelheid Gräfin zu Castell Rüdenhausen (eds.). 1991. *Stadt und Gesundheit. Zum Wandel von 'Volksgesundheit' und kommunaler Gesundheitspolitik im 19. und frühen 20. Jahrhundert*. Stuttgart: Franz Steiner Verlag.

10 Schmauss (1914, 181). Translation by the author.

11 Taut, Bruno. 1919. *Die Stadtkrone*. Jena: Eugen Dieterich. Cited in Troll, Manfred. 1981. *Architektur der Grossstadt. Theoretische Stadtprojekte seit 1900, Entwerfen mit sozioökonomischen Grundlagen*. Berlin: TU Berlin. Translation by the author.

12 Peppler, Albert. 1929. "Das Auto als Hilfsmittel der meteorolgischen Forschung." *Das Wetter* 46: 308. Translation by the author.

13 Cammerer, Joseph Sebastian. 1936. *Die konstruktiven Grundlagen des Wärme- und Kälteschutzes im Wohn- und Industriebau*. Berlin: Julius Springer, p. III. The book is considered one of the founding writings of modern building physics. Translation by the author.

14 Hilberseimer (1944, 49).

15 "Do slums make criminals?" Ibid., 50.

16 Brezina and Schmidt (1937, 183). Translation by the author.

17 Kähler (1985, 21). Translation by the author.

18 Bodenschatz, Harald. 2009. "Städtebau von den neunziger Jahren des 19. Jahrhunderts bis zum ersten Weltkrieg." In *Berlin und seine Bauten. Teil I– Städtebau*, edited by Harald Bodenschatz et al. Berlin: DOM publishers, p. 23.

19 "In the early years of the century, urban planning in Germany was thus faced with the problem of how to channel this growth of the city brought about by private initiatives, how to give it a functioning organization and, in the end, a convincing form. One of the most important questions was, of course, how to provide better housing for the working population". Posener (1981, 70). Translation by the author.

20 On the emergence of the new discipline of urban design, in theory and practice, see Albers, Gerd. 1975. *Entwicklungslinien im Städtebau. Ideen, Thesen, Aussagen 1875–1945: Texte und Interpretationen*. Düsseldorf: Bertelsmann-Fachverlag.

21 Tubbesing, Markus. 2018. *Der Wettbewerb Gross-Berlin 1910. Die Entstehung einer modernen Disziplin Städtebau*. Tübingen: Wasmuth, p. 9. Translation by the author.

22 Bodenschatz (2009, 27). Translation by the author.

23 Kassner (1910).

24 "Meteorology in the Service of Building Technique" is the title of a 1939 paper that in this case referred to "building technique," but has found similar application to various scales of architecture. See Hrudicka, B. 1939. "Meteorologie im Dienste der Bautechnik." *Zeitschrift für angewandte Meteorologie* 54: 37–47. Translation by the author.

25 Le Corbusier. 1925. *Urbanism*. Paris: Editions Crès, p. 158. Translation by the author.

26 Wright, Henry. 1935. *Rehousing Urban America*. New York: Columbia University Press, p. 86.

27 Troll, Manfred. 1981. *Architektur der Grossstadt. Theoretische Stadtprojekte seit 1900, Entwerfen mit sozioökonomischen Grundlagen, TU Berlin* / Banik-Schweitzer, Renate. 1999. "Urban Visions, Plans, and Projects, 1890–1937." In *Shaping the Great City. Modern Architecture in Central Europe, 1890–1937*, edited by Eve Blau and Monika Platzer. Munich: Prestel Verlag, p. 58–72.

28 Taut (1919). Cited in Troll (1981). Translation by the author.

29 Ibid. Translation by the author.

30 Kähler (1985, 20).

31 In his contribution to the CIAM Congress, Le Corbusier interpreted the title as follows: "Should the surface of cities be extended or made smaller?" Le Corbusier. 1931. "Die Bodenaufteilung der Städte." In *Rationelle Bebauungsweisen. Ergebnisse des 3. Internationalen Kongresses für Neues Bauen*, Brussels, November 1930. Frankfurt am Main: Verlag Englert & Schlosser. Reprint 1979, Nendeln: Kraus Reprint, p. 48. Translation by the author; see also the lecture held by Alexander Klein in 1937 on "Horizontal or Vertical Housing?"

32 Kähler (1985, 43).

33 Hilberseimer, Ludwig. 1927. *Grossstadt Architektur*. Stuttgart: Julius Hoffmann, p. 5. Translation by the author.

34 Unwin, Raymond. 1909. *Town Planning in Practice. An Introduction to the Art of Designing Cities and Suburbs*. London: T. F. Unwin, p. 115.

35 Sitte, Camillo. 1889. *Der Städte-Bau nach seinen künstlerischen Grundsätzen: ein Beitrag zur Lösung modernster Fragen der Architektur und monumentalen Plastik unter besonderer Beziehung auf Wien*: Vienna: Verlag von Carl Graeser.

36 Stübben (1890).

37 Hilberseimer, Ludwig. 1924. "Grosstadtarchitektur." *Der Sturm*, 4: 182. Translation by the author.

38 Schmauss (1914, 181). Translation by the author.

39 Brezina and Schmidt (1937, preface). Translation by the author.

40 Ibid., 195. Translation by the author.

41 Hilberseimer (1944, 75).

42 Geiger (1941 [1927], VI). Translation by the author.

43 Fleming and Jankovic (2011, p. 4.) "This is a space in which the 'natural' atmosphere gets entangled with human energy."

44 Geiger (1941 [1927], VI). Translation by the author.

45 Ibid., 347. Translation by the author.

46 Ibid., 348. Translation by the author.

47 Geiger refers to the work of Willy Hellpach, one of the founders of environmental psychology: Hellpach, Willy. 1939. *Geopsyche, Die Menschenseele unterm Einfluss von Wetter und Klima*. Leipzig: Enke. See Geiger (1941 [1927], 347).

48 Geiger (1941 [1927], 349). Translation by the author.

49 "The representatives of the new branch of research speak as much of the microclimate of a city or a meadow, as of that of a single house, a room – or even an anthill." Andreas, W. 1935–1936. "Klima auf Bestellung: eine neue Wissenschaft: Mikroklimatologie – Das künstliche Klima." In *Am häuslichen Herd: schweizerische illustrierte Monatsschrift*, 39: 453. Translation by the author.

50 Kratzer (1937, 65). Translation by the author.

51 Peppler (1929, 308). Translation by the author.

52 Rossler, Gustav (2008), "Kleine Galerie neuer Dingbegriffe: Hybriden, Quasi-Objekte, Grenzobjekte, epistemische Dinge." In *Bruno Latours Kollektive*, edited by Georg Kneer, Markus Schroer und Gerhard Schüttpelz. Frankfurt am Main: Suhrkamp, p. 76–107.

53 "The natural climatic conditions of the area in which a city is located are, of course, of great importance for the development of the internal climate there as well, but they are modified extraordinarily by the urban conditions." Ibid., 183. Translation by the author.

54 Ibid., 183. Translation by the author.

55 Kassner (1910, 14). Translation by the author.

56 In Munich, measurements have been undertaken since 1931. See Kratzer (1937, 62).

57 Schmidt, Wilhelm. 1927. "Die Verteilung des Minimumtemperaturen in der Frostnacht des 12. Mai 1927 im Gemeindegebiet von Wien." *Fortschr. d. Landw.* 21 / Peppler (1929, 305–308).

58 Peppler (1929, 305). Translation by the author.

59 Ibid. Translation by the author.

60 Büdel, A. and J. Wolf. 1933. "Münchner stadtklimatische Studien." *Zeitschrift für angewandte Meteorologie* 50 (1): 6. Translation by the author.

61 Ibid., 4. Translation by the author.

62 Geiger (1941 [1927], 352). Translation by the author.

63 Kratzer (1937, 62). Translation by the author.

64 Ibid., 64. Translation by the author.

65 Büdel and Wolf (1933, 6). Translation by the author.

66 Ibid., 7. Translation by the author.

67 Peppler (1929, 308). Translation by the author.

68 Büdel and Wolf (1933, 8). Translation by the author.

69 Peppler (1929, 308). Translation by the author.

70 Büdel and Wolf (1933, 8). Translation by the author.

71 Ibid., 9. Translation by the author.

72 Peppler (1929, 307). Translation by the author.

73 Büdel and Wolf (1933, 6). Translation by the author.

74 Ibid., 7. Translation by the author.

75 Ibid. Translation by the author.

76 Peppler (1929, 307). Translation by the author.

77 Brezina and Schmidt (1937, 142). Translation by the author.

78 Albrecht, F. and J. Grunow. 1935. "Ein Beitrag zur Frage der vertikalen Luftzirkulation." *Meteorologische Zeitschrift*, Deutsche Meteorologische Gesellschaft, 52: 103. Translation by the author.

79 Ibid., 103. Translation by the author.

80 Ibid., 104. Translation by the author.

81 Ibid. Translation by the author.

82 Ibid. Translation by the author.

83 Ibid., 106. Translation by the author.

84 Ibid., 107. Translation by the author.

85 Ibid., 108. Translation by the author.

86 Stadler, Friedrich. 1997. *Studien zum Wiener Kreis. Ursprung, Entwicklung und Wirkung des Logischen Empirismus im Kontext*. Frankfurt am Main: Springer, p. 134. Translation by the author.

87 Especially in Marxism-Leninism, this question was raised to an important political one. It was about the relation between Marxian epistemology and historical materialism. See Lenin, Vladimir Ilyich. 1909. "Materialismus und Empiriokritizismus. Kritische Bemerkungen über eine reaktionäre Philosophie," in *Werke*, Vol. 14. Berlin: Dietz Verlag 1956–1972 / Wittich, Dieter. 2001. "Lenins Buch 'Materialismus und Empiriokritizismus'. Seine Entstehungsgeschichte sowie progressive und regressive Nutzung." In Gerhardt, Volker and Hans-Christoph Raus (eds.). *Anfänge der DDR-Philosophie. Ansprüche, Ohnmacht, Scheitern*. Berlin: Christoph Links Verlag, p. 160–179.

88 Wittich (2001). Translation by the author.

89 Mach, Ernst. 1886. *Analyse der Empfindungen*. Jena: G. Fischer. Cited in Stadler (1997, 143). Translation by the author.

90 Stadler (1997, 138). Translation by the author.

91 The understanding displayed among German-speaking meteorologists of the interwar period differed fundamentally from the much broader experiential notion of the environment that have been studied in the United States since the 1960s (see chapter 6).

92 Hilberseimer (1924, 177).Translation by the author.

93 Kratzer (1937, 62). Translation by the author.

94 Brezina and Schmidt (1937, 182). Translation by the author.

95 Ibid., 184. Translation by the author.

96 Ibid., 183. Translation by the author.

97 Ibid., 184. Translation by the author.

98 Kratzer (1937, 64). Translation by the author.

99 Brezina and Schmidt (1937, 183). Translation by the author.

100 Poerschke, Ute. 2018. "Data-Driven Design in High Modernism: Ludwig Hilberseimer's Solar Orientation Studies." *ENQ, the ARCC Journal of Architectural Research*. Accessed on August 20, 2019. https://www.arcc-journal.org

Appendices

101 According to Markus Kilian, this went back "to a study by the Bauhaus student Ernst Hegel as part of the Seminar for Housing and Urban Development". Kilian (2002, 91). Translation by the author.

102 Hilberseimer, Ludwig. 1935. "Raumdurchsonnung" *Moderne Bauformen*, 34: 29–36 / Hilberseimer, Ludwig. 1936. "Raumdurchsonnung und Siedlungsdichtigkeit." *Moderne Bauformen*, 35: 69–76.

103 Hilberseimer (1944, 77).

104 Ibid., 85.

105 Ibid., 87.

106 Ibid., 88.

107 Ibid., 89.

108 Ibid.

109 Brezina and Schmidt (1937, 192). Translation by the author.

110 Ibid. Translation by the author.

111 Ibid. Translation by the author.

112 Schmidt, Alfred. 1926. "Die Bedeutung von Sonne und Wind für den Stadtkörper. Ein Beitrag zu zeitgemässem Städtebau." In *Städtebau*. Monatshefte für Stadtbaukunst, Städtisches Verkehrs-, Park- und Siedlungswesen, edited by Werner Hegemann, Verlag Ernst Wasmuth: Berlin, p. 40.

113 Ibid., 40. Translation by the author.

114 Ibid., 39. Translation by the author.

115 Hilberseimer, Ludwig. 1930. "Vorschlag zur Citybebauung." *Die Form. Zeitschrift für gestaltende Arbeit für den Deutschen Werkbund*, 12: 608. Translation by the author.

116 Ibid., 610. Translation by the author.

117 Ibid. Translation by the author.

118 Ibid. Translation by the author.

119 Ibid. Translation by the author.

120 Hilberseimer (1944, 96).

121 Ibid.

122 Ibid., 97.

123 Missenard (1949 [1940], 71). Translation by the author.

124 Ibid., 61. Translation by the author.

125 Ibid. Translation by the author.

126 Ibid. Translation by the author.

127 Ibid., 9. Translation by the author.

128 On Missenard's collaboration with Le Corbusier, see Requena-Ruiz, Ignacio. 2018. "Building a Brazilian Climate. The Case of the House of Brazil in Paris (France)." In *The Urban Microclimate as Artifact*, edited by Sascha Roesler and Madlen Kobi. Basel: Birkhäuser, p. 134–151.

129 Missenard (1949 [1940], 9). Translation by the author.

130 Missenard (1938 [1937], 15). Translation by the author.

131 Missenard (1949 [1940], 71). Translation by the author.

132 Rudder, B. and F. Linke (eds.). 1940. *Biologie der Grossstadt*. Frankfurter Konferenzen für medizinisch-naturwissenschaftliche Zusammenarbeit, IV Konferenz am 9. und 10. Mai. Dresden, Leipzig: Verlag von Theodor Steinkopf.

133 In his closing remarks, grounded in the new German aspirations to global power, F. Linke wrote, "I think we all feel that this conference has created an essential step towards clarifying the big city question. Nothing more could have been expected. The effect will only become apparent in years and decades." Linke, F. 1940. "Schlusswort". In Rudder and Linke (1940, 200). See in a similar vein, in the preface to the 2nd edition of his book, Rudolf Geiger emphasizes the relevance of microclimate research for the Nazi campaign of conquest during the Second World War. The conscious creation of microclimates takes place in the context of the opening up of new *Lebensraum* ("living space") and the construction of new cities. "In the midst of the tremendous struggle for existence to which our fatherland is forced, the book was brought to completion during difficult wartime trimester work. Great tasks lie ahead of us, which the construction and expansion of the newly secured and newly won German living space poses to us. For this the book wants to be ready as a scientific tool for the service." Geiger (1941 [1927], VII).

134 In addition to labor conditions, Friedrich Engels described in detail the housing conditions of the working class. He had thus co-founded a social-scientific empiricism that linked in-situ inspection with an analysis of society as a whole. Revolution or reform – these were the two poles facing the various actors discussing the housing question. See Engels, Friedrich. 1845. "Die Lage der arbeitenden Klasse in England". In *Marx-Engels-Gesamtausgabe* (MEGA) Abt. 1., Band 4. Marx-Engels-Institut. 1932/1970 / Engels, Friedrich. 1872. "Zur Wohnungsfrage." In *Karl Marx/Friedrich Engels – Werke*. Berlin: (Karl) Dietz Verlag. Vol. 18, 5th edition, 1973.

135 On the notion of eugenics, see: Turda, Marius. 2010. *Modernism and Eugenics*, Basingstoke: Palgrave Macmillan / Rosental, Paul-André. 2016. *Destins de l'eugénisme*. Paris: Editions du Seuil.

136 Missenard (1938 [1937], 15). Translation by the author.

137 Carrel, Alexis. Preface. In Missenard (1949 [1940], 9). Translation by the author.

138 Missenard (1938 [1937], 15). Translation by the author.

139 Brezina and Schmidt (1937, 196). Translation by the author.

140 Ibid. Translation by the author.

141 Missenard (1938 [1937], 15). Translation by the author.

142 "In ancient times, the importance of the sun and the wind for the sanitary conditions in the cities was already recognized and rules were established about the cardinal directions of the streets. In the following centuries, traces can be traced from which it can be concluded that when cities were built, the nature of the place and the weather conditions were carefully observed. It was only during the boom of our cities in the 19th century that the old traditions were lost, perhaps believing it was possible to dispense with the 'outdated' rules, then the progress made in technical improvements to urban construction (water pipes, sewers, street paving, etc.). However, this was a mistake that is now taking its toll on us, and we must now laboriously recover, through scientific investigation, what our ancestors possessed in the way of rules of thumb with which they were able to master their tasks. Only in recent times has it been recognized once again that in the construction and expansion of cities it is not only a matter of traffic considerations, but that one must first and foremost be guided by the sun and the wind if healthy living conditions are to be created." Schmidt (1926, 37). Translation by the author.

143 Ibid., 40. Translation by the author.

144 Hilberseimer (1944, 118).

145 Schmidt (1926, 40). Translation by the author.

146 Kassner (1910, 12). Translation by the author.

147 Ibid., 5. Translation by the author.

148 See also Brezina and Schmidt, a quarter of a century later: "Many cities suffer from unfavorable natural climatic conditions because, on the whole, economic and safety reasons, rather than health rea-

sons, are and had to be decisive in their formation and further growth." Brezina and Schmidt (1937, 183). Translation by the author.

149 Kassner (1910, 11). Translation by the author.

150 Ibid. Translation by the author.

151 Ibid. Translation by the author.

152 Landsberg, Helmut. 1958 (1941). *Physical Climatology*. Pennsylvania: Gray Printing, p. 350.

153 Kratzer (1937, 65). Translation by the author.

154 Kassner (1910, 11). Translation by the author.

155 Coen, Deborah R. 2016. "Seeing Planetary Change, Down to the Smalllest Wildflower." In *Climates: Architecture and the Planetary Imaginary*, edited by James Graham et al. Zurich: Lars Müller Publishers, p. 34.

156 Ibid., 35. Coen speaks of "studying the concrete conditions of (economic) unity in (environmental) diversity."

157 Kassner (1910, 10). Translation by the author.

158 Hilberseimer (1944, 85).

159 See on the topic of weathering: Mostafavi, Mohsen and David Leatherbarrow. 1993. *On Weathering. The Life of Buildings in Time*, Cambridge, MA: MIT Press. "The action of the elements leads to the deterioration of the building […]. Weathering does not construct, it destroys. Over time the natural environment acts upon the outer surface of a building in such a way that its underlying materials are broken down. This breakdown, when left to proceed uninterrupted, leads to the failure of materials and the final dissolution of the building itself-ruination". Mostafavi and Leatherbarrow (1993, 5).

160 Döcker, Richard. 1929. *Terrassentyp: Krankenhaus, Erholungsheim, Hotel, Bürohaus, Einfamilienhaus, Siedlungshaus, Miethaus und die Stadt*. Stuttgart: Wedekind.

161 Stieger, Lorenzo. 2018. *Vom Hang zur Schräge. Das Hangterrassenhaus in der Schweiz: Aufstieg und Niedergang einer gefeierten Wohnbautypologie*. Dissertation, ETH Zurich, p. 70.

162 Ibid., 73.

163 Schmidt (1926, 38).

164 Radt, Wolfgang. 1993. "Landscape and Greek Urban Planning: Exemplified by Pergamon and Priene." In *City and Nature: Changing Relations in Time and Space*, edited by T. M. Kristensen et al. Odense: Odense University Press, p. 201–209.

165 Hilberseimer (1944, 76).

166 Socrates is mentioned as an early promoter of planning oriented to the sun. "Soc. 'Do you admit that any one purposing to build a perfect house (13) will plan to make it at once as pleasant and as useful to live in as possible?' and that point being admitted, (14) the next question would be: 'It is pleasant to have one's house cool in summer and warm in winter, is it not?' and this proposition also having obtained assent, 'Now, supposing a house to have a southern aspect, sunshine during winter will steal in under the verandah, (15) but in summer, when the sun traverses a path right over our heads, the roof will afford an agreeable shade, will it not? If, then, such an arrangement is desirable, the southern side of a house should be built higher to catch the rays of the winter sun, and the northern side lower to prevent the cold winds finding ingress; in a word, it is reasonable to suppose that the pleasantest and most beautiful dwelling place will be one in which the owner can at all seasons of the year find the pleasantest retreat, and stow away his goods with the greatest security.'" Xenophon. ~371 BC. *The Memorabilia*. Recollections of Socrates, Book III, VIII. Accessed on February 10, 2022. http://www.gutenberg.org/files/1177/1177-h/1177-h.htm

167 Hilberseimer (1944, 77).

168 Bronin, Sara C. 2009. "Solar rights." *Boston University Law Review*. 89 (4): 1218.

169 Esser (2011, 99).

170 "The influence of the wind on the heating demand is quite insufficiently known. […] It is therefore one of the most important tasks of the urban planner, in cooperation with the meteorologist, to procure the still-missing records, so that some sentimental or slogan-like assertions, which only too often betray an incomprehensible lack of observation of nature, may disappear as soon as possible. South-facing apartments have always been the most sought-after. They are preferred by every tenant, every hygienist, every doctor (sanatorium). In the Urals, in windy and snowy areas, all houses are built east-west with the entrance facing south and the barn facing west. Embankments are heaped up as protection against north winds. The building arrangement is adapted to the particular climatic conditions on the basis of long-term experience. Such experience must be used by the town planner. It would be instructive if, in this respect, studies were also made in our country on buildings which, without being forced to take into account proscribed building

positions, have only sought the best location given the prevailing climatic conditions." Schmitt, Paul. 1930. "Die Besonnungsverhältnisse an Stadtstrassen und die günstigste Blockstellung." In *Zeitschrift für Bauwesen* 5: 124. Translation by the author.

171 Hilberseimer (1927, 5). Translation by the author.

3   Democratizing Urban Nature

1 Hughes, Thomas p. 1987. "The Evolution of Large Technological Systems." In *The Social Construction of Technological Systems. New Directions on the Sociology and History of Technology*, edited by Wiebe E. Bijker, Thomas Hughes and Trevor J. Pinch. Boston: The MIT Press, p. 51–82.

2 Kähler refers to Vienna, Frankfurt and Hamburg. Kähler (1985, 20). Translation by the author.

3 Gandy, Matthew. 2014. "Borrowed Light. Journals Through Weimar Berlin." In *The Fabric of Space. Water, Modernity, and the Urban Imagination.* Boston: The MIT Press, p. 56.

4 Ibid., 56.

5 Julius Posener speaks retrospectively of the "stony" and of the "green" Berlin that had constituted itself at the beginning of the 20th century, also in reference to Werner Hegemann's famous *Das steinerne Berlin. Geschichte der grössten Mietkasernenstadt der Welt* and Martin Wagner's propagation of *Das sanitäre Grün der Städte*. Posener (1981). See Hegemann (1930).

6 Gandy (2014, 59).

7 Seewang, Laila. 2019. *The Scale of Water–Networked Infrastructure and the Making of Municipal Berlin 1872–1900*. PhD Thesis, ETH Zurich, No. 25871.

8 Friedrich Leyden emphasizes the "good" state of climate documentation in the case of Berlin. Leyden, Friedrich (1933), *Gross-Berlin. Geographie der Weltstadt.* Breslau: Ferdinand Hirt, p. 13.

9 Steurer, Hannah. 2017. "'Berlin ist eine Sandwüste. Aber wo sonst findet man Oasen?' Stadtdiskurs als Naturdiskurs in der deutschen und französischen Literatur (1800 bis 1935)." In *Literatur und Ökologie. Neue literatur- und kulturwissenschaftliche Perspektiven,* edited by

10 Ibid., 137. Translation by the author.

11 Ibid., 138. Translation by the author.

12 Kracauer, Siegfried. 2009. *Strassen in Berlin und anderswo*. Frankfurt am Main: Suhrkamp Verlag, p. 244. Citied in Steurer (2017, 139). Translation by the author.

13 Goldmerstein, J. and K. Stodieck. 1931. "Grossstadtsanierung. Gewinnung von Spiel-, Sand- und Grünflächen in Neben- und Seitenstrassen mit Rentabilitätsnachweis." *Deutsche Bauzeitung,* Berlin SW 48.

14 Brezina and Schmidt (1937, 192). Translation by the author.

15 Kratzer (1937, 66). Translation by the author.

16 Ibid., 67. Translation by the author.

17 Stübben, Josef. 1907. *Der Städtebau.* Stuttgart: Kröner, p. 12. Translation by the author.

18 Sitte (1889).

19 Stübben, (1907, 73–74). Translation by the author.

20 Accessed on February 10, 2022. https://www.deutsche-biographie.de/pnd117082066.html

21 General development plan for Greater Berlin (competition for a "basic plan for the development of Greater Berlin").

22 The locomotive manufacturer Borsig, for example, appeared on the scene only at the end of the 19th century—as did the chemical industry in Wedding.

23 Gandy (2014, 57).

24 "The then established boundary of the city area is still valid today". Data according to: Posener, Julius (1981, 70). Translation by the author.

25 The so-called "Special-purpose Association *(Zweckverband)* Gross-Berlin," which came into being during these years, propagated the territory of the capital targeted in the competition.

26 Borsi, Katharina. 2015. "Drawing the region: Hermann Jansen's vision of Greater Berlin in 1910." *Architecture* 20 (1): 50.

27 Ibid., 47.

28 Hertweck, Florian. 2013. "Berliner Ein- und Auswirkungen," in *Die Stadt in der Stadt. Berlin: Ein grünes Archipel. Ein*

Manifest by Oswald Mathias Ungers and Rem Koolhaas with Peter Riemann, Hans Kollhoff and Arthur Ovaska. A critical publication by Florian Hertweck and Sébastien Marot. Baden: Lars Müller Publishers, p. 61. Translation by the author.

29 Jansen, Hermann. 1910. *Vorschlag zu einem Grundplan für Gross-Berlin*. Munich: Callwey, p. 52. Translation by the author.

30 Hertweck (2013, 62). Translation by the author.

31 Jansen (1910). See the map "Forest and meadow belt around Gross-Berlin together with radial connection" in the appendix. Translation by the author.

32 Ibid., 46. Translation by the author.

33 Ibid., 48. Translation by the author.

34 Ibid., 53. Translation by the author.

35 Ibid., 47. Translation by the author.

36 Ibid., 51. Translation by the author.

37 Jansen mentions the "beautiful forest" in the areas of "Wuhlheide, Adlershof, Johannisthal, Alt-Glienicke and Teltow-kanal" as an example. Ibid., 52. Translation by the author.

38 Ibid., 53. Translation by the author.

39 Ibid., 54. Translation by the author.

40 Wagner, Martin. 1915. *Das sanitäre Grün der Städte. Ein Beitrag zur Freiflächentheorie*. Dissertation, TU Berlin, February 27, p. 3. Translation by the author.

41 Hertweck (2013, 65). Translation by the author.

42 Wagner (1915, 1). Translation by the author.

43 On the subject of the ventilated city, see: Goldmerstein, J. and K. Stodieck. 1931. *Wie atmet die Stadt? Neue Feststellungen über die Bedeutung der Parkanlagen für die Lufterneuerung in den Grossstädten*. Berlin: VDI-Verlag GmbH.

44 Wagner, Otto. 1911. *Die Grossstadt*, quoted in: Hackenschmidt, Sebastian and Iris Meder, Ákos Moravánszky. 2018. *Post Otto Wagner. Von der Postsparkasse zur Postmoderne*. Basel: Birkhäuser, p. 86.

45 Kähler (1985, 40).

46 Jansen emphasized the fact that it is by no means the sheer quantity of green spaces that determines the quality of the competition entry but rather their actual feasibility and implementation. Likewise, he emphasized the endangerment of the

existing forests in and around Berlin due to privatization and fiscalization. The green belts of the big city were in need of protection and, accordingly, demanded political regulation.

47 Borsi (2015, 55). See also Borsi, Katharina. 2006. "The Strategies of the Berlin Block." In *Intimate Metropolis: Urban Subjects in the Modern City,* edited by V. Di Palma, D. Periton and M. Lathouri. Abingdon: Routledge, p. 132–152.

48 Borsi (2015, 52 and 55).

49 "A reform [...] will first of all have to endeavor not to repeat the mistakes made earlier in the case of new developments of city districts, housing estates, etc.; and secondly, it will have to examine how, while retaining the existing facilities, it can still be made possible for the population to get out into the open." Flügge, Carl. 1916. *Grossstadtwohnungen und Kleinhaussiedlungen in ihrer Einwirkung auf die Volksgesundheit. A critical discussion for physicians, administrators and builders*. Jena: Fischer, p. 131. Translation by the author.

50 Ibid., 136. Translation by the author.

51 Jansen (1910, 57). Translation by the author.

52 Accessed on February 10, 2022. https://architekturmuseum.ub.tu-berlin.de/images/1600WM/20528.jpg

53 Jansen (1910, 58). Translation by the author.

54 See Development plan for Tempelhofer Feld, Dec. 20, 1910. Museum of Architecture, Technical University of Berlin.

55 Borsi (2015, 52).

56 This consideration of different scales was novel; it anticipated approaches that would be elaborated some 50 years later in an American context by Ian McHarg (see chapter 6). See McHarg, Ian. 1969. *Design with Nature*, New York: Natura History Press.

57 Jansen (1910, 61).

58 Posener (1981, 71).

59 Jansen (1910, 61). Translation by the author.

60 Accessed on August 4, 2020. https://biotope-city.com/de/2019/12/07/das-verhaeltnis-stadt-und-natur-in-der-geschichte-2/

61 "Bornimer Kreis" is not infrequently used as a synonym for the Foerster-Mattern-Hammerbacher working group. The designation encompasses a circle of peo-

Claudia Schmitt and Christiane Sollte-Gresser. Bielefeld: Aisthesis Verlag, p. 136. Translation by the author.

ple who worked closely with Karl Foerster, primarily in the 1920s and 1930s, as well as friends who shared Foerster's natural philosophy and garden ideas. In addition to Foerster, Mattern and Hammerbacher, the core was formed by the garden architects Walter Funcke, Hermann Göritz, Karl-Heinz Hanisch, Richard Hansen, Gottfried Kühn, Alfred Reich and Berthold Körting. The group was expanded to include architect and writer Otto Bartning, as well as his brother, painter Ludwig and his daughter, Esther Bartning, also a painter, architects Hans Poelzig, Hans Scharoun, pianist Wilhelm Kempff, conductor Wilhelm Furtwängler, writer Karla Hoecker, art historian Edwin Redslob, publisher Werner Stichnote, and painter Sigward Sprotte." Translation by the author. Accessed on August 2, 2020. https://karl-foerster-bornimer-kreis.com/2017/05/12/der-bornimer-kreis/

62 Only four of the originally planned eight houses were built. They were built for the Bauhaus supporter Adolf Sommerfeld.

63 Wörner, Martin et al. (eds.). 2013. *Architekturführer Berlin*. Berlin: Reimer, p. 551.

64 "The clients played a decisive role in the aesthetic development of Berlin's industrial architecture. Initially, they placed a great deal of emphasis on representative facades, which were influenced by various styles such as Classicism, Historicism, and Art Nouveau. With the appearance of anonymous joint-stock companies as builders, the extreme interest in the design of facilities with decoration and period architecture receded. However, since industrial construction was not socially recognized, new architectural forms were not yet developed. It was not until the end of the century that engineering companies such as Ludwig Loewe and August Borsig emerged with innovative factory designs. Increasingly, then, artists, architects and industrialists sought the formal interpenetration of building and function. New architectural plans made clear the close connection between construction and production methods and economic conditions. Industrial construction became the experimental field of modern architecture." Accessed on April 26, 2019. http://www.an-morgen-denken.de/tui/99jan/loewe.htm

65 Posener (1981. 73).

66 Wright (1935, 87).

67 City Planning Councilor Martin Wagner and Otto Bartning from the jury opted for a design proposal by Walter Gropius and Stephan Fischer, a row development with ten- to twelve-story high-rise buildings, but the client Gewobag decided for a less radical project, involving various architects.

68 Warhaftig, Rexroth, Oswalt (1994, 52).

69 See the exhibition from Oct. 24, 1984 to Jan. 7, 1985 at the Bauhaus-Archiv, Berlin.

70 Taut, Bruno. 1931. "Der Außenwohnraum". *Gehag-Nachrichten* 2 (1/2): 9.

71 The housing estate is regarded as "one of the most significant architectural achievements of the Weimar Republic." Translation by the author. Wörner (2013, 893).

72 Today, these pines are old and therefore not infrequently have to be felled. Interesting parallels to the forest housing estate Zehlendorf are seen in the forest cemetery Zehlendorf, which was designed by Herta Hammerbacher in 1945.

73 Accessed on February 11, 2022. Cited in www.stadtentwicklung.berlin.de

74 See, for example, the Berlin texts by Walter Benjamin, Franz Hessel, and Siegfried Karacauer. Steurer (2017, 138).

75 Jansen (1910, 47). Translation by the author.

76 Borsi (2015, 55).

77 Hilberseimer, Ludwig. 1967. *Berliner Architektur der 20er Jahre*. Mainz: Florian Kupferberg.

78 Schmidt (1926, 40).

79 Following the state of knowledge of his time, Hilberseimer distinguishes three kinds of solar radiation. "The light-producing rays are directly perceptible, but the invisible rays can be perceived only through their effects. Infra-red rays produce heat; ultra-violet rays exert chemical influences. The effects of the various rays are modified by latitude, by altitude, by season, and by meteorological conditions." Hilberseimer (1944, 81).

80 Heliomorphic design is the counterpart of urban climatic maps. For obvious reasons, heliomorphic design is closer to architects than the physical knowledge, which would require a deeper understanding of physical principles.

81 Behling, Sophia and Stefan Behling. 1996. *Sol Power. The Evolution of Solar Architecture*. Munich, New York: Prestel, p. 167.

82 Ibid., 167.

83 Butti, Ken and John Perlin. 1980. *A Golden Thread. 2500 Years of Solar Architecture and Technology*. Palo Alto: Van Nostrand Reinhold, p. 176.

84 Hebbert, Michael and Webb, Brian. 2012. "Towards a Liveable Urban Climate: Lessons from Stuttgart." In *Isocarp*, 07, p. 124.

85 Ibid.

86 The statement was made in 1931 in the context of a design for an annex building. Data according to Warhaftig, Rexroth, Oswalt (1994, 51).

87 Klein, Alexander. 1927. "Versuch eines Graphischen Verfahrens zur Bewertung von Kleinwohnungsgrundrissen." *Wasmuths Monatshefte für Baukunst und Städtebau* 7. / Lueder, Christoph. 2017. "Evaluator, Choreographer, Ideologue, Catalyst: The Disparate Reception Histories of Alexander Klein's Graphical Method." *Journal of the Society of Architectural Historians* 76 (1): 82–106.

88 Warhaftig et al. (1994, 51). Translation by the author.

89 Alexander Klein. Cited in ibid., 50. Translation by the author.

90 Warhaftig et al. (1994, 50). Translation by the author.

91 Ibid. Translation by the author.

92 Klein, Alexander. 1927. "Untersuchungen zur rationellen Gestaltung von Kleinwohnungsgrundrissen." *Die Baugilde* 9 (22): 1361.

93 Warhaftig et al. (1994, 51). Translation by the author.

94 Klein, Alexander. 1934. *Das Einfamilienhaus. Südtyp, Studien und Entwürfe*. Part of the series: Wohnbau und Städtebau, Beiträge zur Entwurfslehre. Stuttgart: Julius Hoffmann, p. 92.

95 Ibid. Translation by the author.

96 Ibid., 96. Translation by the author.

97 Ibid. Translation by the author.

98 The mentioned study was published in 1942 in Hebrew in *Journal of the Association of Engineers & Architects in Palestine* 2 (5): 2–4. Data according to Warhaftig et al. (1994, 202). Translation by the author.

99 Warhaftig et al. (1994, 54). Translation by the author.

100 No. 578 164, Class 37d, Group 7 01. Date of publication of the grant of the patent: May 18, 1933.

101 De Chiffre, Lorenzo. 2016. *Das Wiener Terrassenhaus. Entwicklungsphasen und Aktualität eines historischen Wohntypus mit Fokus auf den lokalspezifischen architektonischen Diskurs*. Dissertation, University of Vienna.

102 "The most important motivation for the proposal, however, was always the concern for 'air-circulated' outdoor spaces to

promote health, as he paraphrased the modern living conditions in the patent specification and emphasized again in correspondence with his colleague Morton Philip Shand in 1936, while he endeavored to establish contact with the then well-known cardiovascular specialist Professor János Plesch. Plesch cultivated wide-ranging acquaintances (including celebrities from the fields of science, art, music, and theater) and often acted as an intermediary for a variety of companies, from which Behrens probably hoped for support in spreading his idea of the terraced house and the terraced block development." Stieger (2018, 171). Translation by the author.

103 Behrens, Peter. 1930. *Baublock bestehend aus mehr- und vielgeschossigen Einzelhäusern.* Patentiert im Deutschen Reiche *vom* 25. Juni 1930, Nr. 578 164.

104 Ibid. Translation by the author.

105 Ibid. Translation by the author.

106 Ibid. Translation by the author.

107 Thus, Behrens also shows a vacillation between thermal analysis and heliotherapeutic ideology. This lack of definition distinguishes the Modern Movement as a whole, which has largely concealed the massive overheating of apartments at maximum solar radiation.

108 Atkinson works with the modelling of the shadows cast by a cube at different seasons ("Shadows of the cube, autumnal and vernal equinox. Lat. 42°-o' N"). As an illustration of these geometric considerations, he refers to Swiss mountain dwellers who would have recognized the importance of the proper orientation of houses. "The advantage of placing a square building with its diagonal upon the meridian was long ago recognized by the mountain dwellers of Switzerland. [...] It will be observed that the living room is placed in the sunniest corner of the building, with its windows facing southeast and southwest." Atkinson (1912, 23).

109 Ibid., 94.

110 Ibid., 78.

111 Ibid., 87.

112 Ibid., 90.

113 Ibid., 97.

114 Ibid., 88.

115 Butti and Perlin (1980, 178).

116 Atkinson (1912, 94).

117 Ibid., 94.

118 Taylor, Jeremy. 1991. *Hospitals and Asylum Architecture in England 1840–1914. Building for Health Care.* London: Mansell Publishing, p. 8.

119 "Finally as a postscript the *Terrassenbau,* as specialty of Germany and Switzerland for those suffering from tuberculosis. The idea is that patients should be able to profit from fresh air all day long. The south fronts have on all floors terraces and the step back, much as they do in certain flats or tiers of small houses in the fashion of about 1965–70. The idea had first been put forward by a Swiss, Dr. D. Sarasin, in 1901 and was taken up by the architect Richard Döcker of Stuttgart first in 1926 for a sanatorium at Waiblingen." Pevsner, Nikolaus. 1976. *A History of Building Types.* Princeton: Princeton University Press, p. 157.

120 Atkinson (1912, 98).

121 Ibid., 104.

122 Poerschke (2018).

123 In his book *Großstadtarchitektur,* Hilberseimer emphasized the greater significance of "industrial buildings, which are pure works of engineering" compared with the previous design results of architects. Hilberseimer (1927, 92). He criticized, for example, Peter Behrens and his turbine hall for the AEG in Berlin-Moabit. (Ibid., 91).

124 Hilberseimer, Ludwig. 1931. *Hallenbauten. Stadt- und Festhallen, Turn- und Sporthallen, Ausstellungshallen, Ausstellungsanlagen.* Leipzig: J. M. Gebhardt, p. 7.

125 See Elias Canetti, *Crowds and Power* and Siegfried Kracauer's famous 1927 essay "The Mass Ornament". Although Canetti's book did not appear until 1960, it had already taken its starting point as 1922.

126 On the notion of the "heroic style," see Raith, Frank-Bertolt. 1997. *Der heroische Stil. Studien zur Architektur am Ende der Weimarer Republik.* Berlin: Verlag für Bauwesen.

127 Kracauer, Siegfried. 1977 (1927). "The Mass Ornament." In *Weimar Essays,* Harvard University Press, p. 76.

128 Ibid., 51.

129 Hilberseimer (1931, 7).

130 Reich had met Mies in the run-up to the Weissenhof exhibition, where she was responsible for the design and organization of the hall exhibition. This also included the glass room designed with Mies at the Werkbund exhibition in Stuttgart. As a member of the board of the Deutscher Werkbund and experienced exhibition organizer, Reich was an eminent mediator not only of modern crafts-

manship but also between the permanent architecture of exhibition halls and temporary exhibitions to be set up there. See Lange, Christiane. 2011. *Ludwig Mies van der Rohe. Architecture for the Silk Industry.* Berlin: Nicolai, p. 30.

131 The initiators were the Association of German Silk Weavers and the Association of Velvet and Plush Manufacturers. The opening took place on 21 September 1927. In addition to the café, eight booths with showcases were set up in the surrounding gallery of the exhibition hall. The fair was located in the recently completed exhibition hall at the Berlin Radio Tower. See ibid., 71, endnote 10.

132 Ibid., 71.

133 Other colors of the fabrics exhibited in the bunks were "white, red, black, blue, grey, green-yellow." Ibid., 88.

134 Ibid., 73.

135 Deutsche Seide, 1929, p. 216. / Lange (2011, 72).

136 On Gottfried Semper's notion of *Stoffwechsel* (material change), see Roesler, Sascha. 2013. *Weltkonstruktion. Der aussereuropäische Hausbau und die moderne Architektur – ein Wissensinventar.* Berlin: Gebr. Mann Verlag / Moravánszky, Ákos. 2017. *Metamorphism: Material Change in Architecture.* Basel: Birkhäuser.

137 A neologism, combining "gourmet" and "Germania".

138 Wolschke-Bulmahn, Joachim and Peter Fibich. 2004. "Vom Sonnenrund zur Beispiellandschaft. Entwicklungslinien der Landschaftsarchitektur in Deutschland, dargestellt am Werk von Georg Pniower (1896–1960)," Schriftenreihe des Fachbereichs Landschaftsarchitektur und Umweltentwicklung der Universität Hannover, p. 31. Translation by the author.

139 Ibid., 32. Translation by the author.

140 Ibid. Translation by the author.

141 Ibid. See endnote 18. Translation by the author.

142 His work comprised gardens for single-family houses; gardens in new building estates (open space planning); and sample designs for allotment gardens, memorials, cemeteries and public areas. In the Pniower House (c. 1935) at Hochsitzweg 105 in Berlin, a flower window with cacti was installed.

143 Prof. Dr. Reinhold: Report on Prof. Georg Pniower, submitted to the University of Horticulture in Budapest, January 1960, p. 3. Wolschke-Bulmahn and Fibich (2004, 32). Translation by the author.

Endnotes

144 Karl Stodieck was a professor at the Technische Hochschule Berlin-Charlottenburg. Stodieck designed several industrial buildings in Berlin, such as the Fritz-Werner-Werk machine factory in Berlin-Marienfelde (1916–17), the Carl Lorenz AG in Berlin-Tempelhof (1916–18) and the C. J. Vogel Draht- und Kabelwerke AG in Berlin-Köpenick (1916–21). A brief outline of the life and work of Karl Stodieck has been published in an essay by Axel Busch: Luckenwalde. See "Zwei Industriebauten der 1920er Jahre von Karl Stodieck," in *Brandenburgische Denkmalpflege*, Feb. 2019.

145 Goldmerstein, J. and K. Stodieck. 1928. *Thermenpalast. Kur-. Erholungs-, Sport-, Schwimm- und Badeanlage.* Berlin: Wilhelm Ernst & Sohn, p. 4. Translation by the author.

146 Ibid. Translation by the author.

147 The hall should have a diameter of the main hall of 150 m, a ring basin of 6,700 m², a built-up area of about 17,600 m² and an air space of 450,000 m³. See Gröning, Gert and Joachim Wolschke. 1985. "Thermenpalast: ein Projekt aus Weimarer Zeit." *werk, bauen + wohnen* 72 (10): 37. Translation by the author.

148 Goldmerstein and Stodieck. Cited in ibid., 37. Translation by the author.

149 The central hall was surrounded by a three-story ring accommodating the "entrance hall, vestibules, waiting rooms, dressing rooms, drying cells, gymnasiums, shower rooms, tubs and various medical and electric baths, a restaurant and café for 700 people, staff and office rooms, a workshop, a laundry and the necessary lavatories." Goldmerstein and Stodieck (1928, 6). Translation by the author.

150 Ibid. Translation by the author.

151 Ibid. Translation by the author.

152 Ibid., 7. Translation by the author.

153 Ibid., 10. Translation by the author.

154 Ibid., 7. Translation by the author.

155 Ibid., 8. Translation by the author.

156 Ibid. Translation by the author.

157 Expressionist architectural thinking can be understood as pioneering two lines of thought that became influential over the course of the 20th century: A *proto-ecological line of thought* leading from Bruno Taut to Ian McHarg, Anne Whiston Spirn and Ralph Knowles; and a *disurbanist line of thought* leading from Bruno Taut to Mikhail Okhitovich and Moisei

Ginsburg in the early Soviet Union and Kisho Kurokawa in post-war Japan, promoting an agricultural city.

158 Allen, Stan and Marc McQuade (eds.). 2011. *Landform Building: Architecture's New Terrain.* Baden: Lars Müller Publishers, p. 64.

159 Gandy (2014, 59).

160 Ibid., 56.

161 Ibid., 55.

162 Steurer (2017, 137). Translation by the author.

163 Harather, Karin. 1995. *Haus-Kleider. Das Phänomen der Bekleidung in der Architektur.* Vienna: Böhlau.

164 Rodriguez-Lores, Juan. 1988. "Die Zonenplanung – Ein Instrument zur Steuerung des Kleinwohnungsmarktes." In *Die Kleinwohnungsfrage. Zu den Ursprüngen des sozialen Wohnungsbaus in Europa*, edited by Juan Rodriguez-Lores and Gerhard Fehl, Hamburg, pp. 157–207.

165 Aronin, Jeffrey Ellis. 1953. *Climate & Architecture,* New York, p. 17.

166 Hilberseimer (1944, 96).

167 Ibid., 96.

168 "The wind over a city will be one or a combination of the following: the prevailing macroclimatic wind, the microclimatic winds induced by topography, etc., or the wind set up by the presence of the city itself." Aronin (1953, 212).

169 Accessed on February 11, 2022. http://www.staedtebauliche-klimafibel.de

170 According to Aronin (1953, 213).

171 Harbusch, Gregor et al. 2014. "Established Modernists go into Exile, Younger Members go to Athens." In *Atlas of the Functional City. CIAM 4 and Comparative Urban Analysis*, edited by Evelien van Es et al. Bussum: THOTH publishers gta Verlag, p. 171.

172 Van Es, Evelien. 2014. The Exhibition Housing, Working, Traffic, Recreation in the Contemporary City – A Reconstruction, in *Atlas of the Functional City. CIAM 4 and Comparative Urban Analysis*, edited by Evelien van Es et al. Bussum: THOTH publishers gta Verlag, p. 447.

173 Van der Linden, Cornelius, Hubert Hoffmann and Wilhelm Hess. 1932. arbeitsgruppe dessau der int. kongresse f. neues bauen, 11. vor- Nachteil der Lage. Translation by the author.

174 Hilberseimer (1944). *The New City. Principles of Planning,* Chicago, p. 134 and 135.

175 Ibid., 132 and 136.

176 Reuter, Ulrich. 2011. "Implementation of Urban Climatology in City Planning in the City of Stuttgart." In *City Weathers. Meteorology and Urban Design 1950–2010*, edited by Michael Hebbert, Vladimir Jankovic and Brian Webb. Manchester: University of Manchester, p. 157.

177 "As has been seen already, buildings cause eddies in the wind. Thus, there is a braking of friction effect. According to the authorities just mentioned, it is nevertheless difficult to estimate to what degree buildings have influenced the flow of wind in cities. This is because, as the towns were built up, measurements were not usually taken at the same time and the same place. Linke, however, has recorded that there was a slowing down of the mean wind from 5.1 meters to 3.9 meters per second in 10 years in a certain German city. Also, Kratzer's doctoral dissertation noted that in Detroit there was a retardation from 6.5 to 3.8 meters per second in 20 years. The city effect can be noticed too in Stuttgart, where, from 1894 to 1923 the amount of windless days increased from one per cent to twenty-three per cent. It may be determined from this information that buildings, especially when grouped together in small areas, will greatly stop the flow of air. This is not always good; a moving breeze is an excellent thing to prevent the accumulation of foul odors and unsanitary areas." Aronin (1953, 212).

178 Hebbert and Webb (2012, 127).

179 Reuter (2011, 157).

180 "With the beginning of the war in 1939, the municipal meteorologist was appointed to organize measures of air protection. In Stuttgart, these measures were based, among other things, on the idea of depriving attacking bomber planes of the view of the city as a target, which was still necessary at that time, by generating artificial fog–a concept that, admittedly, could only be successful until the extensive introduction of radar in the later years of the war. Smoke screening was carried out from countless points selected according to wind-climatic criteria. This resulted in findings about the air-flow conditions and the ventilation possibilities of the housing-estate areas located in valley locations. It was found that in some areas the artificially generated fog dissipated more quickly than desired, while in other parts of the city it remained for an unusually long time. These facts were correctly associated with the ventilation effect of the many ground-level cold air outflows present in the urban area, which in calm weather conditions appear as nocturnal

slope winds, and in some cases, also as powerful mountain winds. Their topographically preferred routing and drainage paths in the landscape soon became known as the city's fresh air corridors. This involuntary large-scale urban climatic experiment had a considerable impact on urban planning, since it was now important in future urban planning for the city to preserve the function of the fresh air corridors identified in this way, which are important for the urban climate, and to develop them as far as possible." Accessed on July 10, 2019. https://www.stadtklima-stuttgart.de/index.php?service_kontakt_75_jahre_stadtklimatologie. Translation by the author.

181 Schmidt (1926, 40).Translation by the author.

182 Ibid.

183 See Milyutin, N. A. 1930. *Sotzgorod*, Moscow. Mentioned by Hilberseimer in *The New City,* p. 70.

184 Kratzer, Albert (1956 [1937]), *Das Stadtklima*, p. 102.

185 Ibid.

186 May, Ernst. 1931. "Stalingrad." *Frankfurter Zeitung*, January 24. Reprinted in: *Standard Cities. Ernst May in der Sowjetunion 1930–1933. Texte und Dokumente*, edited by Thomas Flierl, Frankfurt am Main, 2012, p. 232. Translation by the author.

187 Bodenschatz, Harald and Christiane Post (eds.). 2003. "Sozialistischer Städtebau im Zeichen der Moderne." In *Städtebau im Schatten Stalins. Die internationale Suche nach der sozialistischen Stadt in der Sowjetunion 1929–1935*. Berlin: Verlagshaus Braun, p. 44. Translation by the author.

188 Accessed on February 11, 2022. http://www.newtowninstitute.org/newtowndata/newtown.php?newtownId=338

189 Bodenschatz and Post (2003, 44).

190 "Within the framework of the industrialization policy of the first five-year plan, a number of city foundations at significant raw material deposits were intended. Magnitogorsk, in connection with the Siberian Kuzneck Basin, was the key to the development of the so-called second metallurgical base. With its location behind the Ural Mountains and thus out of reach of possible attacks on the country from the West, this was to complement the 'first metallurgical base' for the industrialization of the country in the Ukrainian Donetsk Basin." Ibid. Translation by the author.

191 Data according to ibid., 57. Translation by the author.

192 Ibid., 60. Translation by the author.

193 Ibid., 59. Translation by the author.

194 Ibid., 60. Translation by the author.

195 Ibid. Translation by the author.

II Thought Styles of Man-made Climates, *An Introduction*

1 Wagner, Martin. 1951. *Wirtschaftlicher Städtebau.* Stuttgart: Julius Hoffmann, p. 9. Translation by the author.

2 Steinhauser, Ferdinand. 1957. "Klimatologische Sonderbearbeitungen für Zwecke der Bautechnik und der Grossstadthygiene." In *Klima und Bioklima von Wien. Eine Übersicht mit besonderer Berücksichtigung der Bedürfnisse der Stadtplanung und des Bauwesens,* II. Teil, edited by F. Steinhauser, O. Eckel and F. Sauberer. Vienna: Österreichische Gesellschaft für Meteorologie, p. 47. Translation by the author.

3 "Can I plead here for the R.I.B.A. to use its influence in regard to location of pipes?" Manley, Gordon. 1949. "Microclimatology – Local Variations of Climate likely to affect the Design and Siting of Buildings." *The Journal of the Royal Institute of British Architects* 56 (7): 319 (read at meeting of R.I.B.A. on 12 April 1949).

4 Strempler, Werner. 1954. *Klimatologie und Mikroklimatologie in der Stadtplanung. Die Auflockerung der Berliner Innenstadt durch Grünanlagen (Probleme und Vorschläge)*, p. 2. Unpublished manuscript of the *Senats für Bau- und Wohnungswesen*, Berlin. Landesarchiv Berlin, B Rep. 009 Nr. 266 1954–1955, old archive signature: B Rep. 9 Acc. 3851 Nr. 253, Index number: 137. Translation by the author.

5 See in particular the comprehensive overview provided in: Hebbert, Jankovic and Webb (2011).

6 Chandler, Tony. 1970. *Selected Bibliography on Urban Climate*. Secretariat of the World Meteorological Organization / Griffiths, John F. and M. Joan Griffiths. 1969. *A Bibliography of Weather and Architecture*. ESSA Technical Memorandum EDSTM 9, Silver Spring, April.

7 See Bob Frommes' 1978, 1980, and 1983 research reviews: Frommes, Bob. 1978. *Angewandte Klimatologie = Besseres + Billigeres Bauen + Wohnen*. Ständiger Ausschuss Stadt- und Bauklimatologie (IVWSR), Luxemburg; Frommes, Bob. 1980. *Wissensgrundlage der Stadt- und Bauklimatologie*, Ständiger Ausschuss Stadt- und Bauklimatologie (IVWSR), Luxembourg / Frommes, Bob. 1983. *Stadt- und Bauklimatologie für Architekten und Planer Leitfaden für Lehrprogramme*, Ständiger Ausschuss Stadt- und Bauklimatologie (IVWSR), Luxembourg.

8 Böer, Wolfgang. 1954. *Klimaforschung im Dienste des Städtebaus*. Berlin: Deutsche Bauakademie, p. 48. Translation by the author.

9 Olgyay, Victor (with Aladar Olgyay). 1963. *Design with Climate – Bioclimatic Approach to Architectural Regionalism.* Princeton: Princeton University Press, p. 44.

10 Aronin (1953, 214–18).

11 Spirn, Anne Whiston. 1984. *The Granite Garden. Urban Nature and Human Design*. New York: Basic Books, p. 5.

12 Ibid., xiii.

13 Spirn (1984). Cited in Stephen M. Wheeler and Timothy Beatley (eds.). 2004. *The Sustainable Urban Development Reader*. London: Routledge, p. 114.

14 Chang, Jiat-Hwee. 2016. "Tropical Variants of Sustainable Architecture: A Postcolonial Perspective." In *Handbook of Architectural Theory,* edited by Greig Crysler, Stephen Cairns and Hilde Heynen. London, Thousand Oaks, CA: SAGE Publications. p. xix. "Not just environmental discourses but also cultural discourses [are] inextricably intertwined with the identity politics of the post-colonial globalized world."

15 Hebbert and Webb (2012, 123).

16 Ibid., 1542.

17 Ibid.

18 Rittel, Horst W. J. and Melvin M. Webber. 1973. "Dilemmas in a General Theory of Planning." *Policy Sciences* 4. Rittel and Webber relate wicked problems, for instance, to "air pollution" (p. 168) and "the current concerns with environmental quality and with the qualities of urban life" (p. 157).

19 Ibid., 166.

20 Ibid., 162.

21 Ibid., 168.

22 Ibid., 160.

23 Fleck, Ludwig. 1980. Entstehung und Entwicklung einer wissenschaftlichen Tatsache. Einführung in die Lehre vom Denkstil und Denkkollektiv [1935]. Frankfurt am Main: Suhrkamp. Translation by the author.

24 Blumenberg, Hans. 1998 (1960). *Paradigmen zu einer Mataphorologie*. Frankfurt am Main: Suhrkamp Verlag, p. 8. Translation by the author.

25 Ibid., 8. Translation by the author.

26 Ibid., 9. Translation by the author.

27 Ibid., 11. Translation by the author.

28 On the notion of "the metaphor" in architecture and urban design, see: Ungers, Oswald Mathias. 1982. *Morphologie. City Metaphors*, Cologne: W. König / Gerber, Andri. 2012. *Theorie der Städtebaumetaphern – Peter Eisenman und die Stadt als Text*. Zurich: Chronos Verlag / Hnilica, Sonja. 2012. *Metaphern für die Stadt – Zur Bedeutung von Denkmodellen in der Architekturtheorie*. Bielefeld: transcript Verlag.

4 Thermal Heritage

1 Egli, Ernst. 1945. "Vom regionalen Städtebau: der Einfluss von Klima und Landschaft auf die Wohn- und Stadtform." *Werk* 32 (3): 66. Translation by the author.

2 Egli, Ernst. 1951. *Climate and Town Districts. Consequences and Demands*, Erlenbach-Zurich: Verlag für Architektur, p. 10.

3 Roesler, Sascha. 2018. "Jenseits von innen und aussen. Die Stadt als thermischer Innenraum der Gesellschaft." In *archithese* 2: 54–64.

4 Halbwachs, Maurice. 1967 (1950). *Das kollektive Gedächtnis*. Stuttgart: Enke Verlag. French philosopher Halbwachs developed the notion of "collective memory" in the 1920s.

5 Warnke, Martin and Claudia Brink (eds.). 2003. *Aby Warburg. Der Bilderatlas MNEMOSYNE*. Berlin: Akademie Verlag.

6 Kratzer (1956 [1937], IV). Translation by the author. See also Geddes, Arthur. 1946. *Planning and Climate. Climates of Region, Locality and Site*. London: Association for Planning and Regional Reconstruction.

7 Neumann, Erwin. 1954. "Das Stadtklima." In *Die städtische Siedlungsplanung unter besonderer Berücksichtigung der Besonnung*. Stuttgart: Witter, p. 118. Translation by the author.

8 Halbwachs (1967 [1950], 128). Translation by the author.

9 Konau, Elisabeth. 1977. *Raum und soziales Handeln. Studien zu einer vernachlässigten Dimension soziologischer Theoriebildung*. Stuttgart: Enke Verlag, p. 29. Translation by the author.

10 Halbwachs (1967 [1950], 130). Translation by the author.

11 Ibid., 133. Translation by the author.

12 Ibid., 135. Translation by the author.

13 On Kon Wajiro's urban ethnography, see: Fujii, James A. (ed.). 2004. *Text and the City. Essays on Japanese Modernity*. Durham, NC: Duke University Press / Harootunian, Harry. 2000. *Overcome by Modernity. History, Culture, and Community in Interwar Japan*. Princeton: Princeton University Press / Fawcett, Chris. 1980. *The New Japanese House. Ritual and Anti-Ritual, Patterns of Dwelling*. London, New York: Granada.

14 Meanwhile, Chernobyl (Ukraine) is also known for its rich flora and fauna. As a closed "zone" it is reminiscent of Andrei Tarkovsky's film *Stalker*.

15 Lowenhaupt Tsing, Anna. 2015. *The Mushroom at the End of the World. On the Possibility of Life in Capitalist Ruins*. Princeton: Princeton University Press, p. 21.

16 Ibid.

17 See the Film *One with the earth's cycle* on the work of the Japanese architect Hiroshi Sambuichi (Japan), Louisiana Channel 2017.

18 Shitara, Hiroshi. 1957. "Effects on Buildings upon the Winter Temperature in Hiroshima City." *Geographical Review of Japan* 30 (6): 468–82.

19 The architect developed the design for the Orizuro Tower, adjacent to the Hiroshima Peace Park, remembering these natural flows.

20 Accessed on February 11, 2022. http://www.iwk.ac.at/institut/geschichte

21 "Vorträge über Umweltforschung." *Wetter und Leben* 8 (1948): 248.

22 "Aufgabe und Ziel unserer Zeitschrift". *Wetter und Leben* 1 (1948): 1. Translation by the author.

23 Sauberer, Franz. 1948. *Wetter, Klima und Leben. Grundzüge der Bioklimatologie*. Vienna: Hollinek.

24 Already during the Habsburg Empire a so-called "climatography" had developed.

25 Steinhauser, F., O. Eckel and F. Sauberer. 1955. *Klima und Bioklima von Wien. Eine Übersicht mit besonderer Berücksichtigung der Bedürfnisse der Stadtplanung und des Bauwesens*, I. Teil. Vienna: Österreichische Gesellschaft für Meteorologie, p. 3. Translation by the author.

26 Steinhauser, F. 1957. "Säkulare Änderungen der klimatischen Elemente." *Klima und Bioklima von Wien. Eine Übersicht mit besonderer Berücksichtigung der Bedürfnisse der Stadtplanung und des Bauwesens*, II. Teil, edited by F. Steinhauser, O. Eckel and F. Sauberer. Vienna: Österreichische Gesellschaft für Meteorologien, p. 5. Translation by the author.

27 Steinhauser, F., O. Eckel and F. Sauberer. 1959. *Klima und Bioklima von Wien. Eine Übersicht mit besonderer Berücksichtigung der Bedürfnisse der Stadtplanung und des Bauwesens*, III. Teil, edited by F. Steinhauser, O. Eckel and F. Sauberer. Vienna: Österreichische Gesellschaft für Meteorologie, p. 118. Translation by the author.

28 Steinhauser, Eckel and Sauberer (1959, 119). Translation by the author.

29 Ibid., 118. Translation by the author.

30 Ibid., 120. Translation by the author.

31 Ibid. Translation by the author.

32 Ibid., 27. Translation by the author.

33 McHarg (1969, 57).

34 Ibid.

35 Spirn, Anne Whiston. 1993. "Deep Structure: On Process, Form, and Design in the Urban Landscape." In *City and Nature: Changing Relations in Time and Space,* edited by T. M. Kristensen et al. Odense, Denmark: Odense University Press, p. 9.

36 Ibid., 10.

37 Ibid., 11.

38 Konau (1977, 29).

39 Spirn (1993, 11).

40 See also the "High-Deck-Siedlung" in Berlin.

41 The project of the Chimbote Regulatory Plan was presented as one of the themes of CIAM VIII in 1955 in Bergamo (Italy).

42 Whittick, Arnold. 1950. *European architecture in the twentieth century*. London: C. Lockwood. Cited in Aronin (1953, 112).

43 Aronin (1953, 112).

44 Manley (1949, 317).

45 Henschel, Christhardt. 2021. "Kommunizierende Röhren der Singularitäten? Deutsche Debatten um den Holocaust aus Warschauer Perspektive." *Geschichte der Gegenwart*, July 7. Accessed on July 9, 2021. https://geschichtedergegenwart. ch/kommunizierende-roehren-der-singularitaeten-deutsche-debatten-um-den-holocaust-aus-warschauer-perspektive/. Translation by the author.

46 CENTRALA. 2018. *Amplifying Nature. The Planetary Imagination of Architecture in the Anthropocene*, edited by Anna Ptak. Warsaw: Zachęta National Gallery of Art, p. 58.

47 Ibid.

48 Ibid., 59.

49 Ibid., 61.

50 Ibid., 62.

51 Ibid., 58.

52 Ibid., 63.

53 Frampton, Kenneth. 2011. "Megaform as Urban Landscape." In *Landform Building: Architecture's New Terrain*, edited by Stan Allen and Marc McQuade. Baden: Lars Müller Publishers, p. 238.

54 Ibid.

55 Kratzer (1956 [1937], 103).

56 "Magistrate of Berlin, Department of Construction and Housing. The authority was formed on May 17, 1945 as the 'Department of Construction and Housing.' Hans Scharoun was appointed city councilor and head of the department. After the elections of October 1946, Karl Bonatz was entrusted with this function." Accessed on July 9 2020. http://www.content.landesarchiv-berlin.de/php-bestand/. Translation by the author.

57 Selman Selmanagić, Wils Ebert, Peter Friedrich, Sergius Ruegenberg, Herbert Weinberger, Reinhold Lingner (landscape architect). Karl Böttcher (management of the main office for reconstruction implementation).

58 Wolschke-Bulmahn and Fibich (2004, 92).

59 Bodenschatz, Harald et al. 2009. *Berlin und seine Bauten. Teil 1 – Städtebau.* Berlin: DOM publishers, p. 145.

60 The damage mapping had been supervised by Hans Stephan on behalf of Albert Speer since January 28, 1944, as part of the "reconstruction planning" of Berlin; the results of the bombardments were "continuously entered" into the damage plan. Geist, Johann Friedrich and Klaus Küvers. 1995. "Tatort Berlin, Pariser Platz. Die Zerstörung und 'Entjudung' Berlins." In *1945. Krieg – Zerstörung – Aufbau. Architektur und Stadtplanung 1940–1960*, Schriftenreihe der Akademie der Künste, Vol. 23. Berlin: Henschel Verlag, p. 109. Translation by the author.

61 Ibid., 110.

62 Steurer (2017, 137).

63 Go Jeong-Hi. 2006. "Herta Hammerbacher (1900–1985): Virtuosin der Neuen Landschaftlichkeit – der Garten als Paradigma." *Landschaftsentwicklung und Umweltforschung.* Special issue, Vol. 18, Technische Universität Berlin, p. 40.

64 Posener, 1979–1985. Cited in Macieczyk and Syfrig (1985, 9).

65 Düwel, Jörn. 1995. Berlin. "Planen im kalten Krieg." In *1945. Krieg – Zerstörung – Aufbau. Architektur und Stadtplanung 1940–1960*, Schriftenreihe der Akademie der Künste, Vol. 23. Berlin: Henschel, p. 197.

66 Hans Scharoun. Cited in ibid., 198. In May 1949, a summary presentation entitled "Principles of the New Planning of Berlin" *(Grundzüge zur Neuplanung Berlins)* was submitted as the last comprehensive attempt to redesign the entire city after the war. To the East Berlin Magistrate, a planning collective that came from the former Main Office for Planning II submitted "various proposals for a continuation of the previous results of our work." Ibid., 204.

67 Bodenschatz (2009, 145).

68 Wolschke-Bulmahn and Fibich (2004, 92).

69 Düwel (1995, 203).

70 Göderitz, Johannes und Roland Rainer, Hubert Hoffmann. 1957. *Die gegliederte und aufgelockerte Stadt*, Tübingen. The main author of the study, Bauhaus student Hubert Hoffmann, whom Scharoun had since 1934 been associated with in the informal Friday Group, had personally given him a copy. Data according to: Geist and Küvers (1995, 110).

71 Geist and Küvers (1995, 110).

72 Hilberseimer (1944, 96).

73 Ibid., 95.

74 Pfützner was motivated to conduct his research by Georg Pniower. Together with Reinhold Lingner, Pniower is considered the most important landscape planner of the GDR. After Pniower's death in 1960, the dissertation was further supervised by Lingner.

75 Pfützner, Friedrich-Herman. 1962. *Über den Einfluß städtischer Grünanlagen auf die Abkühlungsgröße*. Dissertation, Humboldt-Universiät zu Berlin, p. 97.

76 One should "break up the residential and working areas to such an extent that the microclimate of these areas no longer differs significantly from that of the outskirts." Hassenpflug, Gustav. 1951. "Das Krankenhaus in der Stadtplanung." In *Handbuch für den Neuen Krankenhausbau*, edited by Paul Vogler and Gustav Hassenpflug, Munich, Berlin: Urban & Schwarzenberg, p. 55.

77 Pfützner (1962, 88). Translation by the author.

78 Ibid., 89.

79 Ibid.

80 Ibid., 96.

81 Ibid., 90.

82 Ibid., 93.

83 Ibid., 91.

84 Ibid., 93.

85 Neumann (1954, 118).

86 Pfützner (1962, 90). Translation by the author.

87 Perry, Clarence. 1929. *The Neighbourhood Unit: From the Regional Survey of New York and Its Environs*, Vol. VII, "Neighbourhood and Community Planning." New York: Wiley.

88 Taut, Max. 1946. *Berlin im Aufbau*. Berlin: Aufbau-Verlag.

89 "The surface structure of most modern urban landscapes obscures deep structure with layers of human constructions, many of which are unrelated to deep structure. To most people, the deep structure of a city is invisible, destroyed by the processes of city-building. Although hidden, deep structure continues to exert a powerful influence upon the urban landscape, and should be attended to, especially in the design of the infrastructure that sustains the city and the organism that dwell within it." Spirn (1993, 12).

90 The competition was announced on March 30, 1957. In the invitation, "geological and topographical conditions, climate, living and working, utility networks and greenery are presented in detail." Bodenschatz (2009, 224). Translation by the author.

91 Gandy (2014, 56).

92 Bodenschatz (2009, 231). Translation by the author.

Endnotes

243

93 Förster, Kim. 2019. "The Green IBA. On a Politics of Renewal, Ecology, and Solidarity." *Candide* 11: 9–50. / Bartoli, Sandra et al. (eds.). 2020. *Licht Luft Scheiße. Perspektiven auf Ökologie und Moderne*, Vol. 1, Archäologien der Nachhaltigkeit, Hamburg: Adocs.

94 Kennedy, Margrit. 1984. "Preface." In Fisch, Rose, Inge Maass and Katrin Rating *Der grüne Hof. Grundlagen und Anforderungen an die Hofbegrünung in der Stadterneuerung.* Karlsruhe: C. F. Müller, p. V.

95 Schmalz, Joachim. 1987. *Das Stadtklima. Ein Faktor der Bauwerks- und Städteplanung. Unter besonderer Berücksichtigung der Berliner Verhältnisse, mit Beispielen aus Planungsgebieten der 'Internationalen Bauaustellung Berlin 1987'.* Karlsruhe: C. F. Müller. Translation by the author.

96 Fisch, Rose, Inge Maass and Katrin Rating. 1984. *Der grüne Hof. Grundlagen und Anforderungen an die Hofbegrünung in der Stadterneuerung.* Karlsruhe: C. F. Müller. Translation by the author.

97 Schmalz (1987, 95). Translation by the author.

98 Ibid., 96. Translation by the author.

99 Ibid., 102. Translation by the author.

100 Ibid., 104. Translation by the author.

101 S.T.E.R.N.-Fotobestand, Signatur: 0080/K/16

102 Schmalz (1987, 118). Translation by the author.

103 Fisch, Maass and Rating (1984, 7). Translation by the author.

104 Ibid., 20. Translation by the author.

105 Olgyay (1963, 175).

106 Ibid., 100.

107 One could also refer to Italian architect Gaetano Vinaccia, who, in the context of Italian fascism, worked on a climate-based reformulation of cities in Italy and its colonies. In particular, the first volume, *Come il clima plasma la forma urbane e l'architettura* (1943), examines climatic influences as determinants of urban form and determinants of architecture. Vinaccia's research – which explicitly sought a synthesis of urban climatology, urban planning and architecture – demonstrates the factual difficulties of merging these disciplines; the major qualitative differences between Victor Olgyay's and Gaetano Vinaccia's work reveals the complexity of such an endeavor. The second volume, *La citta e il contado. Gli organismi vitali della città e l'ordine cittadino,* presents in particular the architectural investigations and designs of the author. Here, a schematically elaborated climatic determinism is expressed. Vinaccia, Gaetano. 1943, 1952. *Per la città di domani.* Rome: Fratelli Palombi Editori, two volumes / Chiri, Giovanni and Ilaria Giovagnorio. 2015. "Gaetano Vinaccia's (1881–1971) Theoretical Work on the Relationship between Microclimate and Urban Design." *Sustainability* 7: 4448–4473.

108 Manuscript "Theorie des Städtebaues. 1. Die Stadt und die Erdoberfläche." Lecture series at ETH Zurich, spring semester 1946, ETH Zurich Research Collection, Ernst Egli, Hs. 785:5, p. 3. Translation by the author.

109 Manley, Gordon. 1958. "The Revival of Climatic Determinism." *Geographical Review* 48 (1): 98–105.

110 Egli (1951, 66).

111 Ibid., 81. "Are there different basic structures for the different climatic zones, and what are their characteristics?" Ibid., 72. "Can such basic structures change?" Ibid., 80.

112 Egli wrote an (unpublished) play about the "Seasons," based on antique models, which shows his profound interest in this theme. Egli, Ernst. 1963. *Die Jahreszeiten. Ein Marionettenspiel der Zeit.* Typescript, ETH Zurich Research Collection, Ernst Egli, Hs. 785: 96.

113 Egli (1951, 72). Egli's deductive approach resolutely relies on climatic determinism. His argumentations develop to a large extent without considering novel empirically based urban climatic knowledge as it is present, for example, in the publications of Victor Olgyay.

114 Ibid. 64.

115 Egli (1945, 66). Translation by the author.

116 Ibid., 65. Translation by the author.

117 Ibid. Translation by the author.

118 Ibid., 66. Translation by the author.

119 His book started as "industrial climatology," according to the preface; it is primarily aimed at representatives of the industry.

120 Brooks, Charles Ernest. 1951. *Climate in Everyday Life.* New York: John Wiley & Sons, p. 21.

121 Ellis, F. P. 1953, "Thermal comfort in warm and humid atmospheres." *The Journal of Hygiene* 51 (3): 389.

122 Brooks (1951, 20).

123 Coen (2016).

124 Egli (1951, 72).

125 Ibid., 69

126 Ibid., 82.

127 Ibid.

128 Ibid., 80.

129 Ibid., 74.

130 Ibid., 71.

131 Pfützner (1962, 27). Translation by the author

132 Aronin (1953, 114).

133 The various housing units were designed by Indian architects (with Maxwell Fry and Pierre Jeanneret as resident architects) and Le Corbusier working from his Paris office.

134 Aronin (1953, 214).

135 Ibid.

136 Ibid.

137 Ibid., 112.

138 Ibid., 216.

139 Ibid., 214.

140 Ibid.

141 Ibid., 218.

142 Ibid., 216.

143 Ibid.

144 Ibid., 212.

145 Wedepohl, Claudia. 2005. "Ideengeographie — Ein Versuch zu Aby Warburgs 'Wanderstraßen der Kultur'." In Helga Mitterbauer and Katharina Scherke (eds.). *Entgrenzte Räume – Kulturelle Transfers um 1900 und in der Gegenwart*, Studien zur Moderne 22. Vienna: Passagen-Verlag, p. 253. Translation by the author. See also Wittkower, Rudolf. 1977. *Allegory and the Migration of Symbols.* London: Thames & Hudson.

146 Foucault, Michel. 1969 (1966). *Die Ordnung der Dinge – Eine Archäologie der Humanwissenschaften*, Frankfurt am Main: Suhrkamp, p. 11. Translation by the author.

147 Halbwachs (1967, 1950); Catherin Bull et al. (eds.). 2007. *Cross-Cultural Urban Design. Global or Local Practice?* London: Routledge / Hiroshi Yamamoto and Christine Ivanovic (eds.). 2010. *Übersetzung – Transformation: Umformungs-*

prozesse in/von Texten, Medien, Kulturen. Würzburg: Königshausen & Neumann / Kikuko Kashiwagi-Wetzel and Michael Wetzel (eds.). 2015. *Interkulturelle Schauplätze in der Großstadt. Kulturelle Zwischenräume in amerikanischen, asiatischen und europäischen Metropolen.* Paderborn: Wilhelm Fink.

148 Letter from Fritz Saxl to the publisher B. G. Teubner (Leipzig). Warnke and Brink (2003, XVIII). Translation by the author.

149 See, for instance, the exhibition *Umdenken Umschwenken*, presented in Switzerland, Germany and Austria. See: Roesler, Sascha. 2009. "Nach Sparta – Zwei Ausstellungen als Plädoyers für eine schwach technisierte Lebensweise." *Kunst + Architektur in der Schweiz,* "Lebensstil – Experimente nach 1970" 2: 6–13.

150 "In his book Architecture without Architects, Rudowsky illustrates a design of natural air conditioning device in use for over five hundred years in the lower Sind district of Pakistan." Steadman, Philip. 1975. *Energy, Environment and Building.* Cambridge, MA: Cambridge University Press, p. 32–33.

151 Barthes, Roland. 1989 (1980). *Die helle Kammer. Bemerkungen zur Photographie.* Frankfurt am Main: Suhrkamp, p. 28. "One photo resonates with me [*m'advient*], the other doesn't." Translation by the author.

152 Ibid., 26.

153 The photographic gains of his trip to India have been documented by Hürlimann in five photo albums, of which four are available nowadays in the estate Hürlimann of the Fotostiftung Schweiz. In Album V, which, among other things, is dedicated to the city of Hyderabad, 13 photographs of the windcatchers are gathered.

154 Hürlimann, Martin. 1928. *Indien. Baukunst, Landschaft und Volksleben.* Zurich: Fretz & Wasmuth, p. XXV.

155 In January 1929, Hürlimann published the first issue of the magazine *Atlantis*, showing his photographic works and those of others.

156 From 1925, Galloway began to systematically expand his collection of photographs and to acquire whole archives. In that year, he was able to purchase 8,000 images from Africa and Asia. In the following years, he opened agencies in Chicago, Detroit, Los Angeles, Boston, London, Berlin and Amsterdam.

157 See: Habermas, Jürgen. 1989 (1962). *The Structural Transformation of the Public Sphere: An Inquiry into a Category of Bourgeois Society.* Cambridge, MA: MIT Press.

158 See: Benjamin, Walter. 2008 (1935). *The Work of Art in the Age of Its Technological Reproducibility, and Other Writings on Media.* Cambridge, MA: Belknap Press.

159 The photographic visualization of the world has been continued and brought to fruition after the Second World War by famous stock-photo agencies such as "Magnum." Founded in Paris by Henri Cartier-Bresson and Robert Capa, Magnum was a stock-photo agency with a global reach that retained its importance until the digitalization and the emergence of multinational corporations such as Getty Images in the 1990s.

160 Martin Hürlimann, "Nennen wir das Ding einfach Atlantis!" *Züri Leu* 2, Sept. 1977, p. 13. With the takeover by Conzett und Huber, *Atlantis* was united with the magazine *Du.*

161 Hürlimann (1928, XXIV).

162 The products of Hyderabad were presented at the World Expo 1851 in London. See: Mir Atta Muhammad Talpur. 2007. *The Vanishing Glory of Hyderabad (Sindh, Pakistan)*, in *Webjournal in Cultural Patrimony*, p. 51. Accessed on March 17, 2016. http://www.webjournal.unior.it/dati/19/72/web%20journal%203,%20hyderabad.pdf

163 Fotostiftung Schweiz, Estate Martin Hürlimann, Volume: *Indienreise III,* p. 85–90; Volume: *Indienreise V, Kashmir – Heimreise,* p. 65–70.

164 Sachsse, Rolf. 1984. *Photographie als Medium der Architekturinterpretation: Studien zur Geschichte der deutschen Architekturphotographie im 20. Jh..* Berlin: De Gruyter Saur, p. 91. Cited in Günther, Lutz Philipp. 2009. *Die bildhafte Repräsentation deutscher Städte. Von den Chroniken der Frühen Neuzeit zu den Websites der Gegenwart.* Cologne, Weimar, Vienna: Böhlau, p. 217.

165 Sachsse (1984, 110).

166 Mir Atta Muhammad Talpur (2007, 59).

167 Charles Ernest Brooks mentions the province of Sindh as one of India's regions with "very high temperature" (beside Punjab and the North-west Provinces of India). Brooks (1951, 22).

168 Nova, Alessandro. 2007. *Das Buch des Windes. Das Unsichtbare sichtbar machen.* Munich: Deutscher Kunstverlag, p. 17.

169 Benjamin, Walter. 1978 (1925). *Ursprung des deutschen Trauerspiels.* Frankfurt am Main: Suhrkamp, p. 152. Translation by the author.

170 Ibid., 153. Translation by the author.

171 Ibid., 141. Translation by the author.

172 Zauchtel is today's *Suchdol nad Odrou,* in the eastern Czech Republic.

173 Moravánszky, Ákos (ed.). 2002. *Das entfernte Dorf, Moderne Kunst und ethnischer Artefakt.* Cologne, Weimar, Vienna: Böhlau

174 Wittkower, Rudolf. 1977. "East and West: The Problem of Cultural Exchange." In *Allegory and the Migration of Symbols.* London: Thames & Hudson, p. 10–14.

175 Egli (1951, 106).

176 Ibid., 102.

177 Ibid., 107. Translation by the author.

178 Olgyay (1963, 10).

179 Ibid., 95.

180 Ibid., 94.

181 Rudofsky, Bernard. 1964. *Architecture without Architects. An Introduction to Non-Pedigreed Architecture.* New York: Museum of Modern Art. See images 113, 114 and 115.

182 "These unusual roofscapes are a prominent feature of the lower Sind district in west Pakistan. From April to June, temperatures range above 120° F., lowered by an afternoon breeze to a pleasant 95°. To channel the wind into every building, 'badgir,' wind catchers, are installed on the roofs, one to each room. Since the wind always blows from the same direction, the position of the wind catchers is permanently fixed. In multi-storied houses they reach all the way down, doubling as intramural telephones. Although the origin of this contraption is unknown, it has been in use for at least five hundred years." Ibid., images 113, 114 and 115.

183 Ibid., Preface.

184 Ibid.

185 Ibid., images 113, 114 and 115.

186 Mir Atta Muhammad Talpur (2007, 59).

187 Benjamin (1978 [1925], 145).

188 Atalay Franck, Oya. 2012. *Architektur und Politik: Ernst Egli und die türkische Moderne 1927–1940.* Zurich: gta Verlag.

189 "How will the rooftops of the future look, as we begin to optimize their form in order to aid in the generation of electricity? Will our cities begin to look like Hyderabad, Pakistan, with its wind catchered skyline? From April to June the temperature in Hyderabad can exceed 50° C, but the wind always blows from the same direction, so the position of rooftop wind

Endnotes

245

catchers are fixed and define the image of the city. What are other implications of sculpted rooftops?" Accessed on June 14, 2016. https://thefunctionality.wordpress.com/2009/01/08/mag-lev-turbines-roofs-of-the-future

5 Microclimatic Islands

1 Lenzholzer, Sanda. 2015. *Weather in the City: How Design Shapes the Urban Climate.* Rotterdam: nai010 Publishers.

2 Dahinden, Justus. 1974. *Akro-Polis. Frei-Zeit-Stadt / Leisure City.* Stuttgart: Karl Krämer Verlag, p. 6. Translation by the author.

3 Ovaska, Arthur A. 1978. "The city as a garden, the garden as a city." In The Urban Garden, O. M. Ungers, H. Kollhoff, A. Ovaska, Berlin. NO PAGES.

4 Wagner (1951, 10). Translation by the author.

5 Heschong, Lisa. 1979. *Thermal Delight in Architecture.* Cambridge, MA: MIT Press, p. ix.

6 Ibid., 29.

7 Jiménez Alcalá, Benito. 1999. "Aspectos bioclimaticos de la arquitectura Hispanomusulmana." *Cuadernos de la Alhambra* 35: 13–29 / Petersen, Steen Estvad. 1993. "Desert and City in Islamic Horticulture." In *City and Nature: Changing Relations in Time and Space*, edited by T. M. Kristensen et al. Odense, Denmark: Odense University Press, p. 161–180.

8 "Historical descriptions and recent measurements confirm the pleasant microclimate both in the gardens of the courtyards and inside the buildings." See Hagen, Katrin. 2012. "Mit Sinnen und Verstand. Design principles of Moorish gardens as inspiration for climate-sensitive urban design." *dérive* 48, Stadt KLIMA Wandel, p. 23. Translation by the author. "The microclimatic effect of such a spatial formation can best be explained by the patio-pórtico-torre concept (courtyard-arcade-tower): the lushly landscaped courtyards with their water features can develop a specific microclimate thanks to the insulation provided by the adjacent building facades; the openings of the self-contained buildings are positioned on the courtyard side and in the upper part of the tower-like main room, whereby the cool and humid garden air is sucked through the interiors; the shading by the superior arcades counteracts heating".

Translation by the author. See also her dissertation, assessing comprehensively both historical sources and contemporary measurements in Moorish gardens. Hagen, Katrin. 2011. *Freiraum im Freiraum: Mikroklimatische Ansätze für die städtische Landschaftsarchitektur.* Dissertation, Technical University of Vienna. Translation by the author.

9 Go (2006).

10 Hirschfeld, Christian Cay Lorenz. 1779. *Theorie der Gartenkunst,* Vol. 1, p. 155. Data according to: von Trotha, Hans. 1999. *Angenehme Empfindungen. Medien einer populären Wirkungsästhetik im 18. Jahrhundert vom Landschaftsgarten bis zum Schauerroman.* Munich: Wilhelm Fink Verlag, p. 9. See also Baumgarten, Alexander Gottlieb. 1750–1758. *Ästhetik,* 2 vols. Hamburg, 2007. Translation by the author.

11 Fisher, Mark. 2016. *The Weird and the Eerie.* London: Repeater Books, p. 13.

12 Sirén, Osvald. 1949. *Gardens of China.* New York: The Ronald Press Company, p. 5.

13 Ibid., 6.

14 Chiu Che Bing. 2008. "The Traditional Chinese Garden: A World Apart." In *In the Chinese City. Perspectives on the Transmutations of an Empire,* edited by Frédéric Edelmann, New York: Actar Publications, p. 63.

15 Lovejoy, Arthur O. 1960 (1948). "The Chinese Origin of a Romanticism." *In Essays in the History of Ideas,* New York: Capricorn Books, p. 102.

16 In the sense of Goethe's West-Eastern Divan. Von Goethe, Johann Wolfgang. 1923 (1819). *West-Östlicher Divan. In Goethes sämtliche Werke in sechzehn Bänden,* Vol. 11, Leipzig: Insel. Translation by the author.

17 Ernst Egli, following climate deterministic thinking of his time, conceived the gardens of Baghdad as not only modifying urban microclimates but more generally the "imaginative world" of the residents. "Not much imagination is needed to tell oneself how different life is in the atmosphere of London, compared to that of Baghdad. Any European who has lived for a certain time in the dry zone of the earth knows how every inch of his surroundings will change in importance even for him: the garden, the trees, the water, the land, shade and light. The imaginative world of a person born in that country will be entirely different from our own, for he has quite different needs and desires." Egli (1951, 64).

18 Clément, Gilles. 2015. *"The Planetary Garden" and Other Writings.* Philadelphia: University of Pennsylvania Press.

19 Foucault, Michel. 1984. "Des Espace Autres." *Architecture, Mouvement, Continuite* 5: 46–49.

20 Von Trotha, Hans. 2016. "Hortus conclusus. The Medieval Monastic Garden." In *Gardens of the World. Orte der Sehnsucht und Inspiration,* edited by Albert Lutz (and Hans von Trotha). Cologne: Wienand Verlag, p. 161. Translation by the author.

21 Cited in Bray, David. 2005. *Social Space and Governance in Urban China. The Danwei System from Origins to Reform.* Palo Alto: Stanford University Press, p. 17.

22 Boyd, Andrew. 1962. *Chinese Architecture and Town Planning. 1500 B.C. – A.D. 1911.* Chicago: University of Chicago Press, p. 79.

23 Pavilion and courtyard house form two complementary architectural typologies, which led architect and theorist Werner Blaser to his fundamental distinction between "massive and filigree construction". "The choice of skin-and-skeleton building as a means of joining and disjoining the indoor and outdoor world follows logically from the technological possibilities of combining the two archetypal dwellings: the cave and the tent." Blaser, Werner. 1974. *Chinese Pavillion Architecture. Quality, Design, Structure Exemplified by China.* Niederteufen: Verlag A. Niggli, p. 9.

24 Boyd (1962, 82).

25 Barthes, Roland. 1977. Fragments d'un discours amoureux. Paris: Seuil.

26 Lovejoy (1960 [1948], 101).

27 Glacken (1967, ix).

28 Pevsner (1976, 157).

29 Like Hebenbrand, Hassenpflug spent several years in the Soviet Union (1931–33) with the May group (the hospital-construction group under the leadership of Werner Hebebrand, the so-called "2nd May Group" in Moscow; see for instance the Kurt Liebknecht Hospital for Magnitogorsk since August 1931). Hassenpflug, who had studied at the Bauhaus, showed a great openness to interdisciplinary collaborations, which explains his interest in urban climatology.

30 Hassenpflug (1951, 55). Translation by the author.

31 Ibid. Translation by the author.

32 Ibid. Translation by the author.

33 Knoch, Karl. 1962. *Problematik und Probleme der Kurortklimaforschung als Grundlage der Klimatherapie, Mitteilungen des Deutschen Wetterdienstes,* No. 30, p. 50. Translation by the author. "This visual method certainly includes a strong subjective factor. In order to eliminate it as much as possible, an important meteorological element is measured in all parts of the terrain, namely the incoming solar energy, expressed in cal/cm²/year. It is locally dependent on the exposure and inclination of the irradiated area and is an essential basis for the formation of local climates. [...] The instrument can only provide relative values, but it has the great advantage of being easy to handle in the field. Therefore, it allows to measure a relatively large area in a short time. The first long-term experience was gained in vineyard climatology." Translation by the author.

34 "It is better to use a 5-level evaluation scale. A distinction is made between: 1. normal locations, which correspond to the macroclimate of the wider environment; 2. favorable locations; 3. particularly favorable locations; 4. unfavorable locations; 5. completely unsuitable locations. Favorable and unfavorable locations are to be justified, e.g., sunshine-rich, sunshine-poor (shade location), frost-prone, high wind strength, danger of fog." Ibid., 50. Translation by the author.

35 "In every cure the local climatic facts play a decisive role. They depend on the location of the spa in the terrain (whether in a valley, slope, plateau location, or peak), the formation of the surrounding mountains and their cover (field, meadow, forest). [...] From this arises the necessity to obtain the necessary knowledge about the local climatic differences for the spa. Since appropriate methods have already been tested, this is possible. Their aim is to carry out a mapping of the microclimate in the surroundings of the spa, as far as it serves the spa. In this way, a climate survey is created." Ibid., 49. Translation by the author.

36 Ibid., 53. Translation by the author.

37 Ibid. Translation by the author.

38 Ibid., 54. Translation by the author.

39 Ibid., 52. Translation by the author.

40 Steurer (2017, 129–142). Due to its enclosed character, the young Rem Koolhaas conceived West Berlin as an island and urban laboratory. In his 1972 thesis at the AA in London, Koolhaas explored the potential of the Berlin Wall as infrastructural architecture and the future potential of the death zone after the fall of the Wall.

41 Ovaska (1978).

42 Ibid.

43 Hertweck (2013, 56).

44 The trilogy of speculative architectural projects was elaborated between 1974 and 1978.

45 "For Ungers, the complex around Glienicke Palace thus figured as a miniature archipelago of differentiated architectural events, or, to put it the other way round with Rem Koolhaas: the urban archipelago manifesto made Berlin 'a gigantic enlargement of Schinkel's Glienicke Palace.'" Hertweck (2013, 60). Translation by the author.

46 Ovaska (1978).

47 "The theme, The Ciy as a Garden, is presented in a twoforld intention. First, in the search for a new image of urban structuring after the destruction caused by the poverty of Modern architectural and urban theory, a metaphorical analogy can be seen between the design of gardens and the design of cities. Secondly, the 'garden' analogy can be especially applied to Berlin, as a city of shrinking population and limited boundaries. The model derives as well from: 1.) the richness of Berlin as a 'green' city, 2.) the 'architecturalization of the green or the attempt to insert urban texture and urban elements with minimal means, and 3.) certain moments in the aesthetic history of Berlin, especially the era of Lenné and Schinkel and their Humanist City, which can be re-interpreted and expanded as an urban theory for the future of the city which has, as well, its conceptual roots in tradition." Ibid.

48 The Künstlerhaus Bethanien organized a summer academy in 1977 under the direction of Oswald Mathias Ungers, accompanied by a manifesto by Rem Koolhaas and Ungers.

49 See Ungers, O. M., H. F. Kollhoff and A. A. Ovaska. 1977. *The Urban Villa: a Multi-family Dwelling Type.* Cologne: Studio Press for Architecture.

50 On Ungers in general, see: Bideau, André. 2012. "Elusive Ungers." *AA Files* 64: 3–14 / Bideau, André. 2013. "Housing as Discursive Void: Oswald Mathias Ungers in the 1960s and 1970s." In *Candide* 7.

51 Developed within the framework of a competition for the Landstuhl milking plant. See Ungers, Oswald Mathias. 1980. *5 Energie Häuser.* Cologne: Walther König. Translation by the author.

52 Ibid. Translation by the author.

53 Bettini, Giulio. 2016. *Die "città animata". Mailand und die Architektur von Asnago Vender.* Zurich: gta Verlag, p. 7.

54 Bettini speaks of "following the rules typical of Milanese urban space and at the same time flagrantly transgressing them." Ibid., 7. Translation by the author.

55 Ibid. Translation by the author.

56 Ibid., 29. Translation by the author.

57 Ibid., 25. Translation by the author.

58 Mario Asnago and Claudio Vender in correspondence with the mayor of Milan. Data according to ibid., 40. Translation by the author.

59 Mario Asnago and Claudio Vender. Data according to ibid., 38. Translation by the author.

60 Pero, Elisabetta. 2011. "Environmental Issues as Context." In *Aesthetics of Sustainable Architecture,* edited by Sang Lee, Rotterdam: 010 Publishers, p. 222.

61 Elisabetta Pero continues: "Namely, can the city be regarded an adequate form of contemporary living on which to develop the principles of sustainability? And through this process, should the identity of the city be preserved, and on what criteria should these decisions be based." Ibid., 214.

62 Lovejoy (1960 [1948], 114).

63 Vasilikou, Carolina and Marialena Nikolopoulou. 2014. "Degrees of Environmental Diversity for Pedestrian Thermal Comfort in the Urban Continuum. A New methodological Approach." In *Bridging the Boundaries. Human Experience in the Natural and Built Enviroment and Implications for Research, Policy, and Practice. Advances in People-Enviroment Studies,* Vol. 5. Göttingen: Hogrefe Publishing, p. 103.

64 Ibid., 97.

65 Ibid., 101.

66 Ibid., 106.

67 Ibid., 99.

68 Broadly, sensory walks have been applied in research on urban microclimates by Jean-Paul Thibaud and André Potvin since the early 2000s. Potvin has deployed mobile monitoring methods since his PhD. See André Potvin. 2000. "Assessing the microclimate of urban transitional spaces." In *Proceedings of PLEA2000.* Cambridge: Routledge / Potvin (2004, 121–143). / Thibaud, Jean-Paul. 2007. "La fabrique de la rue en marche: essai sur l'altération des ambiances urbaines." *Flux* 66/67: 111–119.

69 Fleming and Jankovic (2011, 4).

Endnotes

70 Bick Hirsch, Alison. 2014. *City Choreographer: Lawrence Halprin in Urban Renewal America.* Minneapolis: University of Minnesota Press, p. 5.

71 Ibid., 201.

72 Accessed on February 14, 2022. https://time2switch.wordpress.com/2011/08/15/score-quad

73 "In the late 1960s and 70s, Peter Smithson wrote several guides for walkers: 'Bath: Walks within the walls: a study of Bath as a built form taken over by other uses' (Architectural Design, October 1969), followed in 1975 by the second volume, and finally a guide to walks in Oxford and Cambridge ('Oxford and Cambridge Walks,' Architectural Design, June 1976)." See: van den Heuvel, Dirk. 2007. "Situativer Urbanismus." *arch+* 183: 48.

74 Smithson, Peter. 1969. "Bath: Walks within the walls: a study of Bath as a built form taken over by other uses." *Architectural Design,* October: 1.

75 "Daily journeys from the city area to the surrounding heights clearly revealed the susceptibility of the area to conditions of temperature inversion and associated fogs. This and other phenomena [...] suggest that an investigation into local climatic conditions might yield some interesting results." Balchin, W. G. V. and Norman Pye. 1947. "A Microclimatological Investigation of Bath and the Surrounding District." *Quarterly Journal of the Royal Meteorological Society* 73 (317–318): 297–323.

76 In this respect see: Grimmond (2011).

77 Balchin and Pye (1947, 303).

78 See also Schinkel's "View of Park Crescent", 1826; or Hilberseimer's "Bath (UK), view from Hedgemead Park" in *The New City,* 1944.

79 Smithson, Peter (1969, 8).

80 Manley (1949, 319).

81 Wasiuta, Mark and Sarah Herda. "Anna Halprin, Lawrence Halprin. The Halprin Workshops Bay Area, California, USA 1966–1971." Accessed on February 14, 2022. http://radical-pedagogies.com/search-cases/a37-anna-halprin-lawrence-halprin-workshops

82 Bick Hirsch (2014, 200).

83 Ibid., 201.

84 Cited in ibid., 201 and 203. Day one quotations from Halprin, The RSVP Cycles.

85 Cited in ibid., 201.

86 Cited in ibid., 7.

87 Alfred Anton Boeke. Verbatim quotation. In Muren, Zara. 1994. *Dream of the Sea Ranch.* A Documentary Film.

88 Lawrende Halprin. Verbatim quotation in ibid.

89 Zara Muren. Verbatim quotation in ibid.

90 Lawrence Halprin. Verbatim quotation in ibid.

91 Joseph Esherick. Verbatim quotation in ibid.

92 Charles Moore. Verbatim quotation in ibid.

93 Charles Moore. Verbatim quotation in ibid.

94 Habermas, Jürgen. 1981. "Modernity versus Postmodernity." *New German Critique* 22. Special issue on modernism, p. 13. Habermas speaks of "a widespread fear regarding the destruction of the urban and natural environment, and of forms of human sociability." Habermas (1981, 7).

95 Heschong, Lisa. 2018. "Between Laboratory and Sea Ranch. Architecture and the Notion of Microclimate (USA)," Lisa Heschong in conversation with Sascha Roesler. In *The Urban Microclimate as Artifact. Towards an Architectural Theory of Thermal Diversity,* edited by Sascha Roesler and Madlen Kobi. Basel: Birkhäuser, p. 32.

96 Villani, Tiziana. 1995. *Athena cyborg. Per una geografia dell'espressione: corpo, territorio, metropoli.* Sesto San Giovanni: Mimesis, p. 118. Quoted in Gandy, Matthew. 2005. "Cyborg Urbanization: Complexity and Monstrosity in the Contemporary City." *International Journal of Urban and Regional Research* 29.1: 26.

97 Gandy (2005, 33).

98 Ibid., 26.

99 Lange, Torsten. 2019. "Forschungsgeleitetes Entwerfen." In *Architekturpädagogiken. Ein Glossar,* edited by Heike Biechteler and Johannes Käferstein, Zurich: Park Books, p. 57. Translation by the author.

100 Vrachliotis, Georg. 2020. *Geregelte Verhältnisse. Architektur und technisches Denken in der Epoche der Kybernetik.* Basel: Orell Fuessli.

101 Gruen, Victor. 1964. *The Heart of Our Cities: The Urban Crisis; Diagnosis and Cure.* New York: Simon and Schuster / Banham, Peter Reyner. 1976. *Megastructure: Urban Futures of the Recent Past.* New York: Harper & Row.

102 Bélanger, Pierre. 2007. "Underground landscape: The urbanism and infrastructure of Toronto's downtown pedestrian network." *Tunnelling and Underground Space Technology* 22: 279.

103 These cities are part of the so-called winter city movement. See: National Research Council of Canada. *Problems of the North.* Design Manuals, Design Codes, 1950–1980.

104 El-Geneidy, Ahmed et al. 2011. "Montréal's roots. Exploring the growth of Montréal's Indoor City." *The Journal of Transport and Land Use* 4 (2): 33–46.

105 Bélanger (2007, 272).

106 See also Cesar Pelli's 1966 Urban Nucleus P/A proposal at Sunset Mountain Park (Santa Monica Mountain, Los Angeles).

107 Together with her daughter Merete Mattern, Herta Hammerbacher founded the interdisciplinary discussion and working group "Society for Experimental and Applied Ecology" in 1972.

108 Schwagenscheidt, Walter. 1949. *Die Raumstadt. Hausbau und Städtebau für jung und alt, für Laien und was sich Fachleute nennt. Skizzen mit Randbemerkungen zu einem verworrenen Thema.* Heidelberg: Lambert Schneider.

109 Dahinden, Justus. 1971. *Stadtstrukturen für morgen.* Teufen: Niggli, p. 209. Quoted in Stieger (2018, 683). Translation by the author.

110 Dahinden (1974, 10). Translation by the author.

111 Ibid., 28. Translation by the author.

112 Ibid., 32. Translation by the author.

113 Ibid., 16. Translation by the author.

114 Ibid., 8. Translation by the author.

115 Ibid., 21. Translation by the author.

116 Ibid., 10. Translation by the author.

117 Ibid., 8. Translation by the author.

118 Ibid., 46. Translation by the author.

119 Ibid., 28. Translation by the author.

120 Ibid., 16. Translation by the author.

121 Ibid., 46. Translation by the author.

122 Ibid., 32. Translation by the author.

123 Ibid., 13. Translation by the author.

124 Ibid., 10. Translation by the author.

125 Ibid., 16. Translation by the author.

126 Ibid., 12. Translation by the author.

127 Ibid., 20. Translation by the author.

128 Ibid., 10. Translation by the author.

129 Ibid., 6. Translation by the author.

130 Ibid., 13. Translation by the author.

131 Rem Koolhaas speaks of the "direct relationship between repetition and architectural quality: the greater the number of floors stacked around the shaft, the more spontaneously they congeal into a single form". Koolhaas, Rem. 1994 (1978). *Delirious New York. A Retroactive Manifesto for Manhattan.* Rotterdam: 010 Publishers, p. 82.

132 Ibid.

133 Kallipoliti (2018).

134 "it is occasionally proposed that a large bubble be put over the whole city, perhaps a pneumatic structure or one of Buckminster Fuller's domes. This climatic envelope would enable the entire city to be air conditioned, indoors and outdoors. Indeed, 'outdoors' would be a thing of the past." Heschong (1979, 20).

135 Sloterdijk, Peter. 2004. *Sphären. Plurale Sphärologie, Vol. III, Schäume.* Frankfurt am Main: Suhrkamp, p. 472

136 AD. 1972. *Designing for Survival* 7.

137 Lopez, Fanny. 2014. *Le rêve d'une déconnexion. De la maison autonome à la cité auto-énergétique.* Paris: Editions de la Villette, p. 222.

6 Energy-Synergy

1 Buckminster Fuller, Richard. 1975. *Synergetics. Explorations in the Geometry of Thinking.* New York: Macmillan, p. 3.

2 Ibid.

3 Maki, Fumihiko. 1964. *Investigations in Collective Form.* St. Louis: School of Architecture, Washington University.

4 Knowles, Ralph. 1974. *Energy and Form. An Ecological Approach to Urban Growth.* Cambridge, MA: MIT Press.

5 Matus, Vladimir. 1988. *Design for Northern Climates. Cold-Climate Planning and Environmental Design.* New York: Van Nostrand Reinhold, p. 137.

6 Crawford, Margaret. 2016. "Productive Urban Environment." In *Ecological Urbanism,* edited by Mohsen Mostafavi and Gareth Doherty. Zurich: Lars Müller Publishers, p. 148.

7 McHarg (1969, 188).

8 Mumford, Lewis (1969). Introduction in ibid., vi.

9 Hawkes (1996, 115).

10 Boyarsky, Alvin. 1970. "Chicago à la carte. The City as an Energy System." *AD*, Dec., p. 632.

11 Park and his colleagues coined the term "urban ecology" in the field of urban sociology. According to Park's and Burgess's theory of urban ecology, cities are environments like those found in nature.

12 Lynch, Kevin. 1960. *The Image of the City.* Cambridge, MA: MIT Press.

13 Boyarsky (1970, 632).

14 Ibid., 604.

15 Ibid., 612. See also Cronon, William. 1991. *Nature's Metropolis: Chicago and the Great West.* New York: W. W. Norton.

16 Boyarsky (1970, 602).

17 Ibid., 603.

18 Ibid., 612.

19 Ibid., 622.

20 Ibid., 633.

21 Ibid., 612.

22 Lucarelli, Fosco. 2017. Accessed on February 14, 2022. http://socks-studio.com/2017/10/02/alvin-boyarsky-chicago-a-la-carte-the-city-as-an-energy-system-1968

23 Within a few years, Banham published three books on this topic: *The architecture of the well-tempered environment* in 1969, *Los Angeles: The Architecture of Four Ecologies* in 1971 and *Megastructure: Urban Futures of the Recent Past* in 1976.

24 Banham, Peter Reyner. 1971. *Los Angeles. The Architecture of Four Ecologies.* New York: Harper and Row, p. 79.

25 Ibid., ii.

26 Appleyard, Donald. 1964. *The View from the Road.* Cambridge, MA: MIT Press.

27 Boyarsky (1970, 602).

28 Koolhaas, Rem. 1978. *Delirious New York.* New York: Oxford University Press, p. 155.

29 Boyarsky (1970, 632).

30 Ibid., 633.

31 Ibid.

32 Physical disease (heart disease, tuberculosis, diabetes, syphilis, cirrhosis of the liver, amoebic dysentery, bacillary dysentery, salmonellosis); social disease (homicide, suicide, drug addiction, alcoholism, robbery, rape, aggravated assault, juvenile delinquency, infant mortality); mental disease; pollution (suspended dust, settled dust, sulphate index); ethnicity; economic factors (income, poverty, unemployed, housing quality, overcrowding, illiteracy).

33 McHarg (1969, 188).

34 Ibid.

35 Ibid., 193.

36 Ibid.

37 Böer, Wolfgang. 1964. *Technische Meteorologie.* Leipzig: Edition Leipzig, p. 90

38 Kratzer (1937, 10).

39 Kratzer (1956 [1937], 17).

40 "The largest part of the energy quantities in the big cities is provided by coal burning. Unfortunately, there is no compilation of coal consumption in the cities. Only from some cities I could find data on coal consumption". Ibid., 17.

41 Ibid., 19.

42 Ibid., 17.

43 Hilberseimer (1944, 81).

44 Accessed on March 23, 2021. https://siteenvirodesign.com/content/best-products

45 Wines, James and Patricia Phillips. 1982. *Highrise of Homes.* New York: Rizzoli.

46 Koolhaas (1978).

47 The postcard was part of the study "Architettura Riflessa". Superstudio, Cubo di Foresta sul Golden Gate (Cubic Forest on Golden Gate), from L' Architettura Riflessa (Architecture Reflected). Date created: 1972, Medium: collage of cut and printed papers with colored pencil, Dimensions: 75 cm × 107.7 cm.

48 "Since cities represent an extreme density of patterns of energy demand as well as profound heterogeneity in patterns of energy use, they offer great opportunities

for cross-sectoral and temporal integration of energy systems to achieve savings on the demand side as well as the supply side. Traditionally much of urban energy infrastructures and functions (such as heating, lighting, transport, etc.) were designed and operated in relative isolation. Most energy models, for example, focus either on individual buildings or functions within buildings (lighting, air conditioning, hot-water supply, other building services, etc.). The interaction of those functions and other buildings or even different urban activities is rarely reflected." Schulz, Niels et al. 2014. "The SynCity Urban Energy System Model." In Mostafavi and Doherty (2016, 446).

49   The term "energy landscape" comes from the introductory chapter of Hough's *City Form and Natural Process*. See Hough, Michael. 1984. *City Form and Natural Process. Toward a New Urban Vernacular*. New York: Van Nostrand Reinhold, p. 9.

50   Harper, Peter and Godfrey Boyle. 1976. *Radical Technology*. London: Wildwood House, p. 166. See also: *AD*, Jan. 1976, p. 48.

51   Ganser, Karl and Wolfgang Bahr. 1979. "Koordination von Stadtentwicklung und Energieplanung." *Bauwelt Stadtbauwelt* 63: 1489–2510.

52   Ibid.

53   Watts, Michael. 2016. "Oil City: Petrolandscapes and Sustainable Futures." In Mostafavi and Doherty (2016, 451).

54   Ovaska (1978).

55   Matus (1988, 137).

56   Ibid., 138.

57   McHarg (1969, 65).

58   Brownson, Jeffrey R. S. 2013. "Framing the sun and buildings as commons." *Buildings: an open access journal for the built environment* 3 (4): 662.

59   "The literature on urban commons is extremely sparse and no research appears to have been undertaken by landscape designers, urban planners or conservationists into the subject." Laurie, Ian (1979), Urban Commons. In *Nature in Cities. The Natural Environment in the Design and Development of Urban Green Space*, edited by Ian Laurie. Chichester: Wiley, p. 232.

60   Ibid.

61   Urban commons "provide the best and largest range of natural habitats we have in single units to represent much of what is implied in the term 'nature in cities'." Ibid.

62   Ibid., 235.

63   McHarg (1969, 56).

64   Ibid., 65.

65   Hough (1984, 15).

66   Spirn, Anne Whiston. 1984. "City and Nature." In *The Granite Garden: Urban Nature and Human Design*. Cited in *The Sustainable Urban Development Reader*, edited by Stephen M. Wheeler and Timothy Beatley. 2004. London: Routledge, p. 115.

67   Spirn (1984, 5).

68   "The realization that nature is ubiquitous, a whole that embraces the city, has powerful implications for how the city is built and maintained and for the health, safety, and welfare of every resident. Unfortunately, tradition has set the city against nature, and nature against the city. The belief that the city is an entity apart from nature and even antithetical to it has dominated the way in which the city is perceived and continues to affect how it is built." Ibid., 115.

69   Ibid., 4.

70   McHarg (1969, 56).

71   Borasi, Giovanna and Mirko Zardini (eds.). 2007. *Sorry, Out of Gas: Architecture's Response to the 1973 Oil Crisis*. Montréal: CCA Montréal..

72   "It is by distinguishing the resource systems from the resource units, and by describing the linked relationship among clients as appropriators and design / construction teams as provider / producers, that we reveal common challenges in shared governance of the solar and building resources." Brownson (2013, 633).

73   73   Only very recently have new types of microgrids emerged in New York (and elsewhere) that share the power generated by individuals with a larger community, using internet technologies.

74   Hawkes (1996, 113).

75   Schulz, Niels et al. 2014. "The SynCity Urban Energy System Model." in Mostafavi and Doherty (2016, 446). See also Bale, Catherine. 2016. "National Energy Policy, Locally Delivered. The Role of Cities." In *Delivering Energy. Law and Policy in EU and the US. A Reader,* edited by Raphael J. Heffron and Gavin F. M. Little. Edinburgh: Edinburgh University Press, p. 105.

76   Watts (2016, 453).

77   Ibid., 450.

78   Ibid.

79   Moe, Kiel. 2013. "The Formations of Energy in Architecture. An Architectural Agenda for Energy." In *Architecture and Energy Performance and Style*, edited by William W. Braham and Daniel Willis. London: Routledge, p. 154.

80   McHarg (1969, 56).

81   Stieger (2018, 224).

82   Ibid.

83   McHarg (1969, 64).

84   Ibid.

85   The project is one of the outstanding examples of Ticino architecture, later called the "Tendenza" movement. See Steinmann, Martin and Thomas Boga. 1975. *Tendenzen*. Zurich: gta Verlag, p. 33–34.

86   "If we take David Harvey's dictum that 'there is nothing unnatural about New York City' seriously, this impels interrogating the failure of twentieth-century urban social theory to take account of physical or ecological processes. While late-nineteenth-century urban perspectives were acutely sensitive to the ecological imperatives of urbanization, these considerations disappeared almost completely in the decades that followed (with the exception of a thoroughly 'de-natured' Chicago school of urban social ecology). Re-naturing urban theory is, therefore, vital to urban analysis as well as to urban political activism." Kaika and Erik Swyngedouw. 2006. "Urban political ecology, polizicizing the production of urban natures." In *In the Nature of Cities. Urban Political Ecology and the Politics of Urban Metabolism,* edited by Nik Heynen et al. London, New York: Routledge, p. 2.

87   Only recently, architectural historian Margaret Crawford emphasized the character of novelty associated with these projects promoting unprecedented landscapes: "to imagine a variety of new landscapes that might emerge. Supported by a sustainable energy grid that can accommodate and distribute both large- and small-scale energy sources, this would be a green environment, with energy and transportation infrastructures, dwellings and workplaces, agricultural and natural spaces interwoven in new and still to be imagined combinations." Crawford (2016, 149).

88   Branzi, Andrea. 2006. *No-Stop City Archizoom Associati*. Orléans: Hyx, p. 134.

89   Waldheim, Charles (2016) "Weak Work: Andrea Branzi's 'Weak Metropolis' and the Projective Potential of an 'Ecological Urbanism'." In Mostafavi and Doherty (2016, 113).

90   Ibid., 113.

91   Ibid., 114.

Appendices   **250**

92 Accessed on January 27, 2020. https://www.youtube.com/watch?v=1KkTewCUKT8

93 "Some years ago I was asked to advise on which lands in the Philadelphia metropolitan region should be selected for open space. I became clear at the onset that the solution could only be obscured by limiting open space to the arena for organized sweating; it seemed more productive to consider the place of nature in the metropolis. In order to conclude in this place it appeared reasonable to suggest that nature performed work for man without his investment and that such work did represent a value. It also seemed reasonable to conclude that certain areas and natural processes were inhospitable to man – earthquake areas, hurricane paths, floodplains and the like – and that those should be prohibited or regulated to ensure public safety. This might seem a reasonable and prudent approach, but let us recognize that it is a rare one." McHarg (1969, 55).

94 Hough (1984, 15).

95 Spirn (1984, 4).

96 Hughes, Thomas Parke. 2004. *Human-built world: how to think about technology and culture*. Chicago: The University of Chicago Press.

97 Kaika and Swyngedouw (2006, 2).

98 Ibid.

99 "While landscape architects like Olmsted and Howard are often credited with 'creating' urban natural landscapes, the metabolization of urban nature has a history as long as urbanization itself (Olmsted 1895). To this end, Gandy (2002:2) suggests that '[n]ature has a social and cultural history that has enriched countless dimensions of the urban experience. The design, use, and meaning of urban space involve the transformation of nature into a new synthesis.' Still, understanding the politicized and uneven nature of this urban synthesis should be the main task." Ibid., 5.

100 Koolhaas (1978, 155).

101 Hough (1984, 16).

102 Ibid., 12.

103 Matus (1988, 137).

104 Ibid., 138.

105 Maki (1964, 28).

106 Kiss, Daniel and Simon Kretz (eds.). 2021. *Relational Theories of Urban Form. An Anthology*. Basel: Birkhäuser Verlag.

107 Maki (1964, 5).

108 Ibid.

109 Ibid., 27.

110 Waldheim (2016, 113).

111 See Bernoulli (1929), Giedion (1951, 1956).

112 *arch+*. 2020. "Architektur Ethnografie." *arch+* 238.

113 Takamasa Yoshizaka, a former employee of Le Corbusier and influential architectural educator, was also writing on the question of the collective form, highlighting it in his theory of "discontinuous unity." Statement by Manfred Speidel: "As a 'disciple' of Takamasa Yoshizaka, who was himself a disciple of Kon Wajiro, I have always done 'field research' everywhere." See Yoshizaka, Takamasa. 1966. "Group Organization and Physical Structure." *The Japanese Architect*, April: 27.

114 The elaboration of the book (in the second half of the 1950s) was conducted before the so called "urban design" conferences. Gyorgy Kepes was the most important intellectual partner and provider of keywords; they collaborated at MIT during the critical years of the elaboration (see the preface).

115 Smithson, Peter and Alison. 1954. *Doorn Manifesto*. Accessed on February 15, 2022. https://evolutionaryurbanism.com/2017/03/24/the-doorn-manifesto

116 "In the lectures we have established that in architecture, as in science, there are no inventions; that rather every proposal arises from knowledge of the real. From this there is an evident relation between the type of analysis and that of the design that one intends to execute." Rossi, Aldo. 1974. "Bemerkungen zu den Arbeiten des Wintersemsters." In *Aldo Rossi – Vorlesungen, Aufsätze, Entwürfe*, Texte zur Architektur 4, Lehrstuhl Aldo Rossi, Verlag der Fachvereine 1974, ETH Zurich, p. 21.

117 Recalling, for example, the "Vienna Courts" *(Wiener Höfe)*, Rossi notes "how, within a clear typological approach, it becomes possible to articulate the various problems of architecture. The problem of form, seen in this way, represents the possibility of the justification of architecture." Ibid., 23.

118 See Šik, Miroslav. 2011. "Lernen von Rossi." In *Aldo Rossi und die Schweiz. Architektonische Wechselwirkungen*, edited by Ákos Moravánszky and Judith Hopfengärtner. Zurich: gta Verlag, p. 72.

119 This effort has been translated into an 800-page publication. See Rossi, Aldo et al. (eds). 1979. *Costruzione del Territorio e Spazio Urbano nel Cantone Ticino*. Lugano: Fondazione Ticino Nostro. See also Buzzi, Giovanni. 2011. "Costruzione del territorio e spazio urbano nel cantone Ticino. Rossis Beitrag zur Untersuchung der Kulturlandschaft." In Moravánszky and Hopfengärtner (2011, 98).

120 Moravánszky and Hopfengärtner (2011, 3).

121 Ibid.

122 Ibid.

123 Ibid., 28.

124 Ibid., 3.

125 Mumford, Lewis. 1969. Introduction to McHarg (1969, viii).

126 Ibid., iv.

127 "Can we, then, create meaningful group-forms in our society? The answer is not a simple one. It requires a new concept and attitude of design. It also requires the participation of cities, and their social institutions." Maki (1964, 20)

128 "Forms in group-forms have their own built-in link, whether expressed or latent, so that they may grow in a system. They define basic environmental space which also partakes of the quality of systematic linkage. Group-form and its space are indeed prototype elements, and they are prototypes because of implied system and linkage. The element and the growth pattern are reciprocal –both in design and in operation. The element suggests a manner of growth, and that, in turn, demands further development of the elements, in a kind of feedback process." Ibid., 19.

129 Ibid., 13.

130 Ibid., 12.

131 Kurokawa, Kisho. "Agricultural City, 1960." *ArchEyes*, May 7, 2016. Accessed on February 15, 2022. http://archeyes.com/agricultural-city-kurokawa-kisho

132 Maki quotes the book *Communitas.* See Goodman, Percival and Paul. 1947. *Communitas: means of livelihood and ways of life*. Chicago: The University of Chicago Press, p. 13.

133 Tange says, "Reformations of natural topography, dams, harbors, and highways are of a size and scope that involve long cycles of time, and these are the man-made works that tend to divide the overall system of the age. The two tendencies – toward shorter cycles and toward longer cycles – are both necessary to modern life and to humanity itself." Maki (1964, 11).

134 Ibid., 12.

135 Ibid., 8.

Endnotes

251

136 Ibid., 30.

137 Knowles (1974, 111).

138 Ibid.

139 Ibid., vii.

140 Ibid., 111.

141 There are three major influences on Ralph Knowles's work: Eduardo Catalano, Kevin Lynch and D'Arcy Wentworth Thompson (with his "diagram of forces").

142 Knowles, Ralph. 2011. "Solar Aesthetic." In *Aesthetics of Sustainable Architecture*, edited by Sang Lee. Rotterdam: 010 Publishers, p. 50. See also the contemporary experiments under Charles Waldheim at Harvard University. Accessed on February 15, 2022. https://www.gsd.harvard.edu/event/inaugural-conference-of-the-harvard-gsd-office-for-urbanization-heliomorphism. See also Martin, Craig Lee and Greg Keeffe. 2007. "The Biomimetic Solar City: Solar Derived Urban Form using a Forest-growth Inspired Methodology." In *Sun, Wind & Architecture,* Proceedings of the 24th Conference on Passive and Low Energy Architecture, PLEA 2007.

143 Knowles, Ralph. 2003. "The solar envelope: its meaning for energy and buildings." *Energy and Buildings* 35: 15.

144 Matus (1988, 138).

145 He identified three "design purposes". "If each of the three design purposes of economics, community, and aesthetics were restated in terms responsive to the natural environment, the result would be a transformation of our existing cities and a different mode of new growth resulting in the conservation of energy and of our natural environment." Knowles (2003, 1).

146 Stein, Clarence S. 1936. "Henry Wright, 1878–1936." *American Architect and Architecture*, August: 23.

147 Ibid., 24.

148 Wright, Henry Nicolls. 1936. *Solar radiation as related to summer cooling and winter radiation in residences*. New York: John B. Pierce Foundation.

149 "A number of summers he invited a group of younger instructors and advanced students to his simple farm in New Jersey. They worked out projects in community planning and in their spare time remodeled the old mill to serve as drafting room and dormitory. In the evening they discussed endlessly, spurred on by the provocative mind of Henry Wright." Stein (1936, 24).

150 Steadman (1975, 40).

151 "The question of the influence of cities on the climate is therefore the question of the changes that the weather phenomena suffer in the area of cities, and their causes." Kratzer (1956 [1937], 20).

152 Brooks (1951, 56).

153 Sanda Lenzholzer rightly noted "that measurements only reflect existing situations; they do not offer insights on future interventions. And it is this type of information that is crucial for design proposals for spatial planning in cities." Lenzholzer (2015, 96).

154 Knowles, Ralph (2011, 54) / Witt, Andrew and Christopher Reznich. 2018. *The Natural Forces Laboratory: Ralph Knowles and the Instrumentalized Studio*. Montréal and Cambridge: CCA and Harvard University Graduate School of Design. Accessed on February 15, 2022. https://www.cca.qc.ca/en/events/63353/the-natural-forces-laboratory-ralph-knowles-and-the-instrumentalized-studio

155 In a recent interview, Lisa Heschong speaks of a "Berkeley/MIT nexus and a sort of Berkeley/Oregon nexus." Heschong (2018, 31).

156 Roesler and Kobi (2018, 29).

157 Lisa Heschong describes it as follows: "We sent students out to go measure microclimates and report back on how patterns of temperature, humidity, wind speed varied in time and space. We might have them design a pavilion to try to optimize microclimate, especially outdoor environments – how could you take this space and make it a little bit more comfortable? We would have them use the water tables and the wind tunnels to validate their design." Roesler and Kobi (2018, 29).

158 Accessed on April 26, 2018. http://exhibits.ced.berkeley.edu/exhibits/show/new-modernism/college-and-communities/enviro-simulation-lab

159 Since antiquity, cities have been perceived as wind-engineered entities by virtue of their structural layout, resulting in a whole body of treatises on ideal winds and their architectural modeling. These wind theories already suggested that cities can also be assumed to be full-scale wind tunnels.

160 "There is a need for simple regulations, which ensure that society protects the values of natural processes and is itself protected. (…) If so, they would satisfy a double purpose: ensuring the operation of vital natural processes and employing lands unsuited to development in ways that would leave them unharmed by these often violent processes. Presumably, too, development would occur in areas that

were intrinsically suitable, where dangers were absent and natural processes unharmed." McHarg (1969, 55).

161 Brownson (2013, 662).

162 Ibid., 56.

163 McHarg (1969, 65).

164 Brownson (2013, 662).

165 Knowles, Ralph. 1981. *Sun Rhythm Form*. Cambridge, MA: MIT Press, p. x.

166 Ibid.

167 Ibid., 30.

168 Ibid., 25.

169 Coplan, Norman. 1981. "Solar access right." *Progressive Architecture* 4 (81): 191.

170 Knowles (1981, x).

171 Coplan (1981, 191).

172 See Thompson, Grant. 1978. "Solar Energy and the Law: Barriers to a Sunnier Tomorrow." In *Proceedings of the National passive solar conference 2*, p. 850–856 / Morton, David. 1979. "Right to Light." *Progressive Architecture*, April / Bradbrook, A. J. 1988. "Future directions in solar access protection." *Envtl. L.* 19: 167.

173 Jeffrey Brownson summarized the discussion on solar access as follows, emphasizing the tension between private and public interests: "the balance between (1) limited surface area for building systems to deliver solar goods to the occupants and (2) the drive to design for sustainable urban communities will soon drive communities to consider how the solar goods are managed and allocated. However, appropriators of solar energy products, in terms of units of heat / power / illumination / shade / money, find themselves interdependent among each other and tied to network pools such as the power grid and public lands, in addition to the physical resource of the sun." The real challenge is to develop a "shared governance of the solar and building resources". Brownson (2013, 663).

174 It can be called "experimental urban development," which is to be distinguished from traditional urban development without superordinate coordination efforts. "Experimental urban development has grown from the critical reflections that some architects have made, regarding the contemporary urban context as an appropriate structure for sustainable development. This trend is based on the observation that the number of homes using active and passive energy systems should increase […]. However, it remains

difficult to consider which forms, densities, designs, technologies and materials these habitats should take." Pero (2011, 213).

175 The Anglo-Saxon theory of solar rights has recently been strongly influenced by the use of solar energy; in the course of time, the focus has moved away from passive to active use of energy sources. "Common goods and services can be very interesting within the context of society and sustainability. The organization, allocation, and management or governance of a commons can be tied to two concepts: the resource system (a system with stocks) and the resource units that flow from the system. Resource systems can have multiple kinds of stocks, called stock variables, from which a maximum quantity of a flow variable may be collected without harming the particular stock or the resource system as a whole. The broader issues of excludability in the solar field have been addressed in the past through the legal precedent of solar access and solar rights." Brownson (2013, 667).

176 Hough (1984, 28).

177 Ibid., 36.

178 Doug Kelbaugh, reviewing Lisa Heschong's book *Thermal Delight in Architecture*. Kelbaugh, Doug. 1981. "Form and Sense." *Progressive Architecture*, Energy-conscious design 4 (81): 103.

179 Knowles, Ralph. 2000. "The solar envelope. Its meaning for urban growth and form." In *Architecture, City, Environment*, Proceedings of PLEA 2000, Cambridge UK, p. 650.

180 Knowles (1981, xiii).

7 Singapore as a Model?

1 Mumford, Lewis. 1961. *The City in History. Its Origins, Its Transformations, and Its Prospects.* London: Harcourt, Brace & World, p. 286.

2 Villiers, John. 1984, 1965. "Südostasien vor der Kolonialzeit." In *Fischer Weltgeschichte,* Vol. 18. Frankfurt am Main: Fischer Taschenbuch Verlag, p. 13.

3 The territorial expression "Southeast Asia" was used in modern speech for the first time during the Second World War by the Allied troops.

4 The monsoon climate of Southeast Asia is mainly shaped by the El Niño (from the Pacific Ocean) and by the Indian Niño (Indian Ocean Dipole).

5 During his six-year journey between 1854 and 1862 across the Malay Archipelago, Wallace wrote a diary describing, among other things, the territory of the island of Singapore. These descriptions give the contemporary reader an insight into how Singapore's nature and climate once presented themselves to a European traveler. Wallace describes a wooded island that was still largely shaped by its natural parameters. "The island of Singapore consists of a multitude of small hills, three or four hundred feet high, the summits of many of which are still covered with virgin forest. The mission-house at Bukit-tima was surrounded by several of these wood-topped hills, which were much frequented by woodcutters and sawyers, and offered me an excellent collecting ground for insects. [...] Several hours in the middle of every fine day were spent in these patches of forest, which were delightfully cool and shady by contrast with the bare open country we had to walk over to reach them. The vegetation was most luxuriant, comprising enormous forest trees, as well as a variety of ferns, caladiums, and other undergrowth, and abundance of climbing rattan palms." From: Wallace, Alfred Russel. 1869, 2010. *The Malay Archipelago.* Oxford: Oxford University Press, p. 23.

6 Pelzer, who travelled alone through the countries of Southeast Asia between 1963 and 1971 under the most severe socio-political conditions, had in mind no less than a comprehensive visual archive of the indigenous architecture of South Asia. See: Institute of Southeast Asian Studies. 1986. *Southeast Asian Cultural Heritage – Images of Traditional Communities.* Singapore, p. 17.

7 Fleming and Jankovic (2011, 5).

8 Rapoport, Amos. 1969. *House, Form, and Culture.* New York: Pearson.

9 Roesler, Sascha. 2013a. "Climate and Culture." In Climate as a Design Factor, second edition, edited by C. Hönger et al., Luzern: Quart.

10 Pelzer, Dorothy. 1982. *Trek Across Indonesia.* Singapore: Graham Brash Pte Ltd, p. 2.

11 Pelzer describes, for instance, how the central component such as a roof ridge beam was chosen. In order to find the necessary material for the building of the house, the forest had to be searched through. "Ridge: Construction of ridge very important. Judge quality of house by it. From djior wood (Indonesian: djuhar). One piece. Its length limits size of house. Search the forest very long to find the best. Djior wood very long and flexible."

From: Institute for Southeast Asian Studies: Dorothy Pelzer Collection, DP 1b, Common Factors, Singapore.

12 Domenig, Gaudenz. 1980. *Tektonik im primitiven Dachbau.* Zurich: ETH Zurich.

13 The constructional characteristics of this large region characterized by timber construction were summarized by Pelzer as follows: "This is a study of construction without nails. When there are nails, there is already Western influence – except possibly Chinese influence, Chinese Nails." Institute for Southeast Asian Studies: Dorothy Pelzer Collection, DP 1b, Common Factors, Singapore.

14 Chang, Jiat Hwee. 2012. "Tropical Variants of Sustainable Architecture: A Postcolonial Perspective." In *Handbook of Architectural Theory,* edited by Greig Crysler, Stephen Cairns and Hilde Heynen. London, Thousand Oaks, CA: SAGE Publications.

15 Feriadi, Henry. 1999. *Natural Ventilation Characteristics of Courtyard Buildings in Tropical Climate.* Dissertation, National University of Singapore.

16 Fathy, Hassan. 1986. *Natural Energy and Vernacular Architecture.* Chicago: University of Chicago Press, p. 52. See also Ruefenacht, Lea A. and Aurel von Richthofen. 2021. "Mitigate Urban Heat." In *Future Cities Laboratory: Indicia 03,* edited by Stephen Cairns and Devisari Tunas. Zurich: Lars Müller Publishers.

17 Waterson, Roxana. 1990. *The Living House – An Anthropology of Architecture in South-East Asia.* Singapore, Oxford: Ruefenacht.

18 Pandean, Topane-petra. 1983. *Klimaangepasster Wohnungsbau in Indonesien – Ländliche Haus- und Siedlungsformen unter Berücksichtigung natürlicher Klimatisierung und Herstellung in Selbstbauverfahren.* Dissertation, Universität Hannover, p. 168.

19 For a history of knowledge on vernacular architecture see: Roesler (2013b).

20 For the urbanization of Southeast Asia see: Dutt, Ashok K. and Naghun Song. 1994. "Urbanization in Southeast Asia." In *The Asian City: Processes of Development, Characteristics, and Planning,* edited by Ashok K. Dutt et. al. Dordrecht: Springer, p. 159–177.

21 See, for example, the exhibition catalog of DAM (German Architecture Museum) on Indonesian architecture: Sopandi, Setiadi and Avianti Armand (eds.). 2015. *Tropicality: Revisited.* Frankfurt am Main.

22 Medan and Singapore share an entangled colonial past, which involved the continuous supply of labor and other resources for the plantations on Sumatra.

23 Buiskool, Dirk A.. 2004. "Medan. A plantation city on the east coast of Sumatra 1870–1942." Accessed on August 9, 2021. http://archaeologyworld.blogspot.com/2009/04/plantation-city-on-east-coast-of.html

24 See the high-rise conglomerate Pinncale@duxton as an example of this development.

25 Yusuf, A. A. and H. A. Francisco. 2009. "Climate Change Vulnerability Mapping for Southeast Asia." In Economy and Environment Program for Southeast Asia (EEPSEA).

26 In the last 30 years, for example, the island of Sumatra has lost more than 50 percent of its primary forest through deforestation.

27 According to the World Health Organization (WHO), Southeast Asian countries (low and middle income) suffered 3.3 million deaths due to indoor air pollution and 2.6 million deaths due to outdoor air pollution. See: World Health Organization, Burden of Disease from Ambient and Household Air Pollution. Accessed on July 1, 2015. www.who.int/phe/health_topics/outdoorair/databases/en

28 Roesler, Sascha. 2017. "Die Wiederkehr des Klimas. Urbane Kulturen der Belüftung in Südostasien." In arch+ 227, "Vietnam – Die Rückkehr des Klimas."

29 Konersmann, Ralf. 2008. "Unbehagen in der Kultur. Veränderungen des Klimas und der Klimasemantik." In Das Wetter, der Mensch und sein Klima, edited by Petra Lutz and Thomas Macho for the German Museum of Hygiene, Göttingen.

30 Fleming and Jankovic (2011, 4).

31 Ibid., 10.

32 Seng, Eunice. 2014. Habitation and the Invention of a Nation, Singapore 1936–1979. Dissertation, Colombia University, New York 2014, abstract.

33 Gibson, Wiliam. 2012, 1993. "Disneyland with the Death Penalty." In Distrust That Particular Flavor. New York: Putnam Adult.

34 Yuen, Belinda. 2005. "Squatters no more: Singapore Social Housing." In Third Urban Research Symposium: Land Development, Urban Policy and Poverty Reduction, April 4–6 2005, Brasília.

35 Moebius, the book's illustrator, was instrumental in shaping the visual appearance of science-fiction films with the hybrid infra-architecture he developed.

36 Wong, Aline K and Stephen Yeh (ed.). 1985. Housing a Nation. 25 Years of Public Housing in Singapore. Maruzen Asia for Housing & Development Board, p. 136.

37 Liping, Wang and Hien Wong Nyuk. 2006. "The impacts of ventilation strategies and facade on indoor thermal environment for naturally ventilated residential buildings in Singapore." Building and Environment 42 (12): 4006–15.

38 "For example, 86% of the people in Singapore live in HDB (Housing and Development Board) residential buildings, which are designed to be naturally ventilated." Liping, Wang, Hien Wong Nyuk and Shuo Li. 2007. "Facade Design Optimization for Naturally Ventilated Residential Buildings in Singapore." In Energy and Buildings 39 (8): 954–961. Housing & Development Board, General Technical Requirements, HDB(ARCH) 2012.

39 Accessed on February 15, 2022. http://www.teoalida.com/singapore/hdbfloorplans

40 Ibid.

41 Ibid.

42 Ibid.

43 Accessed on February 15, 2022. http://en.wikipedia.org/wiki/Public_housing_in_Singapore

44 Ibid.

45 Liu Thai Ker. 1975. "Design for better Living Conditions." In Public Housing in Singapore. A Multi-disciplinary Study, edited by Stephen H. K. Yeh. Singapore: Singapore University Press (for Housing and Development Board), p. 137.

46 Ibid., 138. / Wong and Yeh (1985, 71).

47 Tenorio, Rosangela. 2007. "Enabling the Hybrid Use of Air Conditioning: A Prototype on Sustainable Housing in Tropical Regions." Building and Environment 42 (2): 605–613.

48 Ibid., 185.

49 George, Cherian. 2000. Singapore. The Air-Conditioned Nation. Singapore: Landmark Books.

50 Lee Kuan Yew. 1999. "Air-con gets my vote." Straits Times, Jan. 19, p. 1, quoted in ibid., 14.

51 Banham (1969, 174).

52 Ibid., 172.

53 I would like to thank Aurel von Richthofen for his numerous comments and astute remarks on Singapore's energy consumption.

54 Standard converter AC systems have not seen major technical changes since the times of Carrier. Centralized cooling systems, on the other hand, would be much more efficient.

55 Tenorio (2007).

56 Arsenault, Raymond. 1984. "The end of the long hot summer: the air conditioner and southern culture." The Journal of Southern History 50 (4): 614.

57 Ibid., 616.

58 Ibid., 625.

59 Shove, E., G. Walker and S. Brown. 2014. "Transnational Transitions: The Diffusion and Integration of Mechanical Cooling," Urban Studies 51 (7): 1506–1519.

60 Tenorio (2007).

61 U-values indicate the thermal transmittance (the rate of transfer of heat) through a structure.

62 "In Singapore, the current façade construction material standard for air-conditioned buildings is envelope thermal transfer value (ETTV), which should not exceed 50W/m². However, for naturally ventilated buildings in Singapore, there are no clear façade design guidelines. Building regulations in Singapore only specify that the U-value of any external wall in [a] non-air conditioned building should not be more than 3.5W/m²K. Wong investigated the effects of U-value of construction materials for naturally ventilated buildings in Singapore. It was recommended that U-value for the east- and west-facing facade should be no more than 2W/m²K and for north and south should be no more than 2.5W/m²K. However, in the study, the effects of WWR on indoor air velocity for naturally ventilated buildings are not considered." Liping and Hien Wong Nyuk (2007). / Building and Construction Authority. Guidelines on envelop thermal transfer value for buildings. Singapore, 2004 / The development & building control division. 1979. Handbook on energy conservation in building and building services. Singapore.

63 Rittel and Webber (1973, 155–169).

64 Arsenault (1984, 597).

65 Rittel and Webber (1973, 161).

66 Thee Kian Wie. 1977. Plantation Agriculture and Export Growth. An economic history of East Sumatra, 1863–1942.

Appendices

Jakarta: National Institute of Economic and Social Research, p.1. See also Geertz, Clifford. 1963. *Agricultural Involution: The Processes of Ecological Change in Indonesia.* Berkeley, CA: University of California Press.

67 Thee Kian Wie (1977, 29).

68 Stoler, Ann Laura. 1995, 1985. *Capitalism and Confrontation in Sumatra's Plantation Belt, 1870–1979.* New Haven, London: Yale University Press, p. 1.

69 Ibid.

70 Ibid., 2.

71 Ibid., 12.

72 Thee Kian Wie (1977, 3).

73 Accessed on February 15, 2022. http://earthenginepartners.appspot.com/science-2013-global-forest

74 Nas, Peter J. M. and Martin van Berkel. 2003. "Small town symbolism – The making of the built environment in Bukittinggi and Payakumbuh." In *Indonesian houses – Tradition and Transformation in Vernacular Architecture,* edited by Reimar Schefold, Gaudenz Domenig and Peter Nas. Leiden: KITLV Press, p. 477.

75 Pelras, Christian. 2003. "Bugis and Makassar houses." In *Indonesian houses – Tradition and Transformation in Vernacular Architecture,* edited by Reimar Schefold, Gaudenz Domenig and Peter Nas. Leiden: KITLV Press, p. 276.

76 Ibid., 278.

77 Soper, Kate. 1995. *What is Nature? Culture, Politics and the Non-human.* Oxford: Blackwell Publishers Limited.

78 Stoler (1995, 1985, 3).

79 As Aurel von Richthofen righty remarks, "acting on this requires to almost rewrite the social contract of Singapore and puts even more burden onto the shoulders of a strong patriarchal government. What if there was a technological disruption that would allow to decouple and decentralise?", private email conversation, 7/8/2021.

80 See Vititneva, Ekaterina, Zhongming Shi, Pieter Herthogs, Reinhard König, Aurel von Richthofen and Sven Schneider. 2021. "Informing the Design of Courtyard Street Blocks Using Solar Energy Models: Generating University Campus Scenarios for Singapore." In *Carbon Neutral Cities – Energy Efficiency & Renewables in the Digital Era.* EPF Lausanne.

81 Latour, Bruno. 2001. *Das Parlament der Dinge.* Frankfurt am Main: Suhrkamp.

82 Giedion, Sigfried. 1948. *Mechanization Takes Command: A Contribution to Anonymous History.* Oxford: Oxford University Press.

83 Sloterdijk (2004, 503).

84 Bachelard, Gaston. 1958, 1994. *The Poetics of Space.* Boston: Beacon Press, p. 211.

85 Ibid., 211.

86 Jean Hyppolite commenting on Sigmund Freud's notion of negation *(Verneinung).*

87 Hyppolite, Jean (1956), "Spoken commentary on the Verneinung (negation) of Freud." *La Psychoanalyse* 1: 35. Citied in Bachelard (1958, 1994, 212).

88 Bachelard (1958, 1994, 211). See also Arnheim, Rudolf, Wolfgang M. Zucker and Joseph Watterson. 1966. "Inside and Outside in Architecture: A Symposium." *The Journal of Aesthetics and Art Criticism* 25 (1): 3–15. / Roesler (2018, 54–64).

89 Chakrabarty, Dipesh. 2009. "The Climate of History: Four Theses." *Critical Inquiry* 35 (2): 197–222.

References

Albers, Gerd. 1975. *Entwicklungslinien im Städtebau. Ideen, Thesen, Aussagen 1875–1945: Texte und Interpretationen.* Düsseldorf: Bertelsmann-Fachverlag.

Albrecht, F. and J. Grunow. 1935. "Ein Beitrag zur Frage der vertikalen Luftzirkulation." *Meteorologische Zeitschrift*, Deutsche Meteorologische Gesellschaft, 52.

Allen, Stan and Marc McQuade (eds.). 2011. *Landform Building: Architecture's New Terrain.* Baden: Lars Müller Publishers.

Althaus, Dirk. 2005. *Bauen heute – Bauen morgen. Architektur an der Schwelle zur postfossilen Zeit,* Berlin: Bauwerk.

Andreas, W. 1935–1936. "Klima auf Bestellung: eine neue Wissenschaft: Mikroklimatologie – Das künstliche Klima." *Am häuslichen Herd: schweizerische illustrierte Monatsschrift*, 39.

Appleyard, Donald. 1964. *The View from the Road.* Cambridge, MA: MIT Press.

Architectural Record. 1957. *Buildings for Industry,* New York: F. W. Dodge Corp.

Arnheim, Rudolf, Wolfgang M. Zucker and Joseph Watterson. 1966. "Inside and Outside in Architecture: A Symposium." *The Journal of Aesthetics and Art Criticism* 25 (1): 3–15.

Aronin, Jeffrey Ellis. 1953. *Climate & Architecture.* New York: Reinhold.

Arsenault, Raymond. 1984. "The end of the long hot summer: The air conditioner and southern culture". In *The Journal of Southern History* 50 (4).

Atalay Franck, Oya. 2012. *Architektur und Politik: Ernst Egli und die türkische Moderne 1927–1940.* Zurich: gta Verlag.

Atkinson, William. 1912. *The Orientation of Buildings or Planning for Sunlight.* New York: J. Wiley & sons.

Bachelard, Gaston. 1958, 1994. *The Poetics of Space.* Boston: Beacon Press.

Baier, Rosemarie. 1982. "Leben in der Mietskaserne. Zum Alltag Berliner Unterschichtsfamilien in den Jahren 1900 bis 1920." In *Hinterhof, Keller und Mansarde. Einblicke in Berliner Wohnungselend 1901–1920,* edited by Gesine Asmus. Reinbek: Rowohlt Taschenbuch Verlag.

Balchin, W. G. V. and Norman Pye. 1947. "A Microclimatological Investigation of Bath and the Surrounding District." In *Quarterly Journal of the Royal Meteorological Society* 73 (317–318): 297–323.

Bale, Catherine. 2016. "National Energy Policy, Locally Delivered. The Role of Cities." In *Delivering Energy. Law and Policy in EU and the US. A Reader,* edited by Raphael J. Heffron and Gavin F. M. Little. Edinburgh: Edinburgh University Press.

Banham, Peter Reyner. 1971. *Los Angeles. The Architecture of Four Ecologies.* New York: Harper and Row.

Banham, Peter Reyner. 1976. *Megastructure: Urban Futures of the Recent Past.* New York: Harper & Row.

Banik-Schweitzer, Renate. 1999. "Urban Visions, Plans, and Projects, 1890–1937." In *Shaping the Great City. Modern Architecture in Central Europe, 1890–1937*, edited by Eve Blau and Monika Platzer. Munich: Prestel.

Barber, Daniel. 2016. *A house in the sun. Modern architecture and solar energy in the Cold War.* New York: Oxford University Press.

Barber, Daniel. 2020. *Modern architecture and climate. Design before air conditioning.* Princeton: Princeton University Press.

Barthes, Roland. 1977. *Fragments d'un Discours Amoureux.* Paris: Seuil.

Barthes, Roland. 1989 (1980). *Die helle Kammer. Bemerkungen zur Photographie.* Frankfurt am Main: Suhrkamp Verlag AG.

Bartoli, Sandra, Marco Clausen, Silvan Linden, Åsa Sonjasdotter, Florian Wüst, Kathrin Grotz and Patricia Rahemipour (eds.). 2020. *Licht Luft Scheiße. Perspektiven auf Ökologie und Moderne*, Band 1, Archäologien der Nachhaltigkeit, Hamburg.

Baumgarten, Alexander Gottlieb. 1750–1758. *Ästhetik*, 2 vols. Hamburg, 2007.

Behling, Sophia and Stefan Behling. 1996. *Sol Power. The Evolution of Solar Architecture.* Munich / New York: Prestel.

Behrens, Peter. 1930. *Baublock bestehend aus mehr- und vielgeschossigen Einzelhäusern.* Patentiert im Deutschen Reiche vom 25. Juni 1930, Nr. 578 164.

Bélanger, Pierre. 2007. "Underground Landscape: The Urbanism and Infrastructure of Toronto's Downtown Pedestrian Network." *Tunnelling and Underground Space Technology*, 22.

Benjamin, Walter. 1978 (1925). *Ursprung des deutschen Trauerspiels*. Frankfurt am Main: Suhrkamp.

Benjamin, Walter. 2008 (1935). *The Work of Art in the Age of Its Technological Reproducibility, and Other Writings on Media.* Cambridge, MA: Belknap Press.

Bettini, Giulio. 2016. *Die "città animata". Milan und die Architektur von Asnago Vender.* Zurich: gta Verlag.

Bick Hirsch, Alison. 2014. *City Choreographer: Lawrence Halprin in Urban Renewal America.* Minneapolis: University of Minnesota Press.

Bideau, André. 2012. "Elusive Ungers." *AA Files*, No. 64, 3–14.

Bideau, André. 2013. "Housing as Discursive Void: Oswald Mathias Ungers in the 1960s and 1970s." *Candide 7.*

Blaser, Werner. 1974. *Chinese Pavillion Architecture. Quality, Design, Structure Exemplified by China.* Niederteufen: Verlag A. Niggli.

Blau, Eve and Ivan Rupnik. 2018. *Baku. Oil and Urbanism.* Zurich: Park Books.

Bodenschatz, Harald. 2009. "Städtebau von den neunziger Jahren des 19. Jahrhunderts bis zum ersten Weltkrieg." In *Berlin und seine Bauten. Teil I – Städtebau*, edited by Harald Bodenschatz et al. Berlin: DOM publishers.

Bodenschatz, Harald and Christiane Post (eds.). 2003. "Sozialistischer Städtebau im Zeichen der Moderne." In *Städtebau im Schatten Stalins. Die internationale Suche nach der sozialistischen Stadt in der Sowjetunion 1929–1935.* Berlin: Verlagshaus Braun.

Böer, Wolfgang. 1964. *Technische Meteorologie.* Leipzig: Edition Leipzig.

Borasi, Giovanna and Mirko Zardini (eds.). 2007. *Sorry, Out of Gas: Architecture's Response to the 1973 Oil Crisis.* Montréal: CCA Montréal.

Borsi, Katharina. 2006. "The Strategies of the Berlin Block." In *Intimate Metropolis: Urban Subjects in the Modern City*, edited by V. Di Palma, D. Periton and M. Lathouri. Abingdon: Routledge.

Borsi, Katharina. 2015. "Drawing the region: Hermann Jansen's vision of Greater Berlin in 1910." In *The Journal of Architecture* 20 (1).

Boyarsky, Alvin. 1970. "Chicago à la carte. The City as an Energy System." *AD*, December.

Boyd, Andrew. 1962. *Chinese Architecture and Town Planning. 1500 B.C. – A.D. 1911.* Chicago: University of Chicago Press.

Bradbrook, A. J. 1988. "Future directions in solar access protection." In *Envtl. L.*, 19.

Branzi, Andrea. 2006. *No-Stop City Archizoom Associati.* Orléans: Hyx.

Bray, David. 2005. *Social Space and Governance in Urban China. The Danwei System from Origins to Reform.* Palo Alto: Stanford University Press.

Brezina, Ernst and Schmidt, Wilhelm. 1937. *Das künstliche Klima in der Umgebung des Menschen.* Stuttgart: Ferdinand Enke Verlag.

Bronin, Sara C. 2009. "Solar rights." *Boston University Law Review* 89 (4): 1218.

Brooks, Charles Ernest. 1951. *Climate in Everyday Life.* New York: Van Nostrand.

Brownson, Jeffrey R. S. 2013. "Framing the sun and buildings as commons." *Buildings: an open access journal for the built environment* 3 (4).

Brüggemeier, Franz and Lutz Niethammer. 1976. "Wie wohnten Arbeiter im Kaiserreich". In *Archiv für Sozialgeschichte*, Jg. 16.

Buckminster Fuller, Richard. 1975. *Synergetics. Explorations in the Geometry of Thinking.* New York: Macmillan.

Büdel, A. and J. Wolf. 1933. "Münchner stadtklimatische Studien." *Zeitschrift für angewandte Meteorologie* 50 (1).

Buiskool, Dirk A.. 2004. "Medan. A plantation city on the east coast of Sumatra 1870–1942." Accessed on August 9, 2021 http://archaeologyworld.blogspot.com/2009/04/plantation-city-on-east-coast-of.html

Bull, Catherin, Davisi Boontharm, Claire Parin, Darko Radovic and Guy Tapie (eds.). 2007. *Cross-Cultural Urban Design. Global or Local Practice?* London: Routledge.

Butti, Ken and John Perlin. 1980. *A Golden Thread. 2500 Years of Solar Architecture and Technology.* Palo Alto: Van Nostrand Reinhold.

Buzzi, Giovanni. 2011. "Costruzione del territorio e spazio urbano nel cantone Ticino. Rossis Beitrag zur Untersuchung der Kulturlandschaft." In Moravánszky and Hopfengärtner (2011, 98)

Cammerer, Joseph Sebastian. 1936. *Die konstruktiven Grundlagen des Wärme- und Kälteschutzes im Wohn- und Industriebau.* Berlin: Julius Springer.

Carrington, Thomas Spees. 1912. *Fresh air and how to use it.* New York: Nat. Assoc. for the Study and Prevention of Tuberculosis.

CENTRALA. 2018. *Amplifying Nature. The Planetary Imagination of Architecture in the Anthropocene,* edited by Anna Ptak. Warsaw: Zachęta National Gallery of Art.

Chakrabarty, Dipesh. 2009. "The Climate of History: Four Theses." In *Critical Inquiry* 35 (2): 197–222.

Chang, Jiat Hwee. 2012. "Tropical Variants of Sustainable Architecture: A Postcolonial Perspective." In Greig Crysler, Stephen Cairns and Hilde Heynen (eds.). *Handbook of Architectural Theory.* London, Thousand Oaks, CA: SAGE Publications.

Chang, Jiat-Hwee. 2016. *A genealogy of tropical architecture. Colonial networks, nature and technoscience.* London, New York: Routledge.

Chiri, Giovanni and Ilaria Giovagnorio. 2015. "Gaetano Vinaccia's (1881–1971) Theoretical Work on the Relationship between Microclimate and Urban Design." In *Sustainability* 77: 4448–4473.

Chiu Che Bing. 2008. "The Traditional Chinese Garden: A World Apart." In *In the Chinese City. Perspectives on the Transmutations of an Empire*, edited by Frédéric Edelmann, New York: Actar Publications.

Clément, Gilles. 2015. *"The Planetary Garden" and Other Writings.* Philadelphia: University of Pennsylvania Press.

Coen, Deborah R. 2016. "Seeing Planetary Change, Down to the Smalllest Wildflower." In *Climates: Architecture and the Planetary Imaginary*, edited by James Graham et al. Zurich: Lars Müller Publishers.

Coplan, Norman. 1981. "Solar access right." *Progressive Architecture* 4 (81).

Corbin, Alain. 1984. *Pesthauch und Blütenduft, eine Geschichte des Geruchs.* Berlin: Wagenbach.

Crawford, Margaret. 2016. "Productive Urban Environment." In *Ecological Urbanism*, edited by Mohsen Mostafavi and Gareth Doherty. Zurich: Lars Müller Publishers.

Cronon, William. 1991. *Nature's Metropolis: Chicago and the Great West.* New York: W. W. Norton.

Dahinden, Justus. 1971. *Stadtstrukturen für morgen.* Teufen: Niggli.

Dahinden, Justus. 1974. *Akro-Polis. Frei-Zeit-Stadt / Leisure City.* Stuttgart: Karl Krämer Verlag.

De Chiffre, Lorenzo. 2016. *Das Wiener Terrassenhaus. Entwicklungsphasen und Aktualität eines historischen Wohntypus mit Fokus auf den lokalspezifischen architektonischen Diskurs.* Dissertation, Vienna: University of Vienna.

De Decker, Kris. 2012. "The solar envelope: how to heat and cool cities without fossil fuels." Accessed on February 9, 2022. At http://www.lowtechmagazine.com/2012/03/solar-oriented-cities-1-the-solar-envelope.html

Denzer, Anthony. 2013. *The Solar House. Pioneering sustainable design.* New York: Rizzoli.

Döcker, Richard. 1929. *Terrassentyp: Krankenhaus, Erholungsheim, Hotel, Bürohaus, Einfamilienhaus, Siedlungshaus, Miethaus und die Stadt.* Stuttgart: Wedekind.

Domenig, Gaudenz. 1980. *Tektonik im primitiven Dachbau.* Zurich: ETH.

Dutt, Ashok K. and Naghun Song. 1994. "Urbanization in Southeast Asia." In *The Asian City: Processes of Development, Characteristics, and Planning*, edited by Ashok K. Dutt, Frank J. Costa, Surinder Aggarwal and Allen G. Noble. Dordrecht: Springer, p. 159–177.

Düwel, Jörn. 1995. "Berlin. Planen im kalten Krieg." In *1945. Krieg – Zerstörung – Aufbau. Architektur und Stadtplanung 1940–1960*, Schriftenreihe der Akademie der Künste, Vol. 23. Berlin: Henschel Verlag.

Egli, Ernst. 1945. "Vom regionalen Städtebau: der Einfluss von Klima und Landschaft auf die Wohn- und Stadtform." *Werk* 32 (3).

Egli, Ernst. 1951. *Climate and Town Districts. Consequences and Demands.* Zurich: Verlag für Architektur Erlenbach.

Egli, Ernst. 1963. *Die Jahreszeiten. Ein Marionettenspiel der Zeit.* Typescript, ETH Zurich Research Collection, Ernst Egli, Hs. 785: 96.

El-Geneidy, Ahmed, Lisa Kastelberger and Hatem T. Abdelhamid. 2011. "Montréal's roots. Exploring the growth of Montréal's Indoor City." *The Journal of Transport and Land Use* 4 (2): 33–46.

Ellis, F. P. 1953, "Thermal comfort in warm and humid atmospheres." *The Journal of Hygiene* 51 (3).

Ellsässer, Karl and Horst Ossenberg. 1954. *Bauten der Lebensmittelindustrie. Anlagen und Arbeitsablauf erläutert an 112 Beispielen des In- und Auslands aus allen Gewerbezweigen.* Stuttgart: Hoffmann.

Engels, Friedrich. 1845. "Die Lage der arbeitenden Klasse in England." In *Marx-Engels-Gesamtausgabe* (MEGA) Abt. 1., Band 4. Marx-Engels-Institut. 1932/1970.

Engels, Friedrich. 1872. "Zur Wohnungsfrage." In *Karl Marx / Friedrich Engels – Werke.* Berlin: (Karl) Dietz Verlag. Vol. 18, 5th edition, 1973.

Esser, Katja. 2011. "City Weathers in Aachen during the 1800s." In *City Weathers. Meteorology and Urban Design 1950–2010*, edited by Michael Hebbert, Vladimir Jankovic and Brian Webb. Manchester: University of Manchester Press.

Farr, William. 1852. *Report on the Mortality of Cholera in England, 1848–1849,* London: W. Clowes and Sons.

Fassbinder, Horant. 1975. *Berliner Arbeiterviertel 1800–1918. Analysen zum Planen und Bauen 2.* West Berlin: Verlag für das Studium der Arbeiterbewegung.

Fathy, Hassan. 1986. *Natural Energy and Vernacular Architecture*. Chicago: University of Chicago Press.

Faust, Bernhard Christoph. 1802 (1792). *Entwurf zu einem Gesundheitskatechismus*. Stuttgart: Franz Steiner Verlag.

Fawcett, Chris. 1980. *The New Japanese House. Ritual and Anti-Ritual, Patterns of Dwelling*. London / New York: Granada.

Feriadi, Henry. 1999. *Natural Ventilation Characteristics of Courtyard Buildings in Tropical Climate*. Dissertation, National University of Singapore.

Fisch, Rose, Inge Maass and Katrin Rating. 1984. *Der grüne Hof. Grundlagen und Anforderungen an die Hofbegrünung in der Stadterneuerung*, Karlsruhe: C. F. Müller.

Fisher, Mark. 2016. *The Weird and the Eerie*. London: Repeater Books.

Fleck, Ludwig. 1980. *Entstehung und Entwicklung einer wissenschaftlichen Tatsache. Einführung in die Lehre vom Denkstil und Denkkollektiv* [1935]. Frankfurt am Main: Suhrkamp.

Fleming, James R. and Vladimir Jankovic. 2011. "Revisiting Klima." In *Klima*, Osiris, Vol. 26.

Flügge, Carl. 1916. *Grossstadtwohnungen und Kleinhaussiedlungen in ihrer Einwirkung auf die Volksgesundheit. Eine kritische Erörterung für Ärzte, Verwaltungsbeamte und Baumeister*. Jena: G. Fischer.

Förster, Kim. 2019. "The Green IBA. On a Politics of Renewal, Ecology, and Solidarity." *Candide* 11: 9–50.

Foucault, Michel. 1966. *Les mots et les choses. Une archéologie des sciences humaines*. Paris: Gallimard.

Foucault, Michel. 1984. "Des Espace Autres." *Architecture, Mouvement, Continuité* 5: 46–49.

Frampton, Kenneth. 2011. "Megaform as Urban Landscape." In *Landform Building: Architecture's New Terrain*, edited by Stan Allen and Marc McQuade. Baden: Lars Müller Publishers.

Fujii, James A. (ed.). 2004. *Text and the City. Essays on Japanese Modernity*. Durham, NC: Duke University Press.

Gandy, Matthew. 2005. "Cyborg Urbanization: Complexity and Monstrosity in the Contemporary City." *International Journal of Urban and Regional Research* 29.1.

Gandy, Matthew. 2014. "Borrowed Light. Journals Through Weimar Berlin." In *The Fabric of Space. Water, Modernity, and the Urban Imagination*. Boston: The MIT Press.

Ganser, Karl and Wolfgang Bahr. 1979. "Koordination von Stadtentwicklung und Energieplanung." *Bauwelt Stadtbauwelt* 63: 1489–2510.

Geddes, Arthur. 1946. *Planning and Climate. Climates of Region, Locality and Site*. London: Association for Planning and Regional Reconstruction.

Geertz, Clifford. 1963. *Agricultural Involution: The Processes of Ecological Change in Indonesia*. Berkeley, CA: University of California Press.

Geiger, Rudolf. 1941 (1927). *Das Klima der bodennahen Luftschicht, ein Lehrbuch der Mikroklimatologie*. Braunschweig: Vieweg & Sohn.

Geist, Johann Friedrich. 1969. *Passagen, ein Bautyp des 19. Jahrhunderts*. Munich: Prestel.

Geist, Johann Friedrich and Klaus Kürvers. 1984. *Das Berliner Mietshaus*, Vol. 2. Munich: Prestel.

Geist, Johann Friedrich and Klaus Küvers. 1995. "Tatort Berlin, Pariser Platz. Die Zerstörung und 'Entjudung' Berlins." In *1945. Krieg – Zerstörung – Aufbau. Architektur und Stadtplanung 1940–1960*, Schriftenreihe der Akademie der Künste, Vol. 23. Berlin: Henschel Verlag.

Genzmer, E. 1911. *Sonder-Katalog für die Gruppe Städtebau der wissenschaftlichen Abteilung der Internationalen Hygiene-Ausstellung*. Berlin: Rudolf Mosse.

George, Cherian. 2000. *Singapore. The Air-Conditioned Nation*. Singapore: Landmark Books.

Gibson, Wiliam. 2012, 1993. "Disneyland with the Death Penalty." In *Distrust That Particular Flavor*. New York: Putnam Adult.

Giedion, Sigfried. 1928. *Bauen in Frankreich, Bauen in Eisen, Bauen in Eisenbeton*. Leipzig, Berlin: Klinkhardt & Biermann.

Giedion, Sigfried. 1941. *Space, Time and Architecture. The Growth of a New Tradition*. Cambridge, MA: The Harvard University Press.

Giedion, Sigfried. 1948. *Mechanization Takes Command: A Contribution to Anonymous History*. Oxford: Oxford University Press.

Glacken, Clarence J. 1967. *Traces on the Rhodian Shore. Nature and Culture in Western Thought from Ancient Times to the End of the Eighteenth Century*. Berkeley, CA: University of California Press.

Goethe, Johann Wolfgang von. 1923 (1819). *West-Östlicher Divan. In Goethes sämtliche Werke in sechzehn Bänden*, Vol. 11, Leipzig: Insel.

Go Jeong-Hi. 2006. "Herta Hammerbacher (1900-1985): Virtuosin der Neuen Landschaftlichkeit — der Garten als Paradigma." *Landschaftsentwicklung und Umweltforschung* Special issue, Vol. 18, Technische Universität Berlin.

Goldmerstein, J. and K. Stodieck. 1928. *Thermenpalast. Kur-, Erholungs-, Sport-, Schwimm- und Badeanlage*. Berlin: Wilhelm Ernst & Sohn.

Goldmerstein, J. and K. Stodieck. 1931a. *Grossstadtsanierung. Gewinnung von Spiel-, Sand- und Grünflächen in Neben- und Seitenstrassen mit Rentabilitätsnachweis*. Deutsche Bauzeitung, Berlin SW 48.

Goldmerstein, J. and K. Stodieck. 1931b. *Wie atmet die Stadt? Neue Feststellungen über die Bedeutung der Parkanlagen für die Lufterneuerung in den Grossstädten*. Berlin: VDI-Verlag GmbH.

Goodman, Percival and Paul. 1947. *Communitas: means of livelihood and ways of life*. Chicago: The University of Chicago Press.

Grämiger, Gregory. 2018. *Baugesetze formen. Architektur und Raumplanung in der Schweiz*. Zurich: gta Verlag.

Grimmond, Sue. 2011. "London's urban climate: historical and contemporary perspectives." In *City Weathers. Meteorology and Urban Design 1950–2010*, edited by Michael Hebbert, Vladimir Jankovic and Brian Webb. Manchester: University of Manchester Press.

Gröning, Gert and Joachim Wolschke. 1985. "Thermenpalast: ein Projekt aus Weimarer Zeit." *werk, bauen + wohnen* 72 (10).

Gruen, Victor. 1964. *The Heart of Our Cities: The Urban Crisis; Diagnosis and Cure*. New York: Simon and Schuster.

Günther, Lutz Philipp. 2009. *Die bildhafte Repräsentation deutscher Städte. Von den Chroniken der Frühen Neuzeit zu den Websites des Gegenwart*. Cologne, Weimar, Vienna: Böhlau.

Habermas, Jürgen. 1981. "Modernity versus Postmodernity." *New German Critique* 22. Special issue on modernism.

Habermas, Jürgen. 1989 (1962). *The Structural Transformation of the Public Sphere: An Inquiry into a Category of Bourgeois Society*. Cambridge, MA: MIT Press.

Hagen, Katrin. 2011. *Freiraum im Freiraum: Mikroklimatische Ansätze für die städtische Landschaftsarchitektur*. Dissertation, Technical University of Vienna.

Hagen, Katrin. 2012. "Mit Sinnen und Verstand. Design principles of Moorish gardens as inspiration for climate-sensitive urban design." *dérive* 48, Stadt KLIMA Wandel.

Halbwachs, Maurice. 1967 (1950). *Das kollektive Gedächtnis.* Stuttgart: Enke Verlag.

Harather, Karin. 1995. *Haus-Kleider. Das Phänomen der Bekleidung in der Architektur.* Vienna: Böhlau.

Harbusch, Gregor et al. 2014. "Established Modernists go into Exile, Younger Members go to Athens." In *Atlas of the Functional City. CIAM 4 and Comparative Urban Analysis*, edited by Evelien van Es, Gregor Harbusch, Bruno Maurer, Muriel Pérez, Kees Somer and Daniel Weiss. Bussum: THOTH Publishers gta Verlag.

Harootunian, Harry. 2000. *Overcome by Modernity. History, Culture, and Community in Interwar Japan.* Princeton: Princeton University Press.

Harper, Peter and Godfrey Boyle. 1976. *Radical Technology.* London: Wildwood House.

Hassenpflug, Gustav. 1951. "Das Krankenhaus in der Stadtplanung." In *Handbuch für den Neuen Krankenhausbau*, edited by Paul Vogler and Gustav Hassenpflug. Munich, Berlin: Urban & Schwarzenberg.

Hawkes, Dean. 1996. "The Cambridge School and the environmental tradition." In *The Environmental Tradition. Studies in the Architecture of Environment,* edited by Dean Hawkes. London: E. & F. N. Spon.

Hebbert, Michael and Brian, Webb. 2012. "Towards a Liveable Urban Climate: Lessons from Stuttgart." *Isocarp* 07.

Hegemann, Werner. 1911. *Der Städtebau nach dern Ergebnisse der allgemeinen Städtebau-ausstellung in Berlin nebst einem Anhang: Die internationale Städtebau-Ausstellung in Düsseldorf*, Erster Teil. Berlin: Verlag Ernst Wasmuth.

Hegemann, Werner. 1930. *Das steinerne Berlin. Geschichte der grössten Mietkasernenstadt der Welt.* Berlin: Vieweg & Sohn.

Heinze, W. 1935. "Welches Klima hat ihr Wohnzimmer? Jede Stadt hat ihr eigenes Klima. – Rivieraluft im Zimmer. – Je nach Wunsch: Arktis oder Tropenhitze." *Die Berner Woche in Wort und Bild: ein Blatt für heimatliche Art und Kunst* 25 (1).

Hellpach, Willy. 1939. *Geopsyche, Die Menschenseele unterm Einfluss von Wetter und Klima.* Leipzig: Enke.

Henschel, Christhardt. 2021. "Kommunizierende Röhren der Singularitäten? Deutsche Debatten um den Holocaust aus Warschauer Perspektive." In *Geschichte der Gegenwart*, July 7.

Hertweck, Florian. 2013. "Berliner Ein- und Auswirkungen," in *Die Stadt in der Stadt. Berlin: Ein grünes Archipel. Ein Manifest* (1977) by Oswald Mathias Ungers and Rem Koolhaas with Peter Riemann, Hans Kollhoff and Arthur Ovaska. A critical publication by Florian Hertweck and Sébastien Marot. Baden: Lars Müller Publishers.

Heschong, Lisa. 1979. *Thermal Delight in Architecture.* Cambridge, MA: MIT Press.

Heschong, Lisa. 2018. "Between Laboratory and Sea Ranch. Architecture and the Notion of Microclimate (USA)," Lisa Heschong in conversation with Sascha Roesler. In *The Urban Microclimate as Artifact. Towards an Architectural Theory of Thermal Diversity,* edited by Sascha Roesler and Madlen Kobi. Basel: Birkhäuser.

Hilberseimer, Ludwig. 1924. "Großstadtarchitektur." *Der Sturm* 4.

Hilberseimer, Ludwig. 1927. *Großstadtarchitektur.* Stuttgart: Julius Hoffmann.

Hilberseimer, Ludwig. 1930. "Vorschlag zur Citybebauung." *Die Form. Zeitschrift für gestaltende Arbeit für den Deutschen Werkbund* 12.

Hilberseimer, Ludwig. 1931. *Hallenbauten. Stadt- und Festhallen, Turn- und Sporthallen, Ausstellungshallen, Ausstellungsanlagen.* Leipzig: J. M. Gebhardt.

Hilberseimer, Ludwig. 1935. "Raumdurchsonnung." *Moderne Bauformen* 34.

Hilberseimer, Ludwig. 1936. "Raumdurchsonnung und Siedlungsdichtigkeit." *Moderne Bauformen*, Jg. 35.

Hilberseimer, Ludwig. 1944. *The New City. Principles of Planning.* Chicago: P. Theobald.

Hilberseimer, Ludwig. 1967. *Berliner Architektur der 20er Jahre.* Mainz: Florian Kupferberg.

Hirschfeld, Christian Cay Lorenz. 1779. *Theorie der Gartenkunst*, Vol. 1.

Horkheimer, Max and Adorno, Theodor. 1988 (1947). *Dialectic of Enlightenment. Philosophische Fragmente.* Frankfurt am Main: S. Fischer Verlag GmbH.

Hough, Michael. 1984. *City Form and Natural Process. Toward a New Urban Vernacular.* New York: Van Nostrand Reinhold.

Hrudicka, B. 1939. "Meteorologie im Dienste der Bautechnik." *Zeitschrift für angewandte Meteorologie* 54: 37–47.

https://geschichtedergegenwart.ch/kommunizierende-roehren-der-singularitaeten-deutsche-debatten-um-den-holocaust-aus-warschauer-perspektive. Accessed on July 9, 2022.

Hughes, Thomas P. 1987. "The Evolution of Large Technological Systems." In *The Social Construction of Technological Systems. New Directions on the Sociology and History of Technology*, edited by Wiebe E. Bijker, Thomas Hughes and Trevor J. Pinch. Boston: The MIT Press.

Hughes, Thomas Parke. 2004. *Human-built world: how to think about technology and culture.* Chicago: The University of Chicago Press.

Hürlimann, Martin. 1928. *Indien. Baukunst, Landschaft und Volksleben.* Zurich: Fretz & Wasmuth.

Hürlimann, Martin. 1977. "Nennen wir das Ding einfach Atlantis!" *Züri Leu*, 2. September 1977.

Institute of Southeast Asian Studies. 1986. *Southeast Asian Cultural Heritage – Images of Traditional Communities.* Singapore.

Jansen, Hermann. 1910. *Vorschlag zu einem Grundplan für Gross-Berlin.* Munich: Callwey.

Jiménez Alcalá, Benito. 1999. "Aspectos bioclimaticos de la arquitectura Hispanomusulmana." In *Cuadernos de la Alhambra* 35: 13–29.

Kähler, Gert. 1985. *Wohnung und Stadt. Hamburg Frankfurt Wien. Modelle sozialen Wohnens in den zwanziger Jahren.* Braunschweig, Wiesbaden: Springer.

Kaika, Maria and Erik Swyngedouw. 2006. "Urban political ecology, polizicizing the production of urban natures." In *In the Nature of Cities. Urban Political Ecology and the Politics of Urban Metabolism*, edited by Nik Heynen et al.. London, New York: Routledge.

Kallipoliti, Lydia. 2018. *The Architecture of Closed Worlds or What is the Power of Shit?* Baden: Lars Müller Publishers.

Kassner, Carl. 1908. *Das Wetter und seine Bedeutung für das praktische Leben.* Leipzig: Quelle & Meyer.

Kassner, Carl. 1910. *Die meteorologischen Grundlagen des Städtebaues*, Städtebauliche Vorträge aus dem Seminar für Städtebau an der Königlichen Technischen Hochschule zu Berlin, Band III, Heft VI, edited by Joseph Brix and Felix Genzmer, Ernst & Sohn, Berlin.

Kelbaugh, Doug. 1981. "Form and Sense." *Progressive Architecture*, Energy-conscious design, 4 (81).

Kennedy, Margrit. 1984. "Preface." In Rose Fisch, Inge Maass and Katrin Rating. *Der grüne Hof. Grundlagen und Anforderungen an die Hofbegrünung in der Stadterneuerung.* Karlsruhe: C. F. Müller, p. V.

Kikuko Kashiwagi-Wetzel and Michael Wetzel (eds.). 2015. *Interkulturelle Schauplätze in der Grossstadt. Kulturell Zwischenräume in amerikanischen, asiatischen und europäischen Metropolen.* Paderborn: Wilhelm Fink.

Kilian, Markus. 2002. *Großstadtarchitektur und New City Eine planungsmethodische Untersuchung der Stadtplanungsmodelle Ludwig Hilberseimers.* Dissertation, Karlsruhe: Technische Hochschule Karlsruhe.

Kiss, Daniel and Simon Kretz (eds.). 2021. *Relational Theories of Urban Form. An Anthology.* Basel: Birkhäuser Verlag.

Klein, Alexander. 1927a. "Untersuchungen zur rationellen Gestaltung von Kleinwohnungsgrundrissen." *Die Baugilde* 9 (22): 1361.

Klein, Alexander. 1927b. "Versuch eines Graphischen Verfahrens zur Bewertung von Kleinwohnungsgrundrissen." In *Wasmuths Monatshefte für Baukunst und Städtebau* 7.

Klein, Alexander. 1934. *Das Einfamilienhaus. Südtyp, Studien und Entwürfe.* Part of the series: Wohnbau und Städtebau, Beiträge zur Entwurfslehre. Stuttgart: Julius Hoffmann.

Knoch, Karl. 1962. *Problematik und Probleme der Kurortklimaforschung als Grundlage der Klimatherapie, Mitteilungen des Deutschen Wetterdienstes.* No. 30, Offenbach a. M.: 50.

Knowles, Ralph. 1974. *Energy and Form. An Ecological Approach to Urban Growth.* Cambridge, MA: MIT Press.

Knowles, Ralph. 1981. *Sun Rhythm Form.* Cambridge, MA: MIT Press.

Knowles, Ralph. 2000. "The solar envelope. Its meaning for urban growth and form." In *Architecture, City, Environment*, Proceedings of PLEA 2000, Cambridge UK.

Knowles, Ralph. 2003. "The solar envelope: its meaning for energy and buildings." *Energy and Buildings* 35.

Knowles, Ralph. 2011. "Solar Aesthetic." In *Aesthetics of Sustainable Architecture*, edited by Sang Lee. Rotterdam: 010 Publishers.

Konau, Elisabeth. 1977. *Raum und soziales Handeln. Studien zu einer vernachlässigten Dimension soziologischer Theoriebildung.* Stuttgart: Enke Verlag.

Konersmann, Ralf. 2008. "Unbehagen in der Kultur. Veränderungen des Klimas und der Klimasemantik." In *Das Wetter, der Mensch und sein Klima*, edited by Petra Lutz and Thomas Macho for the German Museum of Hygiene, Göttingen.

Koolhaas, Rem. 1978. *Delirious New York.* New York: Oxford University Press.

Koolhaas, Rem. 1994 (1978). *Delirious New York.* Rotterdam: 010 Publishers.

Kracauer, Siegfried. 1930. "Abschied von der Lindenpassage." In Mülder-Bach, Inka and Ingrid Belke (eds.). 2011. *Kracauer, Siegfried: Werke*, Vol. 5.3. Berlin: Suhrkamp.

Kracauer, Siegfried. 1977 (1927). "The Mass Ornament." In *Weimar Essays.* Harvard University Press.

Kracauer, Siegfried. 2009. *Straßen in Berlin und anderswo.* Frankfurt am Main: Suhrkamp Verlag.

Kratzer, Albert. 1937. *Das Stadtklima.* Braunschweig: Vieweg & Sohn.

Kurokawa, Kisho. "Agricultural City, 1960." *ArchEyes*, May 7, 2016. Accessed on February 15, 2022. http://archeyes.com/agricultural-city-kurokawa-kisho

Landsberg, Helmut. 1958 (1941). *Physical Climatology.* Pennsylvania: Gray Printing.

Lange, Christiane. 2011. *Ludwig Mies van der Rohe. Architecture for the Silk Industry.* Berlin: Nicolai.

Lange, Torsten. 2019. "Forschungsgeleitetes Entwerfen." In *Architekturpädagogiken. Ein Glossar*, edited by Heike Biechteler and Johannes Käferstein. Zurich: Park Books.

Latour, Bruno. 2001. *Das Parlament der Dinge.* Frankfurt am Main: Suhrkamp Verlag.

Laurie, Ian. 1979. "Urban Commons." In *Nature in Cities. The Natural Environment in the Design and Development of Urban Green Space*, edited by Ian Laurie. Chichester: Wiley.

Le Corbusier. 1925. *Urbanisme.* Paris: Editions Crès.

Le Corbusier. 1931. "Die Bodenaufteilung der Städte." In *Rationelle Bebauungsweisen. Ergebnisse des 3. Internationalen Kongresses für Neues Bauen*, Brussels, November 1930. Frankfurt am Main: Verlag Englert & Schlosser. Reprint 1979, Nendeln: Kraus Reprint.

Lee Kuan Yew. 1999. "Air-con gets my vote." *Straits Times*, Jan. 19.

Lenin, Vladimir Ilyich. 1909. "Materialismus und Empiriokritizismus. Kritische Bemerkungen über eine reaktionäre Philosophie." In *Werke*, Vol. 14. Berlin: Dietz-Verlag.

Lenzholzer, Sanda. 2015. *Weather in the City: How Design Shapes the Urban Climate.* Rotterdam: nai010 Publishers.

Leyden, Friedrich. 1933. *Gross-Berlin. Geographie der Weltstadt.* Breslau: Ferdinand Hirt.

Liping, Wang and Hien Wong Nyuk. 2006. "The impacts of ventilation strategies and facade on indoor thermal environment for naturally ventilated residential buildings in Singapore." *Building and Environment* 42 (12): 4006–4015.

Liping, Wang, Hien Wong Nyuk and Shuo Li. 2007. "Facade Design Optimization for Naturally Ventilated Residential Buildings in Singapore." *Energy and Buildings* 39 (8): 954–961. Housing & Development Board, General Technical Requirements, HDB(ARCH) 2012.

Liu Thai Ker. 1975. "Design for better Living Conditions." In *Public Housing in Singapore. A Multi-disciplinary Study,* edited by Stephen H. K. Yeh. Singapore: Singapore University Press (for Housing and Development Board).

Lopez, Fanny. 2014. *Le rêve d'une déconnexion. De la maison autonome à la cité auto-énergétique.* Paris: Editions de la Villette.

Lovejoy, Arthur O. 1960 (1948). "The Chinese Origin of a Romanticism." In *Essays in the History of Ideas*, New York: Capricorn Books.

Lowenhaupt Tsing, Anna. 2015. *The Mushroom at the End of the World. On the Possibility of Life in Capitalist Ruins.* Princeton: Princeton University Press.

Lueder, Christoph. 2017. "Evaluator, Choreographer, Ideologue, Catalyst: The Disparate Reception Histories of Alexander Klein's Graphical Method." *Journal of the Society of Architectural Historians* 76 (1): 82–106.

Lynch, Kevin. 1960. *The Image of the City.* Cambridge, MA: MIT Press.

Mach, Ernst. 1886. *Analyse der Empfindungen.* Jena: G. Fischer.

Maki, Fumihiko. 1964. *Investigations in Collective Form.* St. Louis: School of Architecture, Washington University.

Manley, Gordon, 1949. "Microclimatology – Local Variations of Climate likely to affect the Design and Siting of Buildings." *The Journal of the Royal Institute of British Architects* 56 (7) May 1949 (read at meeting of RIBA on April 12, 1949).

Manley, Gordon. 1958. "The Revival of Climatic Determinism." In *Geographical Review* 48 (1).

Martin, Craig Lee and Greg Keeffe. 2007. "The Biomimetic Solar City: Solar Derived Urban Form using a Forest-growth Inspired Methodology." In *Sun, Wind & Architecture*, Proceedings of the 24th Conference on Passive and Low Energy Architecture, PLEA 2007.

Matus, Vladimir. 1988. *Design for Northern Climates. Cold-Climate Planning and Environmental Design.* New York: Van Nostrand Reinhold.

May, Ernst. 1931. "Stalingrad." *Frankfurter Zeitung*, January 24. Reprinted in: *Standard Cities. Ernst May in der Sowjetunion 1930–1933. Texte und Dokumente*, edited by Thomas Flierl. Frankfurt am Main, 2012.

McHarg, Ian. 1969. *Design with Nature.* New York: Natural History Press.

Mills, Edward. 1951. *The Modern Factory.* London: Architectural Press.

Mir Atta Muhammad Talpur. 2007. "The Vanishing Glory of Hyderabad (Sindh, Pakistan)." In *Webjournal in Cultural Patrimony.* Accessed on March 17, 2016. http://www.webjournal.unior.it/dati/19/72/web%20journal%203,%20hyderabad.pdf

Missenard, André. 1938 (1937). *Der Mensch und seine klimatische Umwelt.* Stuttgart / Berlin: Deutsche Verlags-Anstalt.

Appendices

260

Missenard, André. 1949 (1940). *Klima und Lebensrhythmus.* Meisenheim am Glan: Westkulturverlag Anton Hain.

Moe, Kiel. 2013. "The Formations of Energy in Architecture. An Architectural Agenda for Energy." In *Architecture and Energy Performance and Style*, edited by William W. Braham and Daniel Willis. London: Routledge.

Moravánszky, Ákos (ed.). 2002. *Das entfernte Dorf, Moderne Kunst und ethnischer Artefakt.* Cologne / Weimar / Vienna: Böhlau.

Moravánszky, Ákos. 2017. *Metamorphism: Material Change in Architecture.* Basel: Birkhäuser.

Morton, David. 1979. "Right to Light." *Progressive Architecture*, April.

Mostafavi, Mohsen and David Leatherbarrow. 1993. *On Weathering. The Life of Buildings in Time.* Cambridge, MA: MIT Press.

Mumford, Lewis. 1961. *The City in History. Its Origins, Its Transformations, and Its Prospects.* London: Harcourt, Brace & World.

Milyutin, N. A. 1930. *Sotzgorod*, Moscow. (Mentioned by Hilberseimer in *The New City*).

Nas, Peter J. M. and Martin van Berkel. 2003. "Small town symbolism – The making of the built environment in Bukittinggi and Payakumbuh." In *Indonesian houses – Tradition and Transformation in Vernacular Architecture*, edited by Reimar Schefold, Gaudenz Domenig and Peter Nas. Leiden: KITLV Press.

Neumann, Erwin. 1954. "Das Stadtklima." In *Die städtische Siedlungsplanung unter besonderer Berücksichtigung der Besonnung.* Stuttgart: Witter.

Nova, Alessandro. 2007. *Das Buch des Windes. Das Unsichtbare sichtbar Machen.* Munich: Deutscher Kunstverlag.

Olgyay, Victor (with Aladar Olgyay). 1963. *Design with Climate – Bioclimatic Approach to Architectural Regionalism.* Princeton: Princeton University Press.

Ovaska, Arthur A. 1978. "The City as a Garden, the Garden as a City." In *The Urban Garden*, O. M. Ungers, H. Kollhoff, A. Ovaska, Berlin.

Pandean, Topane-Petra. 1983. *Klimaangepasster Wohnungsbau in Indonesien – Ländliche Haus- und Siedlungsformen unter Berücksichtigung natürlicher Klimatisierung und Herstellung in Selbstbauverfahren.* Dissertation, Hannover: Universität Hannover.

Pelras, Christian. 2003. "Bugis and Makassar houses." In *Indonesian houses – Tradition and Transformation in Vernacular Architecture*, edited by Reimar Schefold, Gaudenz Domenig and Peter Nas. Leiden: KITLV Press.

Pelzer, Dorothy. 1982. *Trek Across Indonesia.* Singapore: Graham Brash Pte Ltd.

Peppler, Albert. 1929. "Das Auto als Hilfsmittel der meteorolgischen Forschung." *Das Wetter*, 46.

Pero, Elisabetta. 2011. "Environmental Issues as Context." In *Aesthetics of Sustainable Architecture,* edited by Sang Lee. Rotterdam: 010 Publishers.

Perry, Clarence. 1929. *The Neighbourhood Unit: From the Regional Survey of New York and Its Environs*, Vol. VII, "Neighbourhood and Community Planning." New York: Wiley.

Petersen, Steen Estvad. 1993. "Desert and City in Islamic Horticulture." In *City and Nature: Changing Relations in Time and Space,* edited by T. M. Kristensen et al. Odense, Denmark: Odense University Press.

Pevsner, Nikolaus. 1956. "The Geography of Art". In *The Englishness of English Art.* London: The Architectural Press.

Pevsner, Nikolaus. 1976. *A History of Building Types.* Princeton: Princeton University Press.

Pfützner, Friedrich-Herman. 1962. *Über den Einfluss städtischer Grünanlagen auf die Abkühlungsgröße.* Dissertation, Berlin: Humboldt-Universität.

Poerschke, Ute. 2018. "Data-Driven Design in High Modernism: Ludwig Hilberseimer's Solar Orientation Studies." *ENQ, the ARCC Journal of Architectural Research*. Accessed on August 20, 2019. https://www.arcc-journal.org

Posener, Julius. 1981. "Vorlesungen zur Geschichte der Neuen Architektur III. Das Zeitalter Wilhelms II." *arch+ Sonderhefte* 59, October.

Posener, Julius. 1982. "Vorlesungen zur Geschichte der Neuen Architektur IV. Soziale und bautechnische Entwicklungen im 19. Jahrhundert." *arch+ Sonderhefte* 63/64, October.

Potvin, André. 2000. "Assessing the microclimate of urban transitional spaces." In *Proceedings of PLEA2000*. Cambridge UK: Routledge.

Potvin, André. 2004. "Intermediate environments," in Steemers, Koen and Mary Ann Steane (eds.). *Environmental Diversity in Architecture*. London, New York: Routledge.

Prinz, Regina. 2012. "Freyes Licht, Freye Luft, Freyes leben von Pol zu Pol" — Die Idee einer 'Sonnenstadt' von Bernard Christoph Faust und Gustav Vorherr". In *L'Architecture Engagée – Manifeste zur Veränderung der Gesellschaft,* edited by Winfried Nerdinger. Munich: Edition Detail.

Radt, Wolfgang. 1993. "Landscape and Greek Urban Planning: Exemplified by Pergamon and Priene." In *City and Nature: Changing Relations in Time and Space,* edited by T. M. Kristensen, S. E. Larsen, P. G. Moller and S. E. Petersen. Odense, Denmark: Odense University Press.

Raith, Frank-Bertolt. 1997. *Der heroische Stil. Studien zur Architektur am Ende der Weimarer Republik.* Berlin: Verlag für Bauwesen.

Rapoport, Amos. 1969. *House, Form, and Culture.* New York: Pearson.

Requena-Ruiz, Ignacio. 2018. "Building a Brazilian Climate. The Case of the House of Brazil in Paris (France)." In *The Urban Microclimate as Artifact*, edited by Sascha Roesler and Madlen Kobi. Basel: Birkhäuser.

Reulecke, Jürgen and Adelheid Gräfin zu Castell Rüdenhausen (eds.). 1991. *Stadt und Gesundheit. Zum Wandel von 'Volksgesundheit' und kommunaler Gesundheitspolitik im 19. und frühen 20. Jahrhundert.* Stuttgart: Franz Steiner Verlag.

Reuter, Ulrich. 2011. "Implementation of Urban Climatology in City Planning in the City of Stuttgart." In *City Weathers. Meteorology and Urban Design 1950–2010*, edited by Michael Hebbert, Vladimir Jankovic and Brian Webb. Manchester: University of Manchester.

Riemann, Gottfried. 1971. "Frühe englische Ingenieurbauten in der Sicht Karl Friedrich Schinkels. Zu einigen Skizzen und Zeichnungen der englischen Reise von 1816." In *Forschungen und Berichte*, Staatliche Museen zu Berlin, Vol. 13, Berlin: Akademie-Verlag.

Rittel, Horst and Melvin Webber. 1973. "Dilemmas in a General Theory of Planning." *Policy Sciences* 4.

Rodenstein, Marianne. 1988. *"Mehr Licht, mehr Luft". Gesundheitskonzepte im Städtebau seit 1750.* Frankfurt / New York: Campus.

Rodriguez-Lores, Juan. 1988. "Die Zonenplanung – Ein Instrument zur Steuerung des Kleinwohnungsmarktes." In *Die Kleinwohnungsfrage. Zu den Ursprüngen des sozialen Wohnungsbaus in Europa*, edited by Juan Rodriguez-Lores and Gerhard Fehl, Hamburg.

Roesler, Sascha. 2013a. "Climate and Culture." In *Climate as a Design Factor*, second edition, edited by C. Hönger et al., Luzern: Quart.

Roesler, Sascha. 2013b. *Weltkonstruktion. Der aussereuropäische Hausbau und die moderne Architektur – ein Wissensinventar.* Berlin: Gebr. Mann Verlag.

Roesler, Sascha (ed.). 2015. *Natural Ventilation, Revisited. Pioneering a New Climatisation Culture*, Future Cities Magazine (special issue), Singapore-ETH Centre, Singapore.

Roesler, Sascha. 2017. "Die Wiederkehr des Klimas. Urbane Kulturen der Belüftung in Südostasien." *arch+* 227, "Vietnam – Die Ruckkehr des Klimas."

Roesler, Sascha. 2018a. "Allegory of the Sustainable City. The Wind Catchers of Hyderabad (Pakistan) and Their Spread in the Modern Architectural Discourse." In *Quaderni dell'Accademia di architettura*, Università della Svizzera italiana, edited by Sonja Hildebrand et al., p. 126-139.

Roesler, Sascha. 2018b. "Jenseits von innen und aussen. Die Stadt als thermischer Innenraum der Gesellschaft." *archithese* 2: 54–64.

Roesler, Sascha. 2019. Mikroklima. In *Architekturpädagogiken: Ein Glossar*, hrsg. von Heike Biechteler und Johannes Käferstein, Park Books: Zurich, 2019, p. 82–83.

Roesler, Sascha. 2020. On Microclimatic Islands. The Garden as a Place of Intensified Thermal Experience. In *Les Cahiers de la recherche architecturale urbaine et paysagère*, 6, special issue on "Architectures des milieux hyper-conditionnés".

Rosental, Paul-André. 2016. *Destins de l'eugénisme*. Paris: Editions du Seuil.

Rossi, Aldo (1974), "Bemerkungen zu den Arbeiten des Wintersemsters," in *Aldo Rossi — Vorlesungen, Aufsätze, Entwürfe*, Texte zur Architektur 4, Lehrstuhl Aldo Rossi, Verlag der Fachvereine 1974, ETH Zurich.

Rossi, Aldo et al. (eds). 1979. *Costruzione del Territorio e Spazio Urbano nel Cantone Ticino*. Lugano: Fondazione Ticino Nostro.

Rossler, Gustav. 2008. "Kleine Galerie neuer Dingbegriffe: Hybriden, Quasi-Objekte, Grenzobjekte, epistemische Dinge." In *Bruno Latours Kollektive*, edited by Georg Kneer, Markus Schroer und Gerhard Schüttpelz. Frankfurt am Main: Suhrkamp.

Rudder, B. and F. Linke (eds.). 1940. *Biologie der Grossstadt*. Frankfurter Konferenzen für medizinisch-naturwissenschaftliche Zusammenarbeit, IV Konferenz am 9. und 10. Mai. Dresden, Leipzig: Verlag von Theodor Steinkopf.

Rudofsky, Bernard. 1964. *Architecture without Architects. An Introduction to Non-Pedigreed Architecture*. New York: Museum of Modern Art.

Ruefenacht, Lea A. and Aurel von Richthofen. 2021. "Mitigate Urban Heat." In *Future Cities Laboratory: Indicia 03*, edited by Stephen Cairns and Devisari Tunas. Zurich: Lars Müller Publishers.

Sachsse, Rolf. 1984. *Photographie als Medium der Architekturinterpretation: Studien zur Geschichte der deutschen Architekturphotographie im 20. Jh.*. Berlin: De Gruyter Saur.

Salzmann, Heinrich. 1911. *Industrielle und gewerbliche Bauten (Speicher, Lagerhäuser und Fabriken)*, Vol. 1–3, Leipzig: De Gruyter.

Sauberer, Franz. 1948. *Wetter, Klima und Leben. Grundzüge der Bioklimatologie*. Vienna: Hollinek.

Schinkel, Karl Friedrich. 1993. *The English Journey. Journal of a Visit to France and Britain in 1826*, edited by David Bindmann and Gottfried Riemann. New Haven: Yale University Press.

Schmalz, Joachim. 1987. *Das Stadtklima. Ein Faktor der Bauwerks- und Städteplanung. Unter besonderer Berücksichtigung der Berliner Verhältnisse, mit Beispielen aus Planungsgebieten der 'Internationalen Bauausstellung Berlin 1987'*. Karlsruhe: C. F. Müller.

Schmauss, August. 1914. "Meteorologische Grundsätze im Haus- und Städtebau." *Bayerisches Industrie- u. Gewerbeblatt* 19: 181

Schmidt, Alfred. 1926. "Die Bedeutung von Sonne und Wind für den Stadtkörper. Ein Beitrag zu zeitgemässem Städtebau." In *Städtebau*. Monatshefte für Stadtbaukunst, Städtisches Verkehrs-, Park- und Siedlungswesen, edited by Werner Hegemann. Berlin: Verlag Ernst Wasmuth.

Schmidt, Wilhelm. 1927. "Die Verteilung des Minimumtemperaturen in der Frostnacht des 12. Mai 1927 im Gemeindegebiet von Wien." *Fortschr. d. Landw.* 21.

Schmitt, Paul. 1930. "Die Besonnungsverhältnisse an Stadtstraßen und die günstigste Blockstellung." *Zeitschrift für Bauwesen* 5: 124.

Schulz, Niels, Nilay Shah, David Fisk, James Keirstead, Nouri Samsatli, Aruna Sivakumar, Celine Weber and Ellin Saunders. 2014. "The SynCity Urban Energy System Model." In Mostafavi and Doherty (2016, 446).

Schulze-Gävernitz, Gerhard von. 1892. Der Großbetrieb, ein wirtschaftlicher und sozialer Fortschritt. Eine Studie auf dem Gebiete der Baumwollindustrie, Leipzig. Cited in Grundke, Günter. 1955. Die Bedeutung des Klimas für den industriellen Standort. Eine Studie auf dem Gebiete der Technischen Geographie. VEB Geographisch-Kartographische Anstalt Gotha.

Schwagenscheidt, Walter. 1949. *Die Raumstadt. Hausbau und Städtebau für jung und alt, für Laien und was sich Fachleute nennt. Skizzen mit Randbemerkungen zu einem verworrenen Thema*. Heidelberg: Lambert Schneider.

Seewang, Laila. 2019. *The Scale of Water – Networked Infrastructure and the Making of Municipal Berlin 1872–1900*. PhD Thesis, Zurich: ETH, No. 25871.

Seng, Eunice. 2014. *Habitation and the Invention of a Nation, Singapore 1936–1979*. Dissertation, New York: Colombia University, abstract.

Shitara, Hiroshi. 1957. "Effects on Buildings upon the Winter Temperature in Hiroshima City." *Geographical Review of Japan* 30 (6): 468–482.

Shove, E., G. Walker and S. Brown. 2014. "Transnational Transitions: The Diffusion and Integration of Mechanical Cooling". *Urban Studies* 51 (7): 1506–1519.

Šik, Miroslav. 2011. "Lernen von Rossi." In *Aldo Rossi und die Schweiz. Architektonische Wechselwirkungen*, edited by Ákos Moravánszky and Judith Hopfengärtner. Zurich: gta Verlag.

Sirén, Osvald. 1949. *Gardens of China*. New York: The Ronald Press Company.

Sitte, Camillo. 1889. *Der Städte-Bau nach seinen künstlerischen Grundsätzen: ein Beitrag zur Lösung modernster Fragen der Architektur und monumentalen Plastik unter besonderer Beziehung auf Wien*. Vienna: Verlag von Carl Graeser.

Sloterdijk, Peter. 2004. *Sphären. Plurale Sphärologie, Band III, Schäume*. Frankfurt am Main: Suhrkamp Verlag.

Smithson, Peter. 1969. "Bath: Walks within the walls: a study of Bath as a built form taken over by other uses." *Architectural Design*, October.

Smithson, Peter and Alison. 1954. *Doorn Manifesto*. Accessed on February 15, 2022. https://evolutionaryurbanism.com/2017/03/24/the-doorn-manifesto

Sopandi, Setiadi and Avianti Armand (eds.). 2015. *Tropicality: Revisited*. Jakarta: Imaji.

Soper, Kate. 1995. *What is Nature? Culture, Politics and the Non-human*. Oxford: Blackwell Publishers Limited.

Spelsberg, Gerd. 1988 (1984). *Rauch Plage. Zur Geschichte der Luftverschmutzung*. Cologne: Kölner Volksblatt Verlag.

Spirn, Anne Whiston. 1984. "City and Nature." In *The Granite Garden: Urban Nature and Human Design*. Cited in *The Sustainable Urban Development Reader*, edited by Stephen M. Wheeler and Timothy Beatley. 2004, London: Routledge.

Spirn, Anne Whiston. 1993. "Deep Structure: On Process, Form, and Design in the Urban Landscape." In *City and Nature: Changing Relations in Time and Space*, edited by T. M. Kristensen et al. Odense, Denmark: Odense University Press.

Stadler, Friedrich. 1997. *Studien zum Wiener Kreis. Ursprung, Entwicklung und Wirkung des Logischen Empirismus im Kontext*. Frankfurt am Main: Springer.

Steadman, Philip. 1975. *Energy, Environment and Building*. Cambridge, MA: Cambridge University Press.

Stein, Clarence S. 1936. "Henry Wright, 1878–1936." *American Architect and Architecture*, August.

Steinhauser, F. 1957. "Säkulare Änderungen der klimatischen Elemente." In *Klima und Bioklima von Wien. Eine Übersicht mit besonderer Berücksichtigung der Bedürfnisse der Stadtplanung und des Bauwesens*, II. Teil, edited by F. Steinhauser, O. Eckel and F. Sauberer. Vienna: Österreichische Gesellschaft für Meteorologie.

Steinhauser, F., O. Eckel and F. Sauberer. 1955. *Klima und Bioklima von Wien. Eine Übersicht mit besonderer Berücksichtigung der Bedürfnisse der Stadtplanung und des Bauwesens*, I. Teil, Vienna: Österreichische Gesellschaft für Meteorologie.

Steinhauser, F., O. Eckel and F. Sauberer. 1959. *Klima und Bioklima von Wien. Eine Übersicht mit besonderer Berücksichtigung der Bedürfnisse der Stadtplanung und des Bauwesens*, III. Teil, Vienna: Österreichische Gesellschaft für Meteorologie.

Steinmann, Martin and Thomas Boga. 1975. *Tendenzen*. Zurich: gta Verlag.

Steurer, Hannah. 2017. "'Berlin ist eine Sandwüste. Aber wo sonst findet man Oasen?' Stadtdiskurs als Naturdiskurs in der deutschen und französischen Literatur (1800 bis 1935)." In *Literatur und Ökologie. Neue literatur- und kulturwissenschaftliche Perspektiven*, edited by Claudia Schmitt and Christiane Sollte-Gresser. Bielefeld: Aisthesis Verlag.

Stieger, Lorenzo (2018), *Vom Hang zur Schräge. Das Hangterrassenhaus in der Schweiz: Aufstieg und Niedergang einer gefeierten Wohnbautypologie*. Dissertation, Zurich: ETH.

Stolberg, Michael. 1994. *Ein Recht auf saubere Luft? Umweltkonflikte am Beginn des Industriezeitalters*. Erlangen: Harald Fischer Verlag.

Stoler, Ann Laura. 1995, 1985. *Capitalism and Confrontation in Sumatra's Plantation Belt, 1870–1979*. New Haven / London: Yale University Press.

Strauss, Sarah and Ben Orlove (eds.). 2003. *Weather, Climate, Culture*. Oxford, New York: Routledge.

Stübben, Josef. 1890. "Der Städtebau." In *Handbuch der Architektur*, vierter Theil, 9. Halb-Band. Darmstadt: Verlag von Arnold Bergsträsser.

Stübben, Josef. 1907. *Der Städtebau*. Stuttgart: Kröner.

Taut, Bruno. 1919. *Die Stadtkrone*. Jena: Eugen Dieterich.

Taut, Bruno. 1931. "Der Außenwohnraum". *Gehag-Nachrichten* 2 (1/2).

Taut, Max. 1946. *Berlin im Aufbau*. Berlin: Aufbau-Verlag.

Taylor, Jeremy. 1991. *Hospitals and Asylum Architecture in England 1840–1914. Building for Health Care*. London: Mansell Publishing.

Tenorio, Rosangela. 2007. "Enabling the Hybrid Use of Air Conditioning: A Prototype on Sustainable Housing in Tropical Regions." *Building and Environment* 42 (2): 605–613.

Thee, Kian Wie. 1977. *Plantation Agriculture and Export Growth. An economic history of East Sumatra, 1863–1942*. Jakarta: National Institute of Economic and Social Research.

Thibaud, Jean-Paul. 2007. "La fabrique de la rue en marche: essai sur l'altération des ambiances urbaines." *Flux*, 66/67: 111–119.

Thies, Ralf. 2006. *Ethnograph des dunklen Berlin. Hans Ostwald und die "Großstadt-Dokumente"*. Köln: Böhlau Verlag.

Thompson, Grant. 1978. "Solar Energy and the Law: Barriers to a Sunnier Tomorrow." In *Proceedings of the National passive solar conference 2*, p. 850–856.

Thomsen, Christian W. 1985. "Die Hure Babylon und ihre Töchter." *Du, Städtephantasien: Architekturutopien in der Literatur* 45 (2).

Troll, Manfred. 1981. *Architektur der Grossstadt. Theoretische Stadtprojekte seit 1900, Entwerfen mit sozioökonomischen Grundlagen*. Berlin: TU Berlin.

Trotha, Hans von. 1999. *Angenehme Empfindungen. Medien einer populären Wirkungsästhetik im 18. Jahrhundert vom Landschaftsgarten bis zum Schauerroman*. Munich: Wilhelm Fink Verlag.

Trotha, Hans von. 2016. "Hortus conclusus. The Medieval Monastic Garden." In *Gardens of the World. Orte der Sehnsucht und Inspiration*, edited by Albert Lutz (and Hans von Trotha). Cologne: Wienand Verlag.

Tubbesing, Markus. 2018. *Der Wettbewerb Gross-Berlin 1910. Die Entstehung einer modernen Disziplin Städtebau*. Tübingen: Wasmuth.

Turda, Marius. 2010. *Modernism and Eugenics*, Basingstoke: Palgrave Macmillan.

Uekötter, Frank. 2003. *Von der Rauchplage zur ökologischen Revolution. Eine Geschichte der Luftverschmutzung in Deutschland und den USA 1880–1970*. Essen: Klartext Verlag.

Ungers, Oswald Mathias. 1980. *5 Energie Häuser*. Cologne: Walther König.

Ungers, O. M., H. F. Kollhoff and A. A. Ovaska. 1977. *The Urban Villa: a Multi-family Dwelling Type*. Cologne: Studio Press for Architecture.

Unwin, Raymond. 1909. *Town Planning in Practice. An Introduction to the Art of Designing Cities and Suburbs*. London: T. F. Unwin.

Usemann, Klaus W. 1993. *Entwicklung von Heizungs- und Lüftungstechnik zur Wissenschaft. Hermann Rietschel. Leben und Werk*, Munich / Vienna: Springer.

Van den Heuvel, Dirk. 2007. "Situativer Urbanismus." *arch+* 183.

Van der Linden, Cornelius, Hubert Hoffmann and Wilhelm Hess. 1932. Arbeitsgruppe Dessau der Int. Kongresse f. Neues Bauen, 11. vor- Nachteil der Lage.

Van Es, Evelien. 2014. "The Exhibition Housing, Working, Traffic, Recreation in the Contemporary City — A Reconstruction." In *Atlas of the Functional City. CIAM 4 and Comparative Urban Analysis*, edited by Evelien van Es, Gregor Harbusch, Bruno Maurer, Muriel Pérez, Kees Somer and Daniel Weiss. Bussum: THOTH Publishers gta Verlag.

Vasilikou, Carolina and Marialena Nikolopoulou. 2014. "Degrees of Environmental Diversity for Pedestrian Thermal Comfort in the Urban Continuum. A New methodological Approach." In *Bridging the Boundaries. Human Experience in the Natural and Built Enviroment and Implications for Research, Policy, and Practice. Advances in People-Enviroment Studies,* Vol. 5. Göttingen: Hogrefe Publishing.

Villani, Tiziana. 1995. *Athena Cyborg. Per una Geografia dell'Espressione: Corpo, Territorio, Metropoli*. Sesto San Giovanni: Mimesis.

Villiers, John. 1984, 1965. "Südostasien vor der Kolonialzeit". In *Fischer Weltgeschichte*, Vol. 18. Frankfurt am Main: Fischer Taschenbuch Verlag.

Vititneva, Ekaterina, Zhongming Shi, Pieter Herthogs, Reinhard König, Aurel von Richthofen and Sven Schneider. 2021. "Informing the Design of Courtyard Street Blocks Using Solar Energy Models: Generating University Campus Scenarios for Singapore." In *Carbon Neutral Cities - Energy Efficiency & Renewables in the Digital Era*. Lausanne: épfl.

Vögele, Jörg. 2001. *Sozialgeschichte städtischer Gesundheitsverhältnisse während der Urbanisierung*. Berlin: Duncker & Humblot.

Volker, Ullrich. 2007. *Die nervöse Großmacht. Aufstieg und Untergang des deutschen Kaiserreichs 1871–1918*. Frankfurt am Main: Fischer Taschenbuch Verlag.

Vrachliotis, Georg. 2020. *Geregelte Verhältnisse. Architektur und technisches Denken in der Epoche der Kybernetik*, Basel: Orell Fuessli.

Wagner, Martin. 1915. *Das sanitäre Grün der Städte. Ein Beitrag zur Freiflächentheorie*. Dissertation, Berlin: TU. February 27.

Wagner, Martin. 1951. *Wirtschaftlicher Städtebau*. Stuttgart: Julius Hoffmann.

Wagner, Otto. 1911. *Die Großstadt*, quoted in: Hackenschmidt, Sebastian, Iris Medre and Ákos Moravánszky. 2018. *Post Otto Wagner. Von der Postsparkasse zur Postmoderne.* Basel: Birkhäuser.

Waldheim, Charles. 2016. "Weak Work: Andrea Branzi's 'Weak Metropolis' and the Projective Potential of an 'Ecological Urbanism'." In Mostafavi and Doherty (2016, 113).

Wallace, Alfred Russel. 1869, 2010. *The Malay Archipelago.* Oxford: Oxford University Press.

Warhaftig, Myra, Susanne Rexroth and Philipp Oswalt. 1994. "Gebäudeklimatische Studien von A. Klein." In *Wohltemperierte Architektur. Neue Techniken des energiesparenden Bauens*, edited by Philipp Oswalt together with Susanne Rexroth. Heidelberg: Verlag C. F. Müller.

Warnke, Martin and Claudia Brink (eds.). 2000. *Aby Warburg. Der Bilderatlas MNEMOSYNE*. Berlin: Akademie Verlag.

Wasiuta, Mark and Sarah Herda. "Anna Halprin, Lawrence Halprin. The Halprin Workshops Bay Area, California, USA 1966–1971." Accessed on February 14, 2022. http://radical-pedagogies.com/search-cases/a37-anna-halprin-lawrence-halprin-workshops

Waterson, Roxana. 1990. *The Living House – An Anthropology of Architecture in South-East Asia.* Singapore, Oxford: Ruefenacht.

Watts, Michael. 2016. "Oil City: Petro-landscapes and Sustainable Futures." In Mostafavi and Doherty (2016, 451)

Wedepohl, Claudia. 2005. "Ideengeographie – Ein Versuch zu Aby Warburgs 'Wanderstraßen der Kultur'." In Helga Mitterbauer and Katharina Scherke (eds.). *Ent-grenzte Räume – Kulturelle Transfers um 1900 und in der Gegenwart*, Studien zur Moderne 22. Vienna: Passagen-Verlag.

Wetter und Leben. 1948. "Aufgabe und Ziel unserer Zeitschrift." April, Heft 1.

Wetter und Leben. 1948. "Vorträge über Umweltforschung." November, Heft 8.

Whittick, Arnold. 1950. *European architecture in the twentieth century.* London: C. Lockwood. Cited in Aronin, Jeffrey Ellis. 1953. *Climate & Architecture.* New York: Reinhold.

Wines, James and Patricia Phillips. 1982. *Highrise of Homes.* New York: Rizzoli.

Witt, Andrew and Christopher Reznich. 2018. *The Natural Forces Laboratory: Ralph Knowles and the Instrumentalized Studio.* Montréal and Cambridge: CCA and Harvard University Graduate School of Design. Accessed on February 15, 2022. https://www.cca.qc.ca/en/events/63353/the-natural-forces-laboratory-ralph-knowles-and-the-instrumentalized-studio

Wittich, Dieter. 2001. "Lenins Buch 'Materialismus und Empiriokritizismus'. Seine Entstehungsgeschichte sowie progressive und regressive Nutzung." In Gerhardt, Volker and Hans-Christoph Raus (eds.). *Anfänge der DDR-Philosophie. Ansprüche, Ohnmacht, Scheitern.* Berlin: Christoph Links Verlag.

Wittkower, Rudolf. 1977a. *Allegory and the Migration of Symbols.* London: Thames & Hudson.

Wittkower, Rudolf. 1977b. "East and West: The Problem of Cultural Exchange." In *Allegory and the Migration of Symbols.* London: Thames & Hudson.

Wolschke-Bulmahn, Joachim and Peter Fibich. 2004. "Vom Sonnenrund zur Beispiellandschaft. Entwicklungslinien der Landschaftsarchitektur in Deutschland", dargestellt am Werk von Georg Pniower (1896-1960), Schriftenreihe des Fachbereichs Landschaftsarchitektur und Umweltentwicklung der Universität Hannover.

Wong, Aline K and Stephen Yeh (ed.). 1985. *Housing a Nation. 25 Years of Public Housing in Singapore.* Maruzen Asia for Housing & Development Board.

World Health Organization, Burden of Disease from Ambient and Household Air Pollution. Accessed on January 1, 2015. www.who.int/phe/health_topics/outdoorair/databases/en

Wörner, Martin, Paul Sigel, Doris Mollenschott and Karl-Heinz Hüter (eds.). 2013. *Architekturführer Berlin.* Berlin: Reimer.

Wright, Henry. 1935. *Rehousing Urban America.* New York: Columbia University Press.

Wright, Henry Nicolls. 1936. *Solar radiation as related to summer cooling and winter radiation in residences.* New York: John B. Pierce Foundation.

Xenophon. ~371 BC. *The Memorabilia.* Recollections of Socrates, Book III, VIII. Accessed on February 10, 2022. http://www.gutenberg.org/files/1177/1177-h/1177-h.htm

Yamamoto, Hiroshi and Christine Ivanovic (eds.). 2010. *Übersetzung – Transformation. Umformungsprozesse in/von Texten, Medien, Kulturen.* Würzburg: Königshausen & Neumann

Yoshizaka, Takamasa. 1966. "Group Organization and Physical Structure." *The Japanese Architect* April: 27.

Yuen, Belinda. 2005. "Squatters no more: Singapore Social Housing." In *Third Urban Research Symposium: Land Development, Urban Policy and Poverty Reduction,* April 4–6 2005, Brasília.

Yusuf, A. A. and H. A. Francisco. 2009. "Climate Change Vulnerability Mapping for Southeast Asia." In Economy and Environment Program for Southeast Asia (EEPSEA).

Index

Subject Index

adaptation
17, 68, 83, 84, 124, 136, 143, 145, 156, 220, 224, 225

agriculture (agricultural land)
51, 199, 207, 215, 221, 224

air conditioning (air-conditioning devices and cooling)
11, 13, 20, 25, 28, 65, 97–98, 100, 104–105, 140, 153, 162–163, 187, 191, 197, 210, 213, 216–218, 220–222, 224–225

air pollution (air quality)
17, 21–22, 24, 29, 37, 40, 56–57, 70, 100, 106–108, 116, 125–126, 138, 144, 186, 215, 221

Anthropocene
12, 15, 225

apartment
11, 36, 43, 89, 217, 218

apartment building
48, 60, 84, 90–91, 136, 163

arcade (arcade house)
36–40, 42, 70, 89–90

atmosphere
21–22, 104, 156–157, 169, 193, 225

authorities
29, 48, 106, 153

autumn
57, 91

balcony (terrace)
36, 76, 90, 152

bamboo
213–214, 224

bathroom
199

baths, *see swimming pool*

bioclimatic conditions
24, 56–57, 175, 187, 217

bioclimatic design
207, 217, 221

bioclimatic research
125, 225

block (urban block, perimeter block)
33, 35, 37, 42, 47, 49, 62, 82–84, 89–90, 98, 135–139, 146–147, 159, 162–163, 217

body (bodily comfort, body temperature), *see also health*
20, 36, 75, 156, 164, 169, 181, 191

brick
21, 224

bridges (bridge city)
17, 100, 164, 181, 184, 187, 196, 199

buffer zones
37

building law, *see law*

chimney effect, *see stack effect*

choreography
164–165, 168

CIAM (Congrès Internationaux d'Architecture Moderne)
13, 49, 60, 67, 80, 106–107, 126, 130, 145, 147, 198

climate

    climate change
    15, 226

    climate control
    11, 13, 16, 20, 43, 51, 59, 62, 65–66, 71, 75, 88–89, 97, 104, 115, 117, 121, 127, 140, 141, 144, 152, 154, 162–163, 186, 189, 199, 213, 204, 210, 213, 215–217, 220–221, 224–226

    desert climate
    137, 140

    humid climate, *see also humidity*
    28, 213–214

    Mediterranean climate
    70, 194

    subtropical climate
    100, 105, 144

clothing (textiles)
31, 51, 157, 220

cloud (cloud cover, cloud formation)
55, 57

coal, *see also fossil fuels*
21–22, 24, 29, 31, 180, 186

collective form
179, 197–200, 204, 207

collective housing
198

collective memory
121, 124, 126, 148, 198

comfort

    comfort research
    68, 144

comfort thinking/paradigm
11, 17, 25, 65, 68, 119, 153–156, 157–158, 213, 221

    thermal comfort levels/perception
    40, 66, 91, 115, 138, 144, 147, 152, 164, 175, 186, 190, 197, 207, 215–216, 218, 220, 224

commons (common good, energy commons)
188–189, 191, 206–207

concrete (reinforced concrete)
28, 98–99, 128, 130, 162, 174, 196, 199, 224

cooling
28, 36, 40, 43, 50, 56–58, 71, 96, 115, 137, 147, 149, 153–154, 159, 163, 170, 187, 191, 205, 215–218, 220–221, 222, 224–225

cotton mill
20–21, 23, 24

cotton workers/spinners
28

countryside (rural area)
21, 24, 35, 57–58, 62, 68, 131, 136–137, 158, 189

courtyard (yard)
31, 35–36, 41–42, 49, 58, 61–62, 65, 76, 83, 89, 100, 105, 127, 135, 138–139, 147, 157, 162, 169, 197–198, 214, 225, 247, 254, 260, 266

courtyard house
152, 157, 214

cortina, *see also curtain*
162

culture
14–15, 35, 68, 75, 85, 100, 105, 121, 144, 152, 154, 156, 175, 180–182, 191, 197, 213–214, 216–218, 220

    architectural culture
    17

    building culture
    121, 213

    material culture
    37, 164–165, 214

    physical culture
    99, 104

    planning culture
    175

curtain
20, 29, 213

density

    high-density
    30–31, 42, 78, 96, 197, 218

doors
20, 29, 70, 157, 218

265

drought
125, 144, 190, 207, 215

dust
21, 28, 35, 40, 51, 57, 76, 107, 158, 221

    coal dust (soot)
    22, 24, 29, 44, 106–107

ecology, *see also environment*
12, 117–119, 124, 152, 164, 168–169, 179, 181, 186–187, 190, 197–198, 214

    ecological urbanism
    164, 191, 196–197, 207

    ecology movement
    12, 119, 148

    political ecology
    12, 117, 186, 197, 225

electricity
51, 58, 104, 153, 180, 191, 199, 220, 224

energy

    energy consumption
    35, 216–218, 220–221

    energy crisis (oil crisis)
    187

    energy demand
    220

    energy efficiency
    24, 28

    energy hinterland
    131, 191, 193

    energy infrastructures
    24, 179–180, 185–186, 188, 191

    energy landscape
    118, 180–181, 188

    energy saving
    179, 187–188, 207

    renewable energy, *see also solar energy*
    179, 188, 191

environment, *see also ecology*
22, 68, 125, 156, 168–169, 181, 186, 206

ethnography (urban ethnography)
180–181, 198

exterior *see also interior*

    conditions
    15

    space
    115, 163

    walls
    24, 85, 89

facade
37, 40, 42, 47–48, 62, 70, 76, 84–85, 90, 100, 137–139, 162–163, 187, 217–218

factories (industrial plants)
24, 29, 107–108, 162

fieldwork
155

fire
20, 22, 68, 71, 215–216, 218, 232

flooding
106, 190, 199, 206, 215, 224

fog
22, 158, 187, 216,

    artificial fog
    107

    black fog
    21–22

form, *see collective form*

fossil fuels (oil and gas)
24, 188, 191, 220

fountain
100, 185, 191

gardens
48–50, 71, 75, 78, 82–85, 100, 105, 136, 152, 155–159, 162–166, 169, 174, 181, 189–190, 205

    Garden City
    49–50, 85, 108, 112, 131, 164, 175, 196

    gardening
    51, 156, 165

gas (gas lights, gas-burning stoves), *see also fossil fuels*
20, 22, 220

Geddes, Patrick
193, 198

glass, *see glass roof, arcade*

global warming, *see climate change*

golf course
181

governance, *see also thermal governance*
16, 83, 186, 206

greening (greenery)

    green corridor
    79, 130

    sanitary green, *see also hygiene*
    48, 75, 78–79, 82, 159

urban park, *see also gardens*
15, 21, 29, 51, 57, 65, 78–79, 82–83, 85–86, 108, 117, 126, 147, 156, 163, 181, 185, 189, 191

guidelines, *see also regulations, standards*
14, 121, 123, 126–127, 168–169, 221

haze, *see also air pollution*
22, 57, 187, 215–216, 218, 221

HDB (Housing Development Board)
215–221, 224–225

health, *see also body, hygiene*
12, 17, 20, 24, 28–29, 31, 35, 44, 48, 64, 68, 76, 78–79, 82, 85, 96, 104, 106–107, 119, 126, 158–159, 186, 216

heat wave
36

heating
20–22, 24, 31, 35, 40, 66, 70–71, 88, 96, 115, 139, 157, 159, 170, 187

    central heating
    13, 20, 35, 75, 157

    fireplace (brazier)
    157

    floor heating
    153

    heating system
    28, 35, 104, 157, 163

    heating zones
    20

    overheating
    17, 21, 30–31, 35–36, 62, 65, 96, 107, 155, 204, 213

    radiator
    35

high-rise building
47–48, 65, 76, 88–89, 145, 180, 187, 214

high-rise city
45, 48–50, 65

hinterlands, *see also energy hinterland* (outskirts, periphery)
24, 30, 35, 57, 79, 83, 98, 104, 121, 158, 159, 131, 191, 193, 214, 221

humidity
24–25, 38, 31, 57, 127, 144, 149, 193, 213, 220–221

hydrology
130, 190, 213

hydrotherapy
104

hygiene (urban hygiene)
12, 14, 17, 29, 31, 42, 44, 46, 48–49, 66, 71, 85, 89, 97–99, 104, 107, 117–119, 186–187

ice skating (rinks)
174–175

identity (urban identity, place identity)
121, 124, 169, 217

    corporate identity
    84

    identity politics
    118

industrial area (industrial zone)
34, 69, 79, 83, 111, 145, 147

    industrial building
    17, 21, 35, 44

    industrial capitalism
    66, 144

    industrial cities (the industrial city)
    44–45, 48–50, 56, 78, 106–107, 110–11,
    117, 199

    industrial landscape
    17, 21–22, 88

    industrial production
    17, 24–25, 28, 66, 76

    Industrial Revolution
    24, 31

    industrial workforce
    17, 28

    pre-industrial city
    68–69, 71, 74

infra-architecture
188

infrastructure
17, 50–51, 75–78, 191–192, 196–197, 199, 224

    energy infrastructure
    24, 179–180, 185–186, 188, 191

    green infrastructure
    76, 78, 83–85

    heating infrastructure
    31

    housing infrastructure
    215, 221

    rail(way) infrastructure
    24, 98, 108, 147, 180–181, 199

    sewage
    68, 138

    thermal infrastructure
    35, 51, 155

    urban infrastructure
    78, 104, 179–180, 197

    water-related infrastructure
    24, 138, 199

inside-outside relationship/experience
11, 187

insolation
25, 36, 42, 62, 64, 88, 91, 96–97, 158, 162, 201, 204

insulation (insulation layer)
13, 36, 85, 89, 140, 162, 220

interior, *see also exterior*
11, 13, 20, 37, 51, 59–60, 64, 66, 100, 178, 215

    interior climate
    31, 62, 68

    interior/exterior relationship
    11, 15, 30, 37, 99, 104, 112, 115, 139, 175,
    178, 187, 213, 226

    interior space
    82, 90, 97, 115, 174, 206

    interiority
    82, 225

    intermediate zones/environments
    37, 131

    urban interior
    62, 216

intimacy
35, 51, 163–164, 216

kitchens
29, 199

landscape, *see also energy landscape and industrial landscape*
14–15, 21, 29, 75, 85, 97, 104–105, 136, 144, 156–159, 164–165, 168, 170, 175, 188–189, 221

    landscape architecture/architects
    15, 48, 59, 70, 82, 85, 90, 99–100, 116–
    118, 126, 130–131, 136–137, 145, 164–165,
    168, 179, 187–188, 191, 205–206

    urban landscape
    16, 75, 118, 120, 131, 136–138, 159, 162,
    164–165, 170, 179–181, 187–188, 193, 205

laws
29, 70, 106, 206

    building laws
    40–42

    natural laws
    14

light (lighting, natural light)
13, 29–30, 42, 62, 64, 71, 78, 89, 97, 100, 104, 175, 181, 183, 196

    artificial light
    100, 153

sunlight
29, 35, 40, 42, 44, 48, 64–65, 70–71, 84, 90, 92, 96–97, 107, 139, 145, 175, 187, 197, 204

louvers (screens)
29, 175

luxury
37, 163, 189

man-made climate (man-made climate, anthropogenic climate, anthropogenic conditions)
11, 22, 29, 43, 68, 215–216

map (mapping)
46, 77, 106, 125, 131, 155, 160, 164–165, 186, 194, 205, 218–219

memory (collective memory)
121, 126, 130, 148, 198

memory carrier
124

metabolism
12, 16, 51

meteorology
13–14, 16, 44, 51, 70, 106, 125

microclimates

    interior microclimates
    20

    microclimatic conditions
    17, 24–25, 28, 31, 82, 88–89, 98, 125,
    130, 139, 155–156, 158, 163–165, 187, 199

    microclimatic diversity
    78, 105, 128, 130, 158, 162–165

    microclimatic islands
    89, 155–156, 189

    urban microclimates
    12–13, 17, 22, 49, 51, 56–57, 59, 64–66,
    69, 100, 105, 115, 118–120, 136, 144, 155,
    159, 162, 164–165, 170, 178, 206

middle class
85, 224

Milan
37, 40, 161–163

mill (cotton mill)
20, 26, 28

modelling
88, 118, 140, 206

    simulation
    20, 105, 118, 127, 153, 155, 168–169, 175,
    205–209

Modernism (Modern architecture, Modern Movement)
12–15, 17, 30, 43, 49, 50, 62, 65, 84, 106, 138, 140, 144, 148, 153–154, 180, 198

modernity
118, 178, 189, 197, 213, 216

modernization
64, 124, 140, 144, 180, 189, 196, 224

monsoon weather/climate
213–215

monsoon winds
149

NASA (National Aeronautics and Space Administration)
178

nature
12, 31, 75, 85, 104–105, 108, 112, 158, 175, 179–181, 186, 189–190, 194, 196–197, 213–214

indoor nature
98–99

nature and culture
15, 75, 100, 105, 120, 175

urban nature (nature in the city), *see also greening*
49, 75, 84–85, 117, 121, 124, 126, 130–131, 138, 162, 164, 170, 181, 187, 189, 190, 196, 210

neighborhood
17, 24–25, 31, 42, 50, 62, 78, 82, 90, 106, 108, 130, 136, 138–139, 174–175, 188, 196, 217

noise
31, 163, 218

outdoor spaces
13, 22, 35, 82, 84–85, 97, 105, 137, 207, 217, 220

outskirts, *see hinterlands*

overheating
17, 19–21, 30–31, 35, 36, 62, 65, 96, 107, 155, 204, 213,

park, *see urban park*

passive climate control
88–89, 97, 141, 152, 186, 204, 213, 215, 217

building orientation
13, 25, 31, 40, 42, 59, 62, 64–65, 71, 89, 92, 96–98, 112, 127, 145–147, 159, 163, 169, 174–175, 201, 217

cross-ventilation
29–31, 36, 147, 162, 217–218, 224

daylighting
42

shading (shade, shadows)
30, 36, 42, 50, 65, 69–70, 90, 96, 163, 175, 188, 207, 217

solar gain, *see also insolation*
170

periphery, *see hinterlands*

permeability
37, 78, 97, 126, 162–163, 214–215, 220–221, 225

plantation
78, 100, 215, 221, 223–224

plants (planting), *see also greening*
100

pollution, *see air pollution*

population density, *see also density*
44, 62

practices, *see thermal practices*

pre-industrial city, *see also industrial city*
68–69, 71, 74, 144, 157

privacy
136, 215, 217–218, 220

private property (private land)
37, 48, 207

public space
36–37, 40, 84, 163, 170, 175, 189, 191

radiation (solar radiation)
36, 40, 60, 70, 98, 145, 162, 186–187, 205

radiator
35

rain (rainfall)
22, 36, 40, 49, 55, 70, 104, 144, 157, 213,

rattan
213

regulations, *see also standards*
42, 70–71, 82–83, 96, 106–107, 162, 206, 215, 221, 224

renewable energy
179, 188, 191

renovation
175

research

architectural research
42, 49, 163

environmental research
125, 140

urban climate research
12, 14, 116, 190

resources

common-pool resources
189

cooling resources
71

energy resources, *see also energy*
24, 126, 186–187, 190, 207

land resources
189

natural resources
187, 191, 213

rivers
24, 57, 79, 106–108, 124, 130, 139, 147, 149, 157, 165, 181, 196

roof (rooftop, roofscape)
148–153, 154

flat roof
30, 84, 169

glass roof
37–40

pitched roof (gable)
84, 70, 169

roof terrace
85, 97

rural areas (rural communities), *see also countryside*
55–56, 158, 193, 199, 218

rural buildings
62

season (seasonal change, time of year)
25, 70, 84, 165, 213–214, 224

shutters
20, 29, 97, 157

smells (odors)
31, 35, 106, 147–148, 168

smoke
21–22, 24, 29, 31, 36, 40, 44, 106, 147, 187, 209,

snow
70

social class (social status, social stratification, working class)
24, 29, 43, 66, 78–79, 83, 85, 89, 105, 174

socialist (countries, construction, housing, etc.)
116, 136–138, 196

solar energy (solar gain), *see also insolation and renewable energy*
207

sound (formation of)
174

Spa Palace
99, 101–102, 104–105, 175

spectacle
155, 174, 181

Appendices

spring
91, 156–157

stack effect
149

standards (green building standards, standardization)
62, 70, 153, 168

status (socio-economic status), *see also social class*
83

suburban areas (suburban context)
11, 78, 83, 184

suburbanization
136

summer (midsummer)
21, 28, 30–31, 35–36, 40, 51, 56–57, 64–65, 70, 89, 96–97, 104, 108, 126, 131, 137, 139, 147, 149, 156–157, 162, 165, 170, 187, 193, 205, 210, 213

sunbathing
97, 159, 181

swimming (bathing)
75, 104–105

swimming pool
104, 181

synergy
179, 197–198

technologies (technological solutions)
17, 20, 22, 29, 56, 88, 121, 140, 155, 188, 207, 215, 224

temperature
21, 24, 28, 36, 51, 58–59, 61–62, 64, 127, 137, 140, 144, 147, 149, 152, 157, 165, 187, 193, 213, 220–221

   air temperature
   25, 36, 58, 137, 157

   indoor temperature
   36, 42, 66, 68, 89

   outdoor temperature
   31, 43, 56–57, 66, 126, 149, 210

   soil temperature
   67

   temperature conditions
   56

   temperature profiles/patterns
   36, 40, 43, 54–58, 66–68, 70, 129, 149

tenements
30, 35–37, 40, 42, 83, 88, 96, 106

terrace (house, construction)
70, 76, 95, 96, 98, 127, 205

thermal comfort, *see comfort*

thermal delight
155–156, 158, 175, 210

thermal design
90, 137, 165

thermal diversity (thermal variability, thermal variation)
62, 66, 165, 189

thermal governance
16, 83, 206

thermal inertia
36

thermal knowledge
59

thermal measurements
43

thermal perception (thermal experience, thermal sensation)
59, 105, 155, 157–158, 169, 175

thermal practices
22, 155, 217

thermal regime
154, 216

thermal stress
107

thermal structures
217

thermal transition between indoors and outdoors, *see also inside-outside relationship*
99

thermal variation

thermodynamics (thermal flows)
16, 62, 164

time (temporal aspects, dimensions, implications)
40, 42, 51, 59, 62, 64–66, 124–126, 164–165, 180, 210

topography
31, 42, 50, 57, 64, 100, 104–105, 107, 126–128, 130–131, 148, 157–158, 169, 170, 174, 180, 188, 191

   artificial (man-made) topography
   104–105, 126–128, 157

trees
31, 43, 50, 58, 64, 69–70, 78, 83, 85, 100, 124–125, 145, 147, 168, 181, 221

tropics
25, 66, 69

   tropical area (zone, region)
   144, 220

   tropical conditions
   214, 225

tropical disease
31

urban climatology
11–16, 30, 43–44, 48–50, 56, 58–59, 62, 66, 68–70, 76, 88, 98, 106–107, 115–120, 140, 145, 148, 158–159, 170, 175, 186, 188, 205, 226

urban design
12–17, 30, 48–51, 59, 68, 70, 75, 78, 84–85, 96, 106, 115–118, 120–121, 136, 139, 144, 152–153, 155–156, 163–164, 168, 170–171, 179, 187–188, 191, 196, 198, 204, 206–207, 210, 225

urban planning
12, 14–15, 29, 31, 40, 44, 48–49, 51, 65, 68–69, 71, 75–79, 82, 85, 96, 106–108, 112, 115–120, 125–126, 131, 136–138, 140, 145, 147, 153, 158, 163, 174, 179, 196, 198

urban heat island (heat-island effects)
13, 55, 57, 117, 159, 164, 189

urbanization
15–17, 22–22, 25, 44, 49, 56, 69, 88, 106, 108, 136, 140, 145, 152, 157, 181, 191, 193, 196–197, 199, 205–206, 210, 215

   suburbanization
   136

urban-rural relationship, *see also rural areas*
136, 159, 179, 199, 224

U-values
221

ventilation, *see also climate control, passive climate control*

   cross-ventilation
   29–31, 36, 147, 162, 217–218, 224

   natural ventilation
   28, 107

vernacular architecture (vernacular structures)
68, 143, 153–154, 169, 198, 214–215, 216, 225

village
159, 167–168, 198–199, 213–214

water, *see also rain, rivers*

   drinking water
   31

   public bathing
   75, 104–105, 159

   waterways
   24, 83, 164

weather, *see also man-made weather, meteorology*
25, 28, 29, 35, 44, 47, 53, 64, 65, 90, 139, 194, 208

   weather conditions
   57, 69–70, 216

Subject Index

weather measurements
125

winds
15, 22, 31, 35–36, 42, 44, 49–51, 55, 57–58, 65,
69, 71, 76, 83–84, 90, 98, 105–108, 121, 124–
127, 130, 137–138, 144–147, 149, 152, 154–155,
168–169, 188, 197

foehn winds
58

monsoon winds
149, 213–214

prevailing winds
44, 69, 106, 125–126, 145–147

wind corridors
130

wind diagram
106

wind direction
23, 58, 108, 125, 127

wind loads/pressure
35, 83

wind regime
107, 155

wind speeds
58–59, 107, 137–139

wind tunnel
169, 206, 208–209

windcatchers
148–154

windmill
185, 191

windows
64, 85, 90, 97, 157, 221

winter
21, 31, 35, 62, 65, 70, 96–97, 139, 144–145, 149,
157, 162, 163, 165, 170, 186–187, 205, 213

winter garden
163

wool (woolen mill, industry)
28

wood

as building material
157, 189, 196, 214, 224

firewood
20, 22, 29, 31, 189

yard, *see courtyard*

zoning
29, 42, 64–65, 70, 75, 78, 106–108, 126, 204,
207

Index of Names

Alberti, Leon Battista 15
Aronin, Jeffrey 116–117, 127, 140, 145, 147
Archizoom 192–193, 196
Atkinson, William 18, 21, 89, 92–93, 96–98, 205
Banham, Reyner 179–182, 197, 220, 226
Behrens, Peter 84, 90, 95–96, 98
Benjamin, Walter 36–37, 105, 149, 152–153
Boccioni, Umberto 37, 180
Boyarsky, Alvin 180–181, 183, 186, 197
Brezina, Ernst 13, 35, 40, 43, 48, 51, 58, 62, 64, 68, 76, 115, 137
Brooks, Charles Ernest 22, 25, 29, 115, 140–141, 144, 205
Canetti, Elias 99
Carrel, Alexis 68
Carrier, Willis 218
Egli, Ernst 121, 139–145, 152–153
Fechner, Gustav Theodor 59
Flügge, Carl 31, 35–36, 40, 42, 83
Foucault, Michel 15, 156, 214
Galfetti, Aurelio 196
Garnier, Tony 45, 47, 99, 103, 191
Geiger, Rudolf 44, 51, 55–56, 58, 67, 69, 116
Giedion, Sigfried 17, 62
Givoni, Baruch 14
Glacken, Clarence 15, 158
Gropius, Walter 13, 62, 98, 127
Habermas, Jürgen 169
Halprin, Anna 165, 167
Halprin, Lawrence 165, 167–168, 185, 191
Hassenpflug, Gustav 14, 158
Heschong, Lisa 155–156, 165, 169, 175, 205–206
Hilberseimer, Ludwig 14, 24–25, 30, 44–46, 48, 50–51, 55, 59–65, 69–74, 88–89, 98–99, 104, 106–107, 109–111, 134–136, 165, 181, 187, 196
Hough, Michael 118, 188–190, 196–197, 210
Hürlimann, Martin 148–151
Kaika, Maria 12
Kassner, Carl 31, 40, 56, 69–70
Klein, Alexander 84, 89–91, 98
Knowles, Ralph 71, 96, 179, 201, 204–205, 207, 210
Kracauer, Siegfried 36, 40, 75, 99
Kratzer, Albert 13, 43, 48–49, 55, 59, 61, 76, 108, 116, 121, 131, 136, 145, 186–187
Krier, Léon 159
Kurokawa, Kisho 199–200
Landsberg, Helmut 70, 116–117, 147
Latour, Bruno 56, 225
Le Corbusier 44–45, 47–49, 76, 84, 147
Lenin, Vladimir Ilyich 108
Mach, Ernst 59
Maki, Fumihiko 197–200
Manley, Gordon 127, 129–130, 140, 165
Marx, Karl 12
Matus, Vladimir 179, 188, 195, 197, 201, 204
McHarg, Ian 126, 179, 186, 188–191, 193, 196–198, 206
Meyer, Hannes 13, 106
Missenard, André 28, 44, 65–68, 147
Mumford, Lewis 22, 179, 198, 213

Olgyay, Victor and Aladar 116, 139, 141, 152–154, 205
Ovaska, Arthur 159
Pelzer, Dorothy 213–214
Pevsner, Nikolaus 22, 98, 158
Poelzig, Hans 84, 88, 94, 99, 102–104
Posener, Julius 31, 35, 42, 48, 83
Riemann, Gottfried 17
Riemann, Peter 160
Rossi, Aldo 198, 202
Ruchat-Roncati, Flora 196
Rudofsky, Bernard 148, 151–154
Russel Wallace, Alfred 213
Sant'Elia, Antonio 49, 180, 191
Schinkel, Karl Friedrich 17–24, 29, 37–38, 88, 159, 165, 229
Schmidt, Alfred 64, 68–69, 71, 107
Schmidt, Wilhelm 35, 40, 43, 48, 51, 54, 56, 58, 62
Scott Brown, Denise 180
Simmel, Georg 88
SITE 184, 187, 196
Sitte, Camillo 47, 50, 78–79, 163
Smithson, Alison and Peter 138, 164–165
Smithson, Peter 138, 164–166, 198
Spirn, Anne Whiston 117, 125–126, 138, 189–190, 197
Superstudio 184, 187, 192, 196
Swyngedouw, Erik 12
Taut, Bruno 44–45, 49–50, 82–83, 85, 88, 105, 108
Taut, Max 138
Team X 130, 198
Trümpy, Ivo 196
Tsing, Anna 124
Ungers, Oswald Mathias 159–160, 162, 169
Venturi, Robert 180
Wagner, Martin 48, 75, 78–79, 82–83, 87, 115, 131, 136, 155, 159
Wagner, Otto 45, 49, 93, 97,
Waldheim, Charles 179, 196–197
Wittgenstein, Ludwig 59

Geographical Index
(place names, rivers, districts)

Bath 18
Berlin 12, 14, 17, 21, 23–27, 30–38, 40–41, 48–49, 56–59, 61–63, 75–79, 81–83, 85, 95–97, 99–105, 115, 121–122, 131–138, 148, 159–160, 162–163, 175, 186
Bern 80
Birmingham 17, 22
Boston 96–97
Budapest 37, 152
California 96, 165, 168–169
Chandigarh 117, 145–147
Chicago 48, 64, 179–181, 183, 186, 197
Cologne 49, 84
Derby 20
Dudley 21, 23
Edinburgh 17–18
Frankfurt am Main 49–50, 66, 84, 99, 102, 112
Hamburg 49, 62, 94
Hyderabad 148–154
Karlsruhe 43, 56–58
London 17, 22, 37–38, 40, 48, 78, 126, 189, 199
Los Angeles 168, 179–182, 201, 204–205
Manchester 17, 21, 23–25
Milan 37, 40, 161–163
Moscow 37, 64
Munich 30, 52–57
New York 48, 61, 78, 98, 116, 139, 147–148, 175–185, 187, 191, 204–205
Newcastle 21
Oldham 24–25
Paris 37–41, 47–49, 62, 64 79, 162, 164, 199
Stuttgart 49, 56, 71, 84, 96, 98, 106–107, 109, 123, 126
Thames Valley 22
Trieste 35, 69
Vienna 12, 31, 35, 37, 43, 55–56, 59, 75, 79, 82, 90, 93, 96–98, 105–106, 115, 125, 152, 186
Vistula River 130
Zurich 60, 148, 174

About the Author

Sascha Roesler

Prof. Dr. Sascha Roesler is an architect and architectural theorist working at the intersection of architecture, ethnography, and science and technology studies. He is the Associate Professor for Theory of Urbanization and Urban Environments at the Academy of Architecture in Mendrisio, Switzerland (Università della Svizzera Italiana). Between 2013 and 2015, Roesler was a senior researcher at the Future Cities Laboratory (Singapore-ETH Centre for Global Environmental Sustainability), and between 2015 and 2021 he held the position of Swiss National Science Foundation Professor for Architecture and Theory at the Academy of Architecture in Mendrisio, leading a research project on "Architecture and Urban Climates". Roesler has published widely on issues of global architecture, sustainability and environmental technologies. His books include *Weltkonstruktion* (Gebr. Mann, 2013), a global history of architectural ethnography, and *Habitat Marocain Documents* (Park Books, 2015), a volume on the transformation of a colonial settlement in Casablanca.

Acknowledgments

My special thanks go to a number of institutions: first and foremost, the Swiss National Science Foundation, which appointed me as assistant professor in this subject area and thus demonstrated great foresight. This comprehensive publication (with a second volume) has only been made possible by a focus on this single subject over a period of six years. I would also like to mention the Future Cities Laboratory of ETH Zurich, which led me to the topic of urban climate and whose intellectual environment taught me the meaning and purpose of contemporary interdisciplinary research – an experience I would not have missed for the world.

Special thanks also go to the Institute of Technology in Architecture (ITA) at ETH Zurich, where I also benefited from a stimulating environment as an academic guest and visiting professor during the last phase of this research project. In this context, I would like to especially thank Fabio Gramazio and Arno Schlüter. I am also very grateful to the Max Planck Institute for the History of Science in Berlin and its director Dagmar Schäfer. The opportunity to share as a fellow my ideas and insights with experts in science and technology studies has been enormously encouraging.

This book is undoubtedly also based on the insights of outstanding colleagues who have inspired me as reader of their texts, in personal conversations and at professional conferences: Marc Angélil, Tom Avermaete, Daniel Barber, André Bideau, Roger Boltshauser, Jiat-Hwee Chang, Kim Förster, Johan Lagae, Fanny Lopez, Philippe Rahm, Marlyne Sahakian, Elizabeth Shove, Milica Topalović and Aurel von Richthofen. A big thank you also goes to the former members of my research group, Madlen Kobi and Lorenzo Stieger, as well as to all my colleagues at the Academy of Architecture in Mendrisio.

My special thanks go to my great editors Philip Shelley (copy-editing and translations) and Henriette Mueller-Stahl (at Birkhäuser) who carefully guided the publication process. I would like to also warmly thank the team at Birkhäuser, Floyd Schulze and Heike Strempel, for the way they managed the design and production of this book. The Open Access publication costs for this book were covered by the generous funding of the Swiss National Science Foundation.

As always, I am indebted to my partner Katja Jug, who creates open spaces where there are seemingly only impenetrable walls.

Sascha Roesler
December 2024

Illustration Credits

The author would like to thank all institutions and photographers who have made their work available for this publication. The author has made every effort to ensure the accuracy of the figure sources at time of publication and to find the copyrights of external illustrations. In the unlikely event that copyright owners have been overlooked we kindly ask them to let the author know in order to insert the appropriate acknowledgment in any subsequent edition of the book.

Fig. 1, 125
Sascha Roesler

Fig. 2, 3, 7, 9, 28, 78, 85, 102, 110, 111, 113, 186
TU Berlin, Architekturmuseum.

Fig. 4, 34, 96, 99–101
Atkinson (1912).

Fig. 5
Schinkel (1993).

Fig. 6
Ungers (1980).

Fig. 8
The Frick Collection.

Fig. 10
Schneider, Richard (ed.). 2004. Berlin um 1900, Nicolaische Verlagsbuchhandlung, Berlin, p. 111.

Fig. 11
d'Arcet, Jean-Pierre-Joseph. 1843. Des rapports de distances qu'il est utile de maintenir entre les fabriques insalubres et les habitations qui les entourent, Paris.

Fig. 12
Coll. Archives municipales de Mulhouse.

Fig. 13, 15
www.assets.new.siemens.com

Fig. 14
Alte Nationalgalerie,
Staatliche Museen zu Berlin.

Fig. 16, 17
Prinz (2012).

Fig. 18
Heiligenthal (1926).

Fig. 19
Deutsches Historisches Museum.

Fig. 20, 22, 142
Fassbinder (1975).

Fig. 21
Stübben (1907).

Fig. 23
Heiligenthal (1932).

Fig. 24–27, 29–32
Geist (1969).

Fig. 33
Brezina and Schmidt (1937).

Fig. 35
https://www.researchgate.net/profile/
Xavier-Roque-2

Fig. 36
www.krisdedecker.org

Fig. 37, 48
Garnier, Tony. 1917. Une cité industrielle: étude pour la construction des villes, Vincent: Paris.

Fig. 38
Wagner (1911).

Fig. 39, 64–65
Hilberseimer (1927).

Fig. 40
Troll (1981).

Fig. 41, 45, 47
Le Corbusier (1925).

Fig. 42, 52, 55–56
Kratzer (1937).

Fig. 43, 54, 60–61, 73–75, 77, 117, 121, 123–124, 147
Hilberseimer (1944).

Fig. 44
Wright (1935).

Fig. 46
Unwin (1909).

Fig. 49
Stadtarchiv München.
Photographer: Walter Mittelholzer.

Fig. 50, 51, 71
Geiger (1942).

Fig. 53, 57
Büdel and Wolf (1933).

Fig. 58, 59, 70, 79–80, 120, 122, 188
gta Archive, ETH Zurich.

Fig. 62, 66
Kratzer (1956).

Fig. 63, 84
Stübben (1980, 1890).

Fig. 67–69
Hilberseimer (1930).

Fig. 72
Missenard (1949, 1940).

Fig. 76, 119
Döcker (1929).

Fig. 81, 82
Goldmerstein and Stodieck (1931).

Fig. 83, 87–88
Heiligenthal, Roman. 1921. Deutscher Städtebau. Ein Handbuch für Architekten, Ingenieure, Volkswirte und Verwaltungsbeamte. Heidelberg: Carl Winter.

Fig. 86
www.documenta-bauhaus.de

Fig. 89, 90
Archiv GeWoSüd

Fig. 91, 92
Warhaftig et al. (1994).

Fig. 93, 94
Klein (1927a).

Fig. 95
Klein (1934).

Fig. 97
ETH library, image collection.

Fig. 98, 105
Vogler and Hassenpflug (1951).

Fig. 103
pinterest.ch

Fig. 104
Patent Nr. 578 164, Berlin.

Fig. 106
Lustenberger, Kurt. 1994. Adolf Loos, Zurich.

Fig. 107–109
Goldmerstein and Stodieck (1928).

Fig. 112
pinterest.de

Fig. 114–115, 176
pinterest.com

Fig. 116
www.grandlyon.com/fileadmin/user_upload/
media/pdf/institution/archives-
repertoires/20160921_gl_
archives_0001ir029_abattoirs_1886-1978.pdf

Fig. 118
Hebbert and Webb (2012).

Fig. 126–127
Geist and Kürvers (1984).

Fig. 128–129
en.wikipedia.org

Fig. 130–131, 140–141, 143–145
Düwel (1995).

Fig. 132–135
The Warsaw's Academy of Fine Arts archive.

Fig. 136–137, 165–169
Aronin (1953).

Fig. 138–139
Manley (1949).

Fig. 146, 148–152
www.sammlung-online.berlinischegalerie.de

Fig. 153
Fisch, Maass and Rating (1984).

Fig. 154, 157–162
Egli (1951).

Fig. 155
Olgyay (1963).

Fig. 156
Brooks (1951).

Fig. 163–164
ETH Zurich Library,
Research collection Ernst Egli.

Fig. 170
Fotostiftung Schweiz, estate Martin Hürlimann.

Fig. 171
Rudofsky, Bernard. 1977.
The Prodigious Builders, London.

Fig. 172
Peter Riemann, Berlinische Landesgalerie

Fig. 173
Knoch, Karl. 1961. Die Geländeklimaaufnahme
im Rahmen der Planung und des Ausbaus eines
Kur – und Erholungsbezirks. In Wissenschaft-
liche Arbeiten aus dem Burgenland, Vol. 30,
pp. 56–65.

Fig. 174
www.blog.urbanfile.org

Fig. 175
www.archipicture.eu

Fig. 177
Kubrick, Stanley. 1975. Barry Lyndon.

Fig. 178
Balchin and Pye (1947).

Fig. 179–180
Smithson (1969).

Fig. 181–183
Lawrence Halprin Collection,
The University of Pennsylvania.

Fig. 184
Yoko Ono.

Fig. 185
www.documenta14.de (9/10/2019)

Fig. 187
Dahinden (1974).

Fig. 189–190
Juan Navarro Baldeweg.

Fig. 191
Stanford University,
Richard Buckminster Fuller Collection.

Fig. 192–193, 203, 209–210
Lopez (2014).

Fig. 194–195
Banham (1971).

Fig. 196–198
Boyarsky (1970).

Fig. 199
Collection of Jonathan Holtzman.

Fig. 200, 206
Superstudio.

Fig. 201
Collection of The Art Institute of Chicago.

Fig. 202
www.portlandpf.org

Fig. 204
Harper and Boyle (1976).

Fig. 205
Spirn (1984); Hough (1984).

Fig. 207–208
Archizoom.

Fig. 211
Roth, Ueli and Fritz Häubi. 1980. "Wechselwirkungen zwischen Siedlungsstruktur und
Wärmeversorgungssystem", in Schweizer Inge-
nieur und Architekt, Band 98, Heft 29, p. 679.

Fig. 212, 218
Matus (1988).

Fig. 213
Vittorio Gregotti.

Fig. 214–215
Maki (1964).

Fig. 216
Kisho Kurokawa.

Fig. 217
Knowles (1974).

Fig. 219
www.metropolismag.com/profiles/
ralph-knowles-pioneer-solar-design/

Fig. 220
Rossi et al. (1979).

Fig. 221
Massachusetts Institute of Technology. 1976.
Housing for the Pequannock Watershed,
Cambridge (Mass.): MIT.

Fig. 222
www.exhibits.ced.berkeley.edu

Fig. 223
Lenzholzer (2015).

Fig. 224
Eran, Ben–Joseph. 2005. The Code of the City.
Standards and the Hidden Language of Place
Making. Cambridge (Mass.): MIT Press.

Fig. 225
Sascha Roesler, FCL, ETH Zurich,
relying on NASA Satellite Images.

Fig. 226
www.scmp.com/magazines/style/
news-trends/article/3006701/8-great-
attractions-newly-opened-jewel-changi-airport

Fig. 227
Ani Vihervaara / Sascha Roesler,
FCL, ETH Zurich.

Fig. 228, 230
Collection Royal Tropical Institute KIT,
Leiden NL.

Fig. 229
Karoline Kostka / Sascha Roesler,
FCL, ETH Zurich.

Illustration Credits

Imprint

Layout, Cover Design and Typesetting
Floyd Schulze, Berlin

Editorial Supervision and Project Management
Henriette Mueller-Stahl, Berlin

Copy-editing
Philip Shelley, London

Production
Heike Strempel-Bevacqua, Ehrenfriedersdorf (Germany)

Lithography
bildpunkt Druckvorstufen GmbH

Printing
Beltz Grafische Betriebe GmbH

Typeface
Tonal Grotesk by Omnitype

KLIMA POLIS is an international book series on issues of urban climate, edited by Sascha Roesler and published by Birkhäuser.

Library of Congress Control Number: 2024947384

Bibliographic information published by the German National Library
The German National Library lists this publication in the Deutsche Nationalbibliografie; detailed bibliographic data are available on the Internet at http://dnb.dnb.de.

This work is subject to copyright. All rights are reserved, whether the whole or part of the material is concerned, specifically the rights of translation, reprinting, re-use of illustrations, recitation, broadcasting, reproduction on microfilms or in other ways, and storage in databases.
For any kind of use, permission of the copyright owner must be obtained.

ISBN 978-3-0356-2946-0
e-ISBN (PDF) 978-3-0356-2945-3

© 2025 Sascha Roesler
Published by Birkhäuser Verlag GmbH
Im Westfeld 8
4055 Basel
Switzerland

www.birkhauser.com

Questions about General Product Safety Regulation
productsafety@degruyterbrill.com

9 8 7 6 5 4 3 2 1

The first edition of this book was made possible with the support of the Swiss National Science Foundation.